A SOJOURN THROUGH GEOMETRY AND ALGEBRA

Michel Helfgott

Michel Helfgott
Department of Mathematics and Statistics
East Tennessee State University
Johnson City, TN 37604
USA

Library of Congress Control Number 2013919740
ISBN 978-1492798897

About the cover: Geometry plays an important role in the analysis of the phenomenon of reflection of light (p. 30). The photograph of Monet's pond was taken by Edith Seier.

Contents

Preface

This textbook is intended for college juniors or seniors majoring in mathematics, who plan to become high school teachers. It seeks to provide a deeper perspective on secondary mathematics, showing the interplay between plane geometry and algebra. At East Tennessee State University we usually covered the first eight chapters in a one-semester course on geometry and left the next seven chapters for a second semester course on teaching secondary mathematics. The latter was a content course and included, among other topics, some of George Polya's ideas on teaching mathematics.

A distinctive characteristic of the book is the frequent discussion of multiple perspectives on the solution of a problem or the proof of a theorem, as well as the inclusion of trigonometry as a chapter of geometry, keeping in mind that the theory of similarity is what leads to the development of the trigonometry of acute angles.

The title reflects the fact that not every aspect of secondary mathematics is discussed and that neither derivatives nor integrals are utilized. Nonetheless, here and there the idea of limit of a sequence is stressed, especially when dealing with the concepts of length and area of a circle.

Practically none of the topics covered in the book overlap with the content of courses taken by mathematics majors, say real analysis, abstract algebra, differential equations, combinatorics, probability and statistics, number theory, etc. These courses, and several others, provide indispensable mathematical maturity but are rather distant from the core of high school mathematics. Precisely, one of our main objectives is to bridge the gap between the latter and college-level mathematics.

Some ten enrichment notes are included, ranging from Saccheri-Legendre Theorem in neutral geometry to Ferrari's method to solve quartics in a general setting. The decision on whether or not to discuss them in class, or to assign them as collateral reading, depends on how much time is available during the semester. No continuity is lost if the instructor chooses to skip any of the enrichment notes.

I wish to thank George Baloglou, Robert Beeler and Harald Andrés Helfgott for several very useful suggestions, which I have incorporated in the body of the textbook.

As in the past, in matters of grammar and style I had the invaluable advice of Federico Helfgott. To all of them I am deeply grateful.

Chapter 1

Introduction

1.1 Conjectures

Let us try to find a formula for the sum $1 + 2 + 3 + ... + n$, where n is any natural number. We define

$$s_n = 1 + 2 + 3 + ... + n.$$

Obviously $s_n = n + (n-1) + (n-2) + ... + 3 + 2 + 1$, hence $2s_n = (n+1) + (n+1) + ... + (n+1)$. Thus, $2s_n = n(n+1)$ and consequently

$$s_n = \frac{n(n+1)}{2}. \tag{1.1}$$

Having achieved this, we might think about the possibility of finding a formula for

$$1^2 + 2^2 + 3^2 + ... + n^2.$$

This is an important task because it will allow us to calculate the area under a parabola in a later chapter, as well as to find a formula for the volume of a cone and the volume of a sphere. Table 1.1 does not seem to provide a pattern, so let us build another table (Table 1.2). Since $1 = 3/3$, $14/6 = 7/3$, $30/10 = 3 = 9/3$, $55/15 = 11/3$, $91/21 = 13/3$, $140/28 = 35/7 = 5 = 15/3$, we can examine Table 1.3 and observe that there is indeed a pattern.

We conjecture that

$$\frac{\Sigma_{i=1}^n i^2}{\Sigma_{i=1}^n i} = \frac{2n+1}{3}$$

for all n. That is,

$$\Sigma_{i=1}^n i^2 = \frac{n}{6}(n+1)(2n+1).$$

Table 1.1: Looking for a pattern

n	$\Sigma_{i=1}^{n} i$	$\Sigma_{i=1}^{n} i^2$
1	1	1
2	3	5
3	6	14
4	10	30
5	15	55
6	21	91
7	28	140

Table 1.2: A pattern emerges

n	$\Sigma_{i=1}^{n} i^2 / \Sigma_{i=1}^{n} i$
1	1
2	5/3
3	14/6
4	30/10
5	55/15
6	91/21
7	140/28

Table 1.3: A pattern is found

n	$\Sigma_{i=1}^{n} i^2 / \Sigma_{i=1}^{n} i$
1	3/3
2	5/3
3	7/3
4	9/3
5	11/3
6	13/3
7	15/3

So far this is a conjecture, thus we need to provide a proof. The principle of mathematical induction, a basic postulate of mathematics, is ideally suited for this purpose. When $n = 1$ it is obviously true that

$$\Sigma_{i=1}^{n} i^2 = \tfrac{1}{6}(1+1)(2+1)$$

since $\Sigma_{i=1}^{1} i^2 = 1^2 = 1$, while $\frac{1}{6}(1+1)(2+1) = \frac{6}{6} = 1$. Next let us assume that the identity is true for n and we wish to prove that it is true for $n+1$. Indeed,

$$\Sigma_{i=1}^{n+1} i^2 = \Sigma_{i=1}^{n} i^2 + (n+1)^2 = \tfrac{n}{6}(n+1)(2n+1) + (n+1)^2 =$$

$$(n+1)[\tfrac{n}{6}(2n+1) + (n+1)] = \tfrac{n+1}{6}(2n^2+7n+6) = \tfrac{n+1}{6}(n+2)(2n+3).$$

The proof is complete. We went from a conjecture to a convincing mathematical proof.

A different approach can be used. Indeed,

$$n^3 = \Sigma_{i=1}^{n} i^3 - \Sigma_{i=1}^{n}(i-1)^3 = \Sigma_{i=1}^{n}(i^3 - (i-1)^3) =$$

$$\Sigma_{i=1}^{n}(i^3 - (i^3 - 3i^2 + 3i - 1)) = 3\Sigma_{i=1}^{n} i^2 - 3\Sigma_{i=1}^{n} i + \Sigma_{i=1}^{n} 1.$$

Thus, $3\Sigma_{i=1}^{n} i^2 = n^3 + \frac{3n(n+1)}{2} - n$. That is,

$$3\Sigma_{i=1}^{n} i^2 = \tfrac{1}{2}(2n^3 + 3n^2 + 3n - 2n) = \tfrac{1}{2}(2n^3 + 3n^2 + n) = \tfrac{n}{2}(n+1)(2n+1).$$

Consequently,

$$\Sigma_{i=1}^{n} i^2 = \tfrac{n}{6}(n+1)(2n+1).$$

The second approach is very ‘sleek’, does not require the use of mathematical induction, and is the technique of choice if we wish to find compact formulas for $\Sigma_{i=1}^{n} i^k$, $k \geq 3$.

The art of making conjectures plays an important role in the process of discovery that often precedes a mathematical proof. For instance, we may observe that $37+73 = 110$ is a multiple of 11 and that $42 + 24 = 66$ is also a multiple of 11. So we might conjecture that $ab + ba$ is always a multiple of 11 (a and b are any digits, 1 through 9). Indeed,

$$ab + ba = 10a + b + 10b + a = 11a + 11b = 11(a+b).$$

Hence $ab + ba$ is a multiple of 11. Having been successful with two-digit numbers we might think that perhaps $abc+cba$ is always a multiple of 11. Although $123+321 = 444$ is a multiple of 11, $137 + 731 = 868$ is not a multiple of 11. We have found a counterexample to the statement that $abc + cba$ is always a multiple of 11. Thus, this assertion is not true for every three-digit number.

How about working with four-digit numbers? We realize that $1234 + 4321 = 5555$ is a multiple of 11 while $3212+2123 = 5335$ is also a multiple of 11. Maybe $abcd+dcba$ is always a multiple of 11? The answer is yes:

$$abcd + dcba = 1000a + 100b + 10c + d + 1000d + 100c + 10b + a = \\ 1001a + 110b + 110c + 1001d = 11(91a + 10b + 10c + 91d).$$

Hence $abcd + dcba$ is always a multiple of 11. It can be proven that the statement is true when the number of digits is even, but that it is not always true when the number of digits is odd.

1.2 Generalizations

In the previous section we were able to find a compact formula for $\Sigma_{i=1}^n i^2$ from the observation that $n^3 = \Sigma_{i=1}^n i^3 - \Sigma_{i=1}^n (i-1)^3$. Maybe this idea works in general, allowing us to find compact formulas for $\Sigma_{i=1}^n i^k$, $k \geq 3$? Indeed

$$\Sigma_{i=1}^n i^4 = \Sigma_{i=1}^n (i-1)^4 + n^4 = \Sigma_{i=1}^n i^4 - 4i^3 + 6i^2 - 4i + 1) + n^4 = \\ \Sigma_{i=1}^n i^4 - 4\Sigma_{i=1}^n i^3 + 6\Sigma_{i=1}^n i^2 - 4\Sigma_{i=1}^n i + n + n^4.$$

Hence

$$4\Sigma_{i=1}^n i^3 = n(n+1)(2n+1) - 4\tfrac{n(n+1)}{2} + n + n^4 = (2n^2 - n)(n+1) + n + n^4 = \\ n^4 + n^2 + 2n^3 = (n^2 + n)^2.$$

Thus

$$\Sigma_{i=1}^n i^3 = (\tfrac{n^2+n}{2})^2 = (\tfrac{n(n+1)}{2})^2.$$

The reader can surmise that this approach will work for $\Sigma_{i=1}^n i^k$. $k \geq 4$. For large values of k it requires a lot of computations, but in principle it can be done.

Another interesting generalization process has to do with the arithmetic-geometric mean inequality (AGM). When $n = 2$ it states that given any $a, b > 0$

1. $\sqrt{ab} \leq \frac{a+b}{2}$.

2. $\sqrt{ab} = \frac{a+b}{2}$ if and only if $a = b$.

The proof of (i) is pretty simple because

$$\sqrt{ab} \leq \tfrac{a+b}{2} \Leftrightarrow ab \leq \tfrac{(a+b)^2}{4} \Leftrightarrow 4ab \leq a^2 + 2ab + b^2 \Leftrightarrow 0 \leq (a-b)^2.$$

The last inequality is certainly true, hence (i) has to be true. In a similar fashion, (ii) can be proven:

$$\sqrt{ab} = \tfrac{a+b}{2} \Leftrightarrow ab = \tfrac{(a+b)^2}{4} \Leftrightarrow 4ab = a^2 + 2ab + b^2 \Leftrightarrow 0 = (a-b)^2 \Leftrightarrow a = b.$$

For applications of AGM ($n = 2$) see the list of exercises at the end of the chapter. When $n = 3$ the AGM inequality asserts that, given $a, b, c > 0$

1. $(abc)^{1/3} \leq \frac{a+b+c}{3}$.

2. $(abc)^{1/3} = \frac{a+b+c}{3} \Leftrightarrow a = b = c$.

Surprisingly, a proof of the case $n = 3$ is much more difficult than the case $n = 2$ (see appendix B at the end of the book). AGM ($n = 3$) plays a vital role in the solution of the isoperimetric problem for triangles, that is, among all triangles of given perimeter p, which one encloses the greatest area? This problem is analyzed in detail in chapter 4, right after Heron's formula is discussed.

AGM is valid for any n, but we will not use this inequality when $n \geq 4$.

1.3 Indirect proofs

The proofs that we have used so far are all direct proofs. That is to say, we start under certain hypotheses and reach the conclusion through a collection of valid statements. But sometimes it is not easy to find a simple direct proof, so we add, as an extra hypothesis, the negation of what we wish to prove. If a contradiction of any sort is reached, we can claim that the proof has come to a conclusion. Indirect proofs are also called 'proofs by contradiction'.

Perhaps the simplest indirect proof has to do with the irrationality of $\sqrt{2}$. Assume that $\sqrt{2}$ is rational, that is,

$$\sqrt{2} = \tfrac{m}{n}$$

where m and n are integers. If m and n were to have factors in common we erase them from the numerator and the denominator. Then $2 = m^2/n^2$, hence $m^2 = 2n^2$. Thus m^2 is even, which in turn implies that m is even (if m were to be odd, say $m = 2a + 1$, then $m^2 = 4a^2 + 4a + 1 = 2(a^2 + 2a) + 1$ is odd). Say $m = 2p$ for a certain integer p. Therefore $4p^2 = 2n^2$, which in turn leads to $n^2 = 2p^2$. Since n^2 is even we must conclude that n is even, say $n = 2q$ for some integer q. We have reached a contradiction since m and n have the factor 2 in common, a possibility that was discarded at the beginning when we erased any possible factors (different from 1) that m and n could have in common. The reader must have heard that there are several other proofs of the irrationality of $\sqrt{2}$. But probably none is as elementary as the one portrayed above, which goes back to Euclid.

A direct proof is preferable, nonetheless on occasion no direct proof can be found, or it is too complicated, and therefore we have to settle for a proof by contradiction. In the realm of plane geometry, as we will illustrate in the next two chapters, sometimes indirect proofs are needed. Indirect proofs should not be confused with the use of the 'logical contrapositive'. The latter asserts that $p \to q$ is logically equivalent to $\neg q \to \neg p$ (not q then not p). Under these circumstances we assume $\neg q$ as a hypothesis and try to reach the thesis $\neg p$. The use of the logical contrapositive sometimes simplifies an argument, for instance when discussing the Basic Inequality (section 2.7) or the proof of the fact that congruence of a pair of alternating interior angles implies parallelism among the corresponding lines (section 3.1). In contrast, to prove $p \to q$ by contradiction means that we assume p and $\neg q$ as hypotheses and try to reach a stage where some statement r and its negation ($\neg r$) are true simultaneously. This is a flagrant impossibility, so q must hold; hence $p \to q$ has been proven.

Exercises for chapter 1

1. Prove that a three-digit number abc is a multiple of 2 if and only if c is even.

2. Prove that a four-digit number $abcd$ is a multiple of 5 if and only if $d = 0$ or $d = 5$.

3. Prove that a three-digit number abc is a multiple of 3 if and only if $a + b + c$ is a multiple of 3.

4. Show that $\sqrt{3}$ is an irrational number.

5. Find a compact formula for $\Sigma_{i=1}^{n} i^4$.

6. Using AGM ($n = 2$), show that among all rectangles of fixed perimeter p, the square encloses the greatest area.

7. Given positive constants A, B define

$$f(x) = Ax + \tfrac{B}{x},\ x > 0.$$

   Show that the function $f(x)$ attains its minimum at $x = \sqrt{\frac{B}{A}}$ (Hint: Start by noticing that, thanks to AGM ($n = 2$),

$$2\sqrt{Ax\frac{B}{x}} \leq Ax + \frac{B}{x}$$

for any $x > 0$.)

8. Among all right triangles whose hypotenuse passes through $(2, 5)$, with legs on the coordinate axes (Figure 1.1), which one has minimal area? (Hint: Observe that area $= \frac{1}{2}(p + 2)(q + 5)$ and $\frac{q}{2} = \frac{5}{p}$. Thereafter use the preceding exercise.)

9. Use mathematical induction to show that $2^n < n!$, $n \geq 4$ ($n!$ is the factorial of the natural number n; that is, $n! = n(n-1)(n-2)...2$).

10. Use mathematical induction to show that $1^3 + 2^3 + ... + n^3 = \frac{n^2(n+1)^2}{4}$.

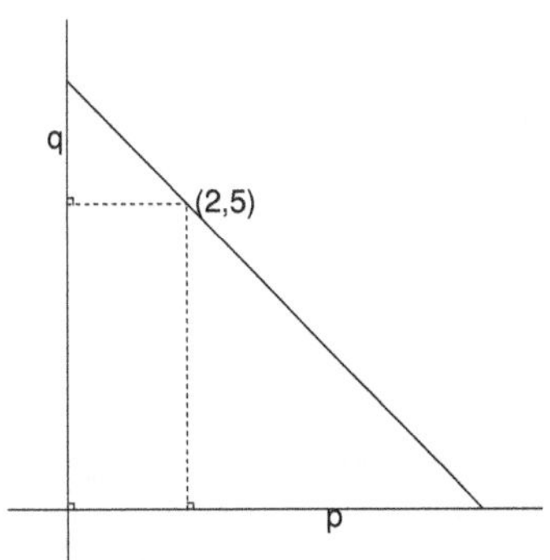

Figure 1.1: Minimization problem

Chapter 2

Some Basic Ideas of Geometry

2.1 SAS and other axioms

Euclid wrote 'The Elements', some 2,300 years ago, using an axiomatic approach. His work had a lasting impact on mathematics and related sciences, so much so that until the 19th century it was used on a regular basis in many countries. In classic Greek mathematics (600 BCE-300 BCE) geometric constructions were those that could be built employing only a straightedge (a ruler without marks on it) and a compass. Thus, Euclid's work reflects this approach to geometry. We will adopt a less restrictive point of view, accepting the use of a common ruler and a protractor, as well as a compass.

Edwin Moise (1918-1998) developed an axiomatic system (Moise, 1963) which uses the real numbers and incorporates the intuitive idea that the reals can be put into a one-to-one correspondence with any line. Moreover, Moise assumed mathematical notions compatible with the fact that a ruler is commonly used to measure the distance between any two points, and that a protractor is designed to measure angles[1]. In a rather informal way we will follow Moise's approach without being too rigorous, especially at the beginning, since our main purpose is to prove non-evident propositions rather than dwell on them with extreme rigor.

Introductory plane geometry deals mostly with lines, rays, segments, angles, triangles, and circles. A **line** that passes through points A and B is denoted $\overleftrightarrow{AB}$ while $\overrightarrow{OA}$ is the **ray** that starts at O, passes through A and extends indefinitely as a line from A onwards. That is to say, a ray $\overrightarrow{AB}$ is a sort of 'half line' with initial point A, which passes through B and extends indefinitely[2] beyond B. For the purposes that

[1]Moise's axiomatic system was preceded by work of George D. Birkhoff (1884-1944), a well-known American mathematician (Birkhoff, 1932).

[2]In Appendix A.2 the reader can find a rigorous definition of a ray and other basic concepts.

we have in mind, it is enough to keep an intuitive notion of a ray.

Two lines L_1 and L_2 are said to be *parallel*, $L_1 \parallel L_2$, if they have no points in common, that is, they never intersect. The **segment** with initial point A and endpoint B is the set that contains A, B and every point in between, and is represented by the symbol $\overline{AB}$; AB denotes its length. As expected, two segments are said to be parallel if the lines that contain them are parallel. The **angle** formed by the rays $\overrightarrow{OA}$ and $\overrightarrow{OB}$, with vertex at O, is represented by the symbol $\angle AOB$ (or $\angle BOA$). It should be noted that given any $\angle AOB$ there is a real number α, $0 < \alpha < 180$, called the measure of the angle in degrees and denoted $m\angle AOB$. An angle with measure less than 90 is called *acute* while an angle with measure bigger than 90 is called *obtuse*. When the measure of the angle is 90 we will say that it is a *right* angle.

There is a long tradition of referring to the measure α as α^o ('α degrees'). Although the concept of 'degree' is not needed from a mathematical point of view, when using diagrams it is often convenient to write, say, 30° instead of simply 30 to emphasize that we are dealing with the measure of an angle rather than the length of a side (section 6.7). We will also use the symbol α^o when it is necessary to stress the distinction between degrees and radians (section 7.11).

Finally, the symbol ΔABC represents the triangle with vertices A, B, and C, sides $\overline{AB}$, $\overline{BC}$, and $\overline{AC}$, and angles[3] $\angle ABC$, $\angle BCA$, and $\angle CAB$. A triangle is said to be an **acute triangle** if all angles are acute while ΔABC is an **obtuse triangle** if one angle is obtuse[4]. If one of the angles has measure 90 we will say that we are dealing with a **right triangle**.

Every line divides the plane in two half-planes, usually denoted H_1 and H_2. Given any number α, $0 < \alpha < 180$ and given any ray $\overrightarrow{OA}$ there exists a unique ray $\overrightarrow{OB_1}$, lying on the half-plane H_1 determined by $\overleftrightarrow{OA}$, such that $m\angle AOB_1 = \alpha$ (Figure 2.1). Similarly, there exists a unique ray $\overrightarrow{OB_2}$, lying on the half-plane H_2 determined by $\overleftrightarrow{OA}$, such that $m\angle AOB_2 = \alpha$. This is the **construction of angles** property.

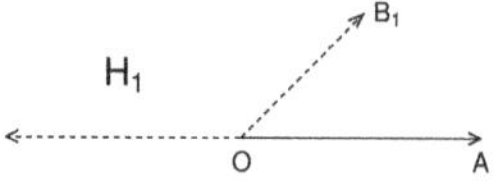

Figure 2.1: Construction of angles

Moreover, the **addition of angles** property asserts that if C lies in the interior

[3]Sometimes called *internal angles* of the triangle.

[4]As a consequence of the Exterior Angle Theorem, to be studied in section 2.7, and the linear pair property (see top of next page), the sum of any two internal angles of a triangle is less than 180; thus, no triangle can have two or more obtuse internal angles.

of $\angle AOB$ then

$$m\angle AOB = m\angle AOC + m\angle COB.$$

We rely on the reader's intuition with regard to the concept of interior of an angle. On the other hand, if $\angle AOB$ and $\angle BOC$ form a 'linear pair', that is, if A, O, and C lie on a line, with O between A and C, and a ray $\overrightarrow{OB}$ is given (Figure 2.2), then

$$m\angle AOB + m\angle BOC = 180,$$

This is the **linear pair** property of angles.

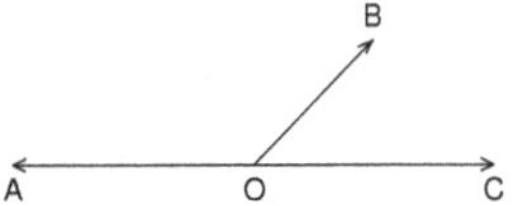

Figure 2.2: A linear pair

Two triangles are **congruent** if their corresponding three sides have the same length and their corresponding three angles have the same measure. In symbols, $\Delta ABC \cong \Delta DEF$ if and only if $AB = DE$, $BC = EF$, $AC = DF$, $m\angle ABC = m\angle DEF$, $m\angle BCA = m\angle EFD$, and $m\angle CAB = m\angle FDE$. We could also write, instead of the last three equalities, $\angle ABC \cong \angle DEF$, $\angle BCA \cong \angle EFD$, and $\angle CAB \cong \angle FDE$. Often we will say that two triangles are congruent if their corresponding sides are congruent (that is, have the same length) and their corresponding three angles are congruent (that is, have the same measure). Intuitively speaking, two triangles are congruent if we can rotate and translate one of them through the plane and make it overlap perfectly on the other triangle. We should add that the **angle bisector** of $\angle AOB$ is the ray $\overrightarrow{OC}$ that divides the angle in two congruent angles, that is, $\angle AOC \cong \angle BOC$.

In **neutral geometry** we do not use the Axiom of Parallels, which states, in one of its versions, that given a line and a point outside the line, there is at most one line that passes through the point and is parallel to the given line. However, in neutral geometry, also called 'absolute geometry', we do have the **SAS congruence** property (side-angle-side congruence), which states that $\Delta ABC \cong \Delta DEF$ provided that $AB = DE$, $BC = EF$ and $\angle ABC \cong \angle DEF$ (Figure 2.3).

Figure 2.3: SAS congruence

It should be noted that any result from neutral geometry is valid both in **Euclidean geometry**, where we add the Axiom of Parallels, and in hyperbolic geometry. In the latter, a member of the family of non-Euclidean geometries, the negation of the axiom of parallels is added to the list of axioms of neutral geometry. The word 'neutral' was probably chosen by mathematicians to emphasize the fact that there is a body of geometrical results that can be accepted by those that work in Euclidean geometry or hyperbolic geometry.

A detailed list of axioms for neutral geometry, in two dimensions, can be found in Appendix A2. For the purposes we have in mind in this chapter, it is only necessary to use, time and again, the above-mentioned four properties[5]: construction of angles, angle addition, linear pair, and SAS congruence. For instance, let us prove the well-known *Vertical Angle Theorem*, **VAT** for short, which states that any two opposite angles have the same measure: $m\angle AOB = m\angle COD$ (Figure 2.4). Indeed

$$m\angle AOB + m\angle BOC = 180 = m\angle BOC + m\angle COD$$

thanks to the linear pair property (applied twice). Therefore $m\angle AOB = m\angle COD$, as we wished to prove. In the same fashion, it can be proven that $m\angle AOD = m\angle BOC$.

We should also add another well-known definition: Two lines, L_1 and L_2, are *perpendicular*, $L_1 \perp L_2$, if they intersect and make a right angle at the point of intersection. Thanks to the linear pair property and VAT, if one of the angles at the point of intersection has a measure of 90 then the other three angles also have a measure of 90. Furthermore, it is to be understood that two segments are perpendicular if the lines that contain them are perpendicular.

Remark

In the future, especially when proving the converse of the Isosceles Triangle Theorem and ASA congruence, we will have to deal with the following situation: Let $\angle AOC$ be given and let B be an interior point of it. Then

[5]These four properties, as well as others, are really axioms. We could rather call them *principles* instead of *properties*.

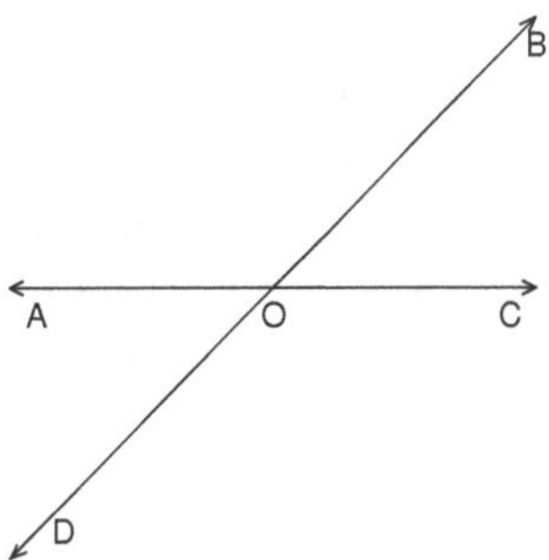

Figure 2.4: Vertical Angle Theorem

$$m\angle AOB < m\angle AOC.$$

This fact seems pretty evident. Indeed, $m\angle AOC = m\angle AOB + m\angle BOC$ by the addition of angles property. Since the three numbers are positive we can conclude that $m\angle AOB < m\angle AOC$.

2.2 Isosceles Triangle Theorem (ITT)

Let us consider[6] ΔABC with $AB = BC$. Then $\angle A \cong \angle C$.

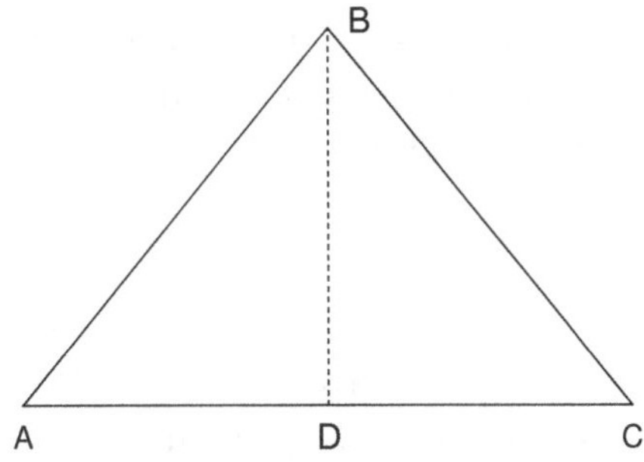

Figure 2.5: A first attempt at proving ITT

Is there a simple proof? Let us start by building the angle bisector of $\angle ABC$, which meets $\overline{AC}$ at D (Figure 2.5). Using SAS we can conclude that $\Delta ABD \cong \Delta CBD$. In

[6]Such a triangle is called 'isosceles'.

particular $\angle A \cong \angle C$. QED

This is not Euclid's proof, which involves a more extensive argument to be discussed soon. How could Euclid have missed the proof that uses the angle bisector? There is a reason behind Euclid's avoidance of the above-mentioned proof. In classic Greek mathematics one had to prove that it is possible to build the angle bisector of any angle using only a straightedge and a compass. As we will discuss in section 2.5, to justify such a construction it is necessary to utilize the SSS property of congruence. But to prove SSS we will require ITT, a fact that will become evident shortly. Of course, it is important to avoid circular arguments.

Commentaries

From a modern perspective, which is not constrained by constructions with straightedge and compass, there is no problem in building the bisector of $\angle ABC$. We would just have to use the axiom of construction of angles, choosing $\alpha = \frac{1}{2}m\angle B$. However, there is a subtle difficulty: Are we sure that the angle bisector will eventually cross the segment $\overline{AC}$? It does, but we would need to prove first the Crossing-The-Bar Theorem (Crossbar Theorem for short). The latter requires some delicate arguments that involve Axiom 4 from the appendix, the axiom of separation of the plane (Moise, 1963). Of course, if a High School student is starting to study geometry, for pedagogical reasons a teacher can well prove ITT using the angle bisector approach, shown above, for this purpose. A beginner might well find it hard to believe that the Crossbar Theorem requires a proof! In any case, our purpose in this book is not to build a 'perfect' edifice of plane geometry, where everything is proved from the axioms, included evident propositions at the beginning, but to develop the corpus of geometry in a sensible and balanced way.

Yet another approach to a proof is to choose the midpoint of $\overline{AC}$, call it E, and then claim that $\Delta ABE \cong \Delta CBE$ by SSS, which in particular implies that $\angle BAE \cong \angle BCE$. But we must avoid using SSS at this stage because we have not proven it yet. It happens that our proof of SSS will involve ITT, so we cannot use the former to prove the latter because we have to avoid the type of circular arguments mentioned before. Nonetheless, from the point of view of pedagogy we might include SSS as an axiom in a first course on geometry at the High School level. However, there are some dangers with regard to piling up results as axioms; not only does our axiomatic system lose independence but a student could get the wrong impression that it is possible to enlarge the list of axioms as much as we wish.

Euclid's proof of ITT

It is worthwhile to study Euclid's original approach. Let us start by prolonging $\overline{AB}$ until we reach a point D. It does not matter how big AD is. Then prolong $\overline{BC}$ up to a point E such that $CE = AD$, and build $\overline{DC}$ and $\overline{AE}$ (Figure 2.6). Keep in mind that we wish to show that $\alpha = \beta$, where $\alpha = m\angle BAC$ and $\beta = m\angle BCA$. Let us pay

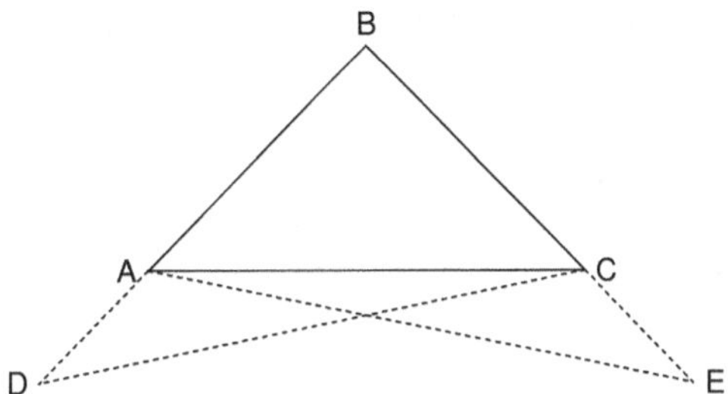

Figure 2.6: Euclid's proof of ITT

attention to ΔABE and ΔCBD. Their angles at the top are the same and we notice that $AB = BC$ and $BE = BC + CE = AB + AD = BD$. Then, SAS implies that

$$\Delta ABE \cong \Delta CBD. \tag{2.1}$$

Are the two triangles at the bottom, namely ΔACD and ΔCAE, congruent? Since $\Delta ABE \cong \Delta CBD$ we can assert that $\angle BDC \cong \angle BEA$ and $DC = AE$. Then SAS implies that $\Delta ACD \cong \Delta CAE$; thus, in particular, $m\angle ACD = m\angle EAC$. Let γ be their common measure. From (2.1) we get that $\angle BAE \cong \angle BCD$, therefore $\alpha + \gamma = \beta + \gamma$. Hence $\alpha = \beta$, as we wished to prove. QED

Yet another proof of ITT

Let us recall that $\Delta ABC \cong \Delta DEF$ if and only if there is a one-to-one correspondence $ABC \leftrightarrow DFE$, between the vertices of both triangles, such that $\overline{AB} \cong \overline{DF}$, $\overline{BC} \cong \overline{EF}$, $\overline{AC} \cong \overline{DF}$, $\angle ABC \cong \angle DFE$, $\angle BCA \cong \angle FED$, $\angle CAB \cong \angle DEF$.

Assume that ΔABC is isosceles with $AB = BC$ and consider the correspondence $ABC \leftrightarrow CBA$. Since $AB = BC$ we will obviously have the equalities $AB = CB$ and $BC = BA$. Moreover, we note that $\angle ABC \cong \angle CBA$. Then, by SAS, $\Delta ABC \cong \Delta CBA$. In particular, $\angle BAC \cong \angle BCA$; that is, the base angles of ΔABC are congruent. QED

Remark

The last proof is short but rather 'abstract', so from a pedagogical point of view it might not be a proof of choice if intended for an audience of students with limited mathematical experience. Yes, Euclid's proof is a little long. However, the techniques that it involves, basically using SAS several times, provide valuable experience with congruence arguments.

Converse of ITT

Let ΔABC be given, with their base angles congruent; that is, $\angle BAC \cong \angle BCA$.Then $AB = BC$.

Figure 2.7: Proving the converse of ITT

Proof

Proceeding by contradiction assume that $AB \neq BC$. Without loss of generality suppose that $BC < AB$. Start by choosing D on $\overline{AB}$ such that $AD = BC$ and build $\overline{CD}$. Let us pay close attention to ΔDAC and ΔBCA (Figure 2.7). By SAS we can conclude that $DAC \cong \Delta BCA$ so, in particular, $\angle DCA \cong \angle BAC$. But $\angle BAC \cong \angle BCA$ by hypothesis, consequently $\angle DCA \cong \angle BCA$. We have reached an impossibility since D is an interior point of $\angle BCA$. QED

2.3 ASA and SSS congruence

Theorem (ASA congruence)

Let us assume that ΔABC and ΔDEF are such that $\angle BAC \cong \angle EDF$, $\angle BCA \cong \angle EFD$, and AC=DF (Figure 2.8). Then $\Delta ABC \cong \Delta DEF$.

Proof

If $AB = DE$ the proof comes to a conclusion thanks to SAS. Suppose that $AB \neq DE$, say $DE < AB$. Choose G on $\overline{AB}$ such that $AG = DE$. Then $\Delta GAC \cong \Delta EDF$

by SAS; in particular $\angle GCA \cong \angle EFD$. But $\angle EFD \cong \angle BCA$ by hypothesis, thus $\angle GCA \cong \angle BCA$. We have reached an impossibility since G is an interior point of $\angle BCA$. This impossible situation was reached when we assumed that $AB \neq DE$. Therefore $AB = DE$, which in turn brings the proof to completion due to SAS. QED

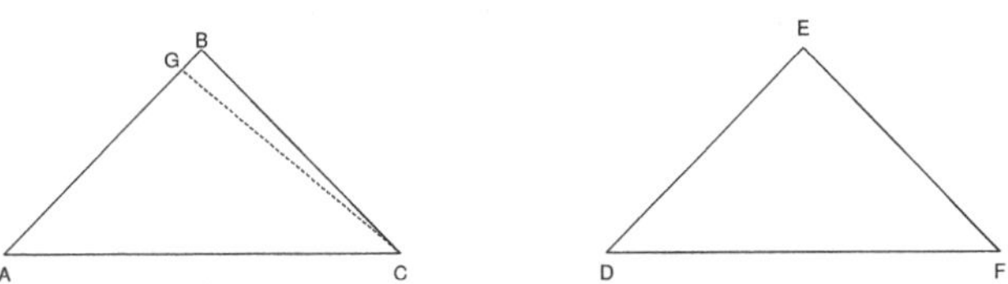

Figure 2.8: Proving ASA congruence

Using ASA to prove the converse of ITT

Let ΔABC be a triangle where $\angle BAC \cong \angle CBA$ and let us consider the correspondence $ABC \leftrightarrow CBA$. Since $\angle BAC \cong \angle BCA$ and $AC = CA$, using ASA we get $\Delta ABC \cong \Delta CBA$. In particular $AB = CB$. QED

The measurement of the width of a river

A person, standing at a point A, wishes to determine the width d of a river. With a pair of binoculars she chooses a point B on the opposite side of the river; it could be the base of a tree, a rock, or any discernible object from the landscape (Figure 2.9). Then, with a transit of height h set perpendicular to the terrain, she points toward B and measures $\angle ACB$. Next she carefully rotates the transit, keeping without change the perpendicularity with the terrain and $\angle ACB$, and locates the point D on the side of the river where she stands (Figure 2.9). Thereafter she asks a helper to set a marker at D and measure AD. We observe that, due to ASA, $\Delta CAB \cong \Delta CAD$, consequently $d = AB = AD$.

Theorem (SSS Congruence)

Let ΔABC and ΔDEF be triangles such that $AB = DE$, $AC = DF$, and $BC = EF$. Under these circumstances $\Delta ABC \cong \Delta DEF$.

Proof

1. Suppose that the triangle under consideration is acute; that is, all the internal angles are less than 90. Start by building a segment $\overline{AG}$ in such a way that

Figure 2.9: Measuring the width of a river

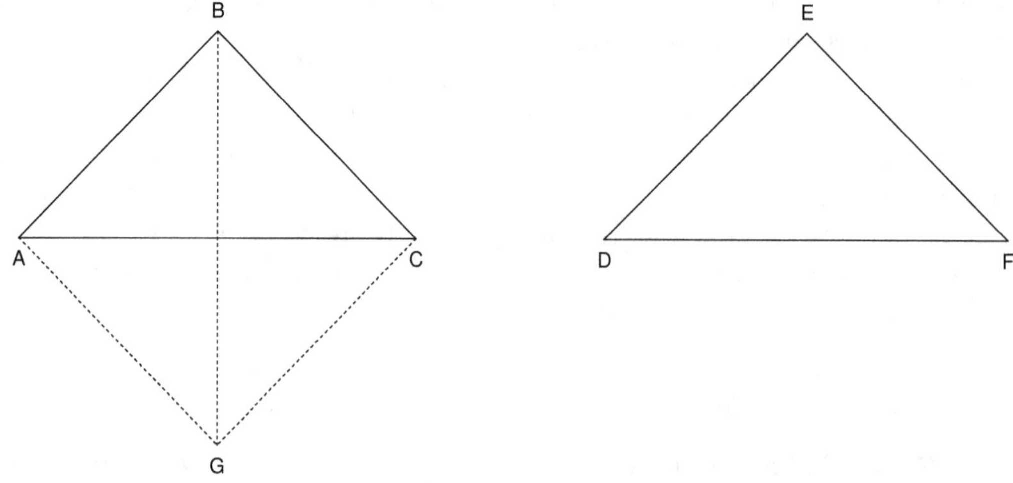

Figure 2.10: Proving SSS congruence for acute triangles

$AG = DE$ and $\angle CAG \cong \angle EDF$. Then join points C and G (Figure 2.10). Thanks to the SAS axiom we have

$$\Delta AGC \cong \Delta DEF. \tag{2.2}$$

In particular, $CG = FE$. We notice that ΔGAB is isosceles, then by ITT $\angle ABG \cong \angle AGB$. Similarly, $CBG \cong \angle BGC$ since ΔBCG is isosceles, therefore $\angle ABC \cong \angle AGC$. Hence, by SAS,

$$\Delta ABC \cong \Delta AGC. \tag{2.3}$$

From (2.2) and (2.3) we can conclude that $\Delta ABC \cong \Delta DEF$.

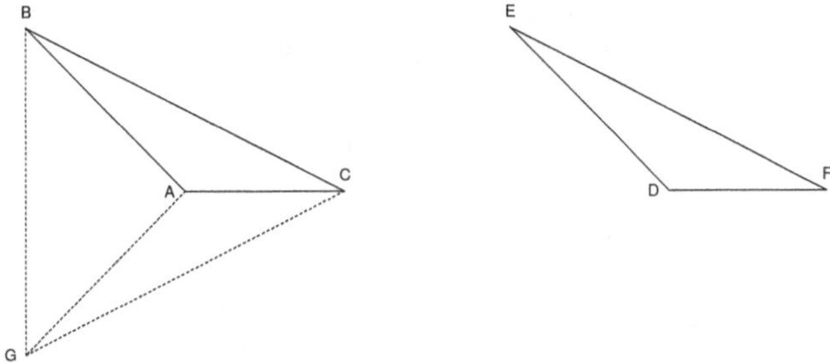

Figure 2.11: Proving SSS congruence for obtuse triangles

2. Assume that the triangle is obtuse (Figure 2.11). Let us build $\overline{AG}$ so that $AG = DE$ and $\angle CAG \cong \angle FDE$. Then , by SAS,

$$\Delta CAG \cong \Delta FDE. \tag{2.4}$$

In particular, $GC = FE$. Since ΔBAG is isosceles we can conclude, thanks to ITT, that the base angles are congruent (with common measure α). Similarly, since ΔBCG is isosceles we can conclude that its base angles are congruent (with common measure β). Then $m\angle ABC = m\angle AGC = \beta - \alpha$. Using SAS it follows that

$$\Delta CAG \cong \Delta CAB. \tag{2.5}$$

Finally, (2.4) and (2.5) imply that $\Delta CAB \cong \Delta FDE$ or, what is the same, $\Delta ABC \cong \Delta DEF$.

3. If one of the triangles is a right triangle, say with right angle at A, the proof closely resembles the steps taken in case 1 when the triangles under consideration are acute (exercise 18 at the end of the chapter). Obviously, if both triangles are right triangles then SAS leads to congruence between them immediately. QED

2.4 Building proofs

When faced with the challenge of proving some geometrical proposition, we often have to start by constructing something. We do not know in advance whether our construction will work, but at least we can try to solve the puzzle; this is part of the beauty of mathematics. Let us discuss two problems that exemplify quite well the difficulties that we have to overcome.

1. Suppose that $\overline{BD}$ is both an angle bisector and a median of ΔABC; that is, $\angle ABD \cong \angle CBD$ and $AD = DC$ (Figure 2.12). Show that ΔABC is isosceles.

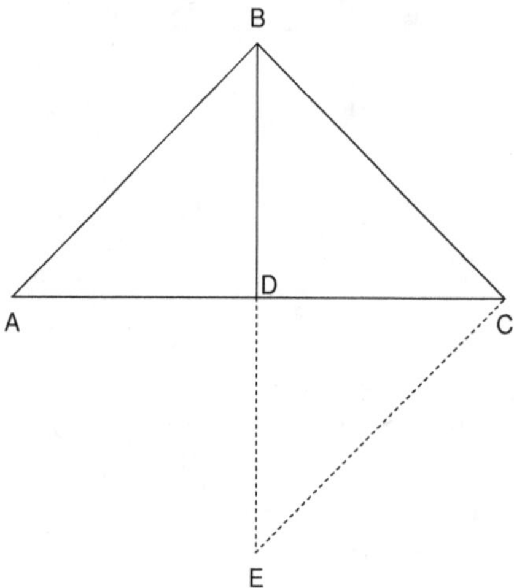

Figure 2.12: Using a construction to solve a problem

Right away we cannot apply SAS, ITT (or its converse), ASA, or SSS. So let us extend $\overline{BD}$ in such a way that $DE = BD$ and build $\overline{CE}$ (Figure 2.12). Since $\angle ADB \cong \angle CDE$ (VAT proposition) we can conclude, thanks to SAS, that $\Delta ADB \cong \Delta CDE$. In particular $\angle ABD \cong \angle CED$. Then the converse of ITT implies that $BC = CE$. But $AB = CE$ (recall that $\Delta ADB \cong \Delta CDE$), consequently $AB = BC$. The reader can check that the proof does not depend on the fact that the triangle is acute (exercise 17 at the end of the chapter).

2. Let $ABCD$ be a quadrilateral (a simple polygon of four sides[7]) with $\angle BAD \cong \angle CDA$ and $AB = CD$ (Figure 2.13). Show that $\angle ABC \cong \angle DCB$.

[7]A more formal definition of a simple quadrilateral can be found in section 3.2.

We start by choosing the midpoint M of $\overline{AD}$ and then build the segments $\overline{BM}$ and $\overline{CM}$. Using SAS we get $\Delta BAM \cong \Delta CDM$, so $BM = CM$ and

$$\angle ABM \cong \angle DCM. \tag{2.6}$$

Then ITT implies that

$$\angle MBC \cong \angle MCB. \tag{2.7}$$

From (2.6) and (2.7) we can conclude that $\angle ABC \cong \angle DCB$, precisely what we wanted to prove.

Sometimes there is more than one way of proving a result in plane geometry. Let us analyze a different proof for the second problem:

Build $\overline{AC}$ and $\overline{BD}$ (Figure 2.13). By SAS $\Delta BAD \cong \Delta CDA$, hence $BD = CA$ and

$$\angle ABD \cong \angle DCA. \tag{2.8}$$

Next let us pay attention to ΔABC and ΔDCB. Since both triangles share the segment $\overline{BC}$, SSS implies that $\Delta ABC \cong \Delta DCB$. In particular,

$$\angle DBC \cong \angle ACB. \tag{2.9}$$

From (2.8) and (2.9) we can conclude that $\angle ABC \cong \angle DCB$.

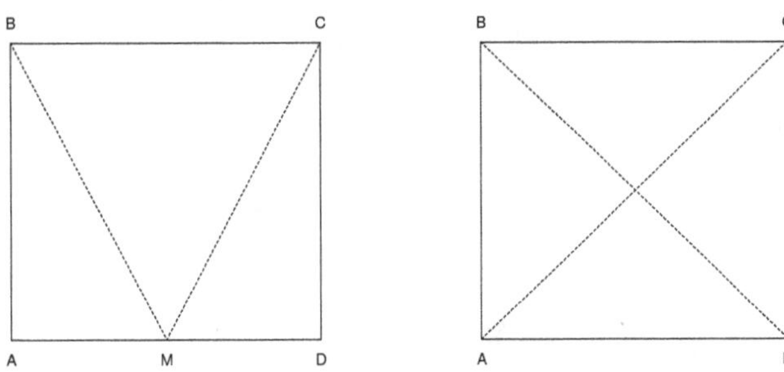

Figure 2.13: Two ways of solving a problem

It should be noted that when $m\angle BAD = m\angle CDA = 90$ and $AB = CD$, the quadrilateral under discussion is called a **Saccheri quadrilateral**. Obviously, the angles at the top of any Saccheri quadrilateral are congruent.

2.5 Three constructions with straightedge and compass

In classic Greek mathematics constructions could be done only by using a straightedge (a ruler without marks) and a compass. They are usually very appealing and simple. Let us discuss three of the most important ones.

1. Let us start with the construction of the bisector of an arbitrary angle, $\angle COD$ (Figure 2.14). With the compass centered at O choose a point A on $\overrightarrow{OC}$ and a point B on $\overrightarrow{OD}$ in such a way that $OA = OB$. Then, with center at A, open the compass until it touches point B. Thereafter, without changing the opening of the compass, make a circle. With the same opening, but with center at B, make another circle. Both circles intersect at a point P. Finally, draw the ray $\overrightarrow{OP}$. We claim that $\overrightarrow{OP}$ is the bisector of $\angle COD$. The proof of this fact is pretty easy: By the construction itself we have $OA = OB$ and $AP = BP$. Then SSS implies that $\Delta OAP \cong \Delta OBP$. In particular $\angle AOP \cong \angle BOP$.

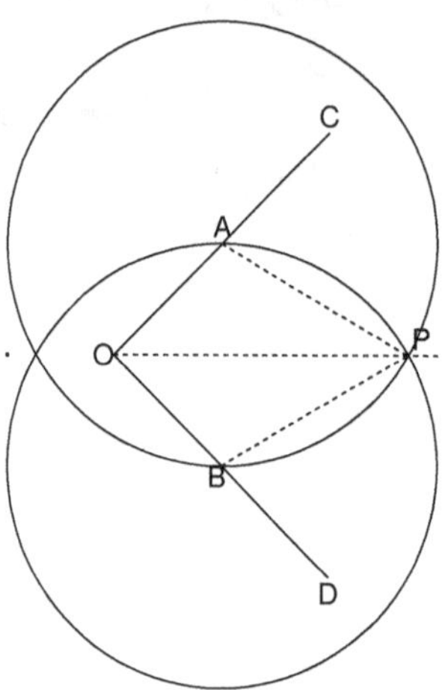

Figure 2.14: Bisecting an angle

2. Let us analyze how we can construct the perpendicular bisector of a given segment $\overline{AB}$ (Figure 2.15). Start with the compass centered at A and draw a circle with radius AB, then draw another circle with the same radius and centered at B. Both circles meet at a point P above the segment and a point Q below the segment. We claim that the line $\overleftrightarrow{PQ}$, which we draw using the straightedge and which meets $\overline{AB}$ at R, is the perpendicular bisector of $\overline{AB}$.

Let us provide a proof: Thanks to ITT, the two base angles of ΔAPB are congruent and the two base angles of ΔAQB are congruent. Then, by SAS, $\Delta PAQ \cong \Delta PBQ$. In particular, $\angle APQ \cong \angle BPQ$. Thus, again by SAS, it follows that $\Delta APR \cong \Delta BPR$. Hence $AR = BR$ and $\angle ARP \cong \angle BRP$. But $\angle ARP$ and $\angle BRP$ form a linear pair, consequently

$$m\angle ARP + m\angle BRP = 180.$$

Therefore $m\angle ARP = 90$. We have just shown that $\overleftrightarrow{PQ}$ is the perpendicular bisector of $\overline{AB}$. By the way, it is evident that the preceding construction can be used to find the midpoint of any segment. Moreover, given a point P on a line L, it is possible to construct a line perpendicular to L that passes through P. Indeed, with a compass label a point R to the left of P and a point S to the right of P such that $RP = PS$, and then construct the perpendicular bisector of RS; this line is perpendicular to L and passes through P.

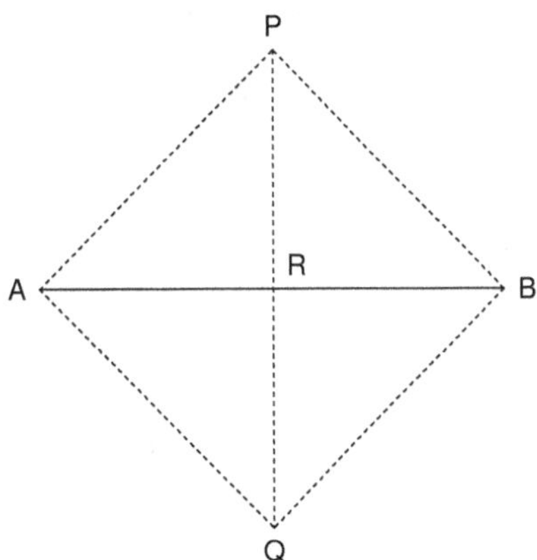

Figure 2.15: Constructing the perpendicular bisector

3. The third construction deals with the task of building a perpendicular line from a point P to a given line L, where P does not belong to L. With the compass centered at P, we touch the line L at a point A and then build a circle, with radius PA, that touches the line at another point B (Figure 2.16). So far we have that $PA = PB$. Then, using the compass, we build a circle centered at A and with a radius AB, then another circle with the same radius but centered at B. Both circles meet at a point Q, 'below' the line L. We claim that

$\overleftrightarrow{PQ}$, which crosses L at the point R, is a line that passes through P and is perpendicular to L. The underlying proof is quite similar to the one we used in the preceding construction: Using ITT it follows that $\angle PAR \cong \angle PBR$ and $\angle QAR \cong \angle QBR$. Then $\Delta PAQ \cong \Delta PBQ$ by SAS; thus, $\angle APQ \cong \angle BPQ$. Then, by SAS, $\Delta APR \cong \Delta BPR$. Therefore $\angle PRA \cong \angle PRB$. Since they form a linear pair we can conclude that $m\angle PRA = 90$.

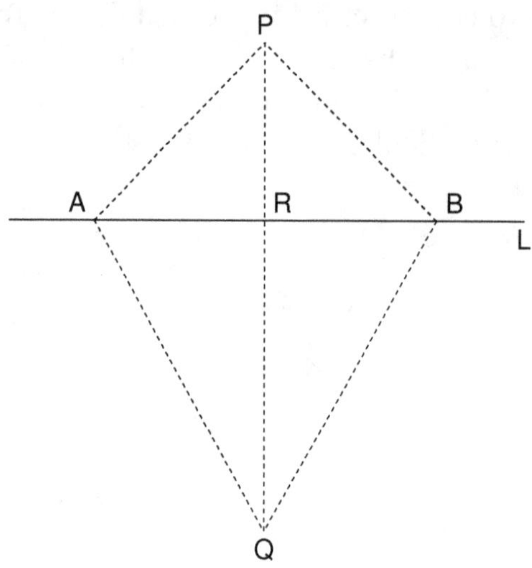

Figure 2.16: The third construction

2.6 Characterization of the perpendicular bisector

We have already dealt with the perpendicular bisector of a segment in the section on constructions with straightedge and compass. Now it is time to pay a closer look to this important notion. Let us recall that, given a segment $\overline{AB}$, a line L is the perpendicular bisector of the segment if it touches $\overline{AB}$ at a point O in such a way that $AO = BO$ and if L is perpendicular to $\overline{AB}$, that is, $m\angle BOC = 90$ where C is any point on the line L.

Theorem

Let L be the perpendicular bisector of $\overline{AB}$. Then L is the set of all points P such that $PA = PB$. That is to say, $L = \{P : PA = PB\}$.

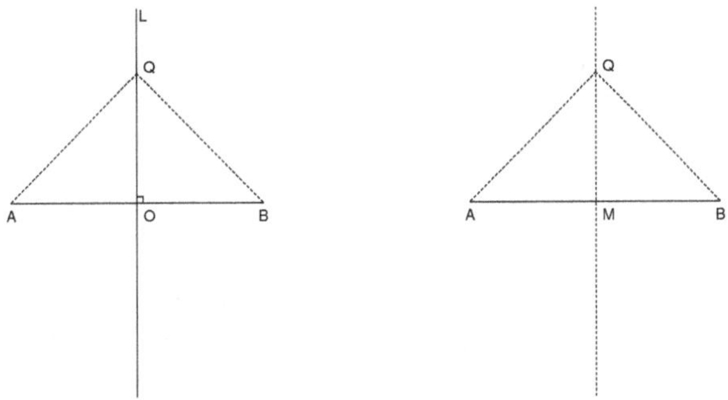

Figure 2.17: Proving the characterization property of perpendicular bisectors

Proof

We need to show that one set is a subset of the other. Thus, the proof will be divided in two parts.

1. Let us try to prove that $L \subseteq \{P : PA = PB\}$. Assume that Q belongs to L and build segments $\overline{AQ}$, $\overline{BQ}$ (Figure 2.17). Since $AO = BO$, using SAS we can assert that $\Delta AOQ \cong \Delta BOQ$. In particular, $QA = QB$, thus Q belongs to the set $\{P : PA = PB\}$.

2. Next we need to prove that $\{P : PA = PB\} \subseteq L$. With this purpose in mind let us assume that Q belongs to $\{P : PA = PB\}$; thus, $QA = QB$. Then choose the midpoint M of $\overline{AB}$ and build the segments $\overline{QA}$, $\overline{QB}$, and $\overline{QM}$ (Figure 2.17). Using SSS we can assert that $\Delta QMA \cong \Delta QMB$. Hence, in particular, $\angle QMA \cong \angle QMB$. Since both angles form a linear pair, and are congruent, we can conclude that

$$m\angle QMA = m\angle QMB = 90.$$

Therefore $\overleftrightarrow{QM} = L$, so Q belongs to the line L. QED

Proposition

The three perpendicular bisectors of any triangle meet at a point, called the **circumcenter**[8].

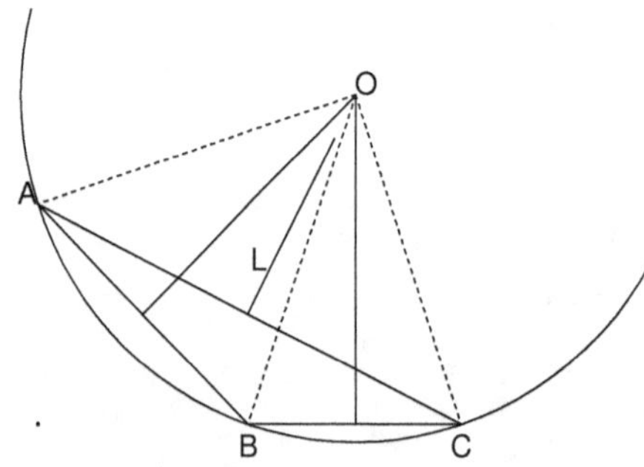

Figure 2.18: Circumcenter of a triangle

Proof

Let ΔABC be an arbitrary triangle. The perpendicular bisectors of $\overline{AB}$ and $\overline{BC}$ meet at a point O; otherwise, if they were to be parallel, $\overline{AB}$ and $\overline{BC}$ would be lying on the same line and consequently we would not be dealing with a triangle. The outstanding question is whether L, the perpendicular bisector of $\overline{AC}$, passes through O (Figure 2.18).

We notice that $OA = OB$ and $OB = OC$ due to the characterization theorem, hence $OA = OC$. Once again we apply the characterization theorem and conclude that O belongs to L, consequently L passes through O too. QED

2.7 Exterior Angle Theorem (EAT)

As we will see, EAT involves inequalities. This result has an astonishing number of consequences, from the triangle inequality to AAS and SsA congruence, and many other important propositions of plane geometry.

Theorem

Let ΔABC be an arbitrary triangle and let DCB be an *exterior angle* of the given triangle (Figure 2.19). Then $m\angle DCB > m\angle ABC$ and $m\angle DCB > m\angle BAC$. That is

[8]The name 'circumcenter' stems from the fact that it is the center of the circle that circumscribes the given triangle.

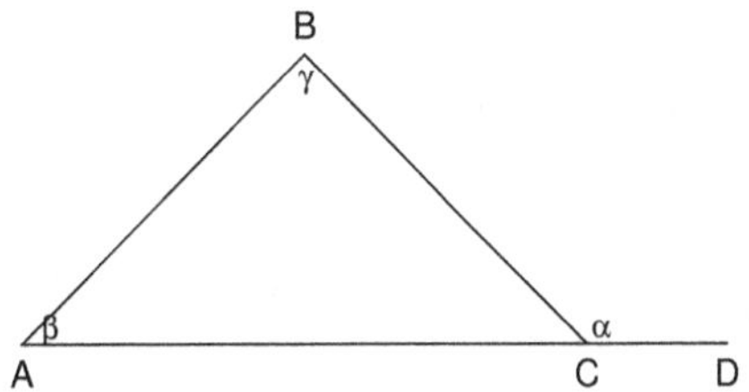

Figure 2.19: The Exterior Angle Theorem

to say, the measure of any exterior angle is strictly larger than the measure of each of the two non-adjacent internal angles: $\alpha > \beta$ and $\alpha > \gamma$.

Proof

Let $\alpha = m\angle DCB$, $\beta = m\angle BAC$, and $\gamma = m\angle ABC$ and let M be the midpoint of $\overline{BC}$. Then prolong $\overline{AM}$ so that $MN = AM$ and build $\overline{NC}$ (Figure 2.20). By SAS we get that $\Delta BMA \cong \Delta CMN$. In particular, $\angle ABM \cong \angle NCM$, thus $m\angle NCM = \gamma$. Since N is an interior point of $\angle BCD$ we can conclude that $\gamma < \alpha$.

Our next task is to prove the inequality $\beta < \alpha$. The proof will resemble the one described in the previous paragraph. Let M be the midpoint of $\overline{AC}$ and extend $\overline{BM}$ so that $MN = BM$ (Figure 2.20). Then, thanks to SAS, $\Delta AMB \cong \Delta CMN$. In particular, $\angle MCN \cong \angle MAB$; thus, $m\angle MCN = \beta$. Now choose point E on the line $\overleftrightarrow{NC}$, with E lying in the interior of $\angle BCD$. On the other hand, the Vertical Angle Theorem implies that $m\angle ECD = \beta$. Since E lies in the interior of $\angle BCD$ we can conclude that $m\angle ECD < m\angle BCD$, that is, $\beta < \alpha$. QED

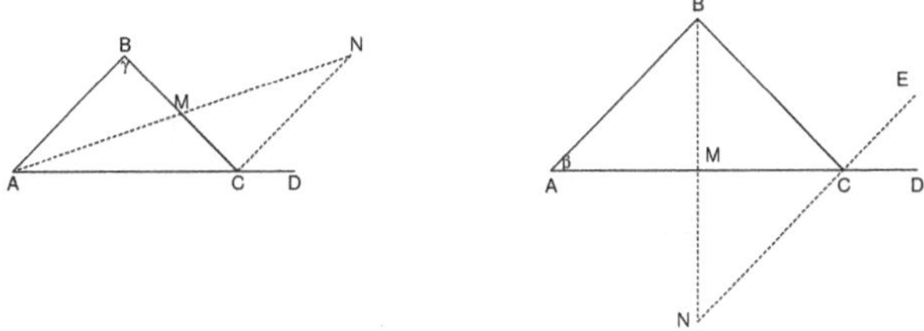

Figure 2.20: Proving the Exterior Angle Theorem

Remarks

1. In section 2.5 we learned how to build a segment that starts at a point P, on the outside of a line L, and is perpendicular to it. Could there be another segment that starts at P and is perpendicular to L ? EAT provides a negative answer (Figure 2.21). Then we can define the **distance between P and the line L, $d(P, L)$**, simply as PT where T is the point where the unique perpendicular from P to L intersects L.

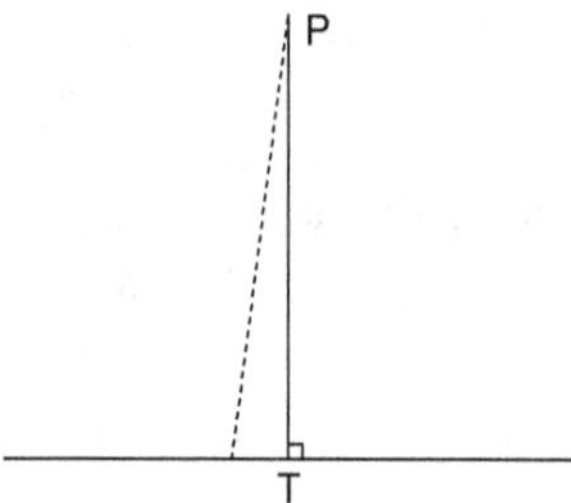

Figure 2.21: The perpendicular from a point to a line

2. Given an arbitrary triangle, the segment that starts at a vertex and is perpendicular to the opposite side is called a **height**.

3. In every triangle, the sum of any two internal angles is less than 180. Why is this so? Let α and β be the measures of any two internal angles of a triangle ΔABC (Figure 2.22) and extend $\overline{AC}$ letting $\gamma = m < BCD$. By EAT we will have $\gamma > \alpha$, so $\beta + \gamma > \alpha + \beta$. But $\beta + \gamma = 180$, consequently $\alpha + \beta < 180$.

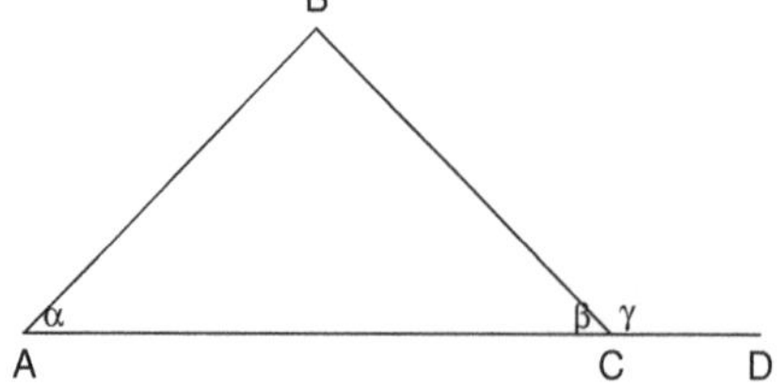

Figure 2.22: The sum of two internal angles

Basic Inequality (BI)

Given any triangle ABC, $a < b$ if and only if $\alpha < \beta$, where a is the length of the side opposite the angle with measure α and b is the length of the side opposite the angle with measure β (Figure 2.23).

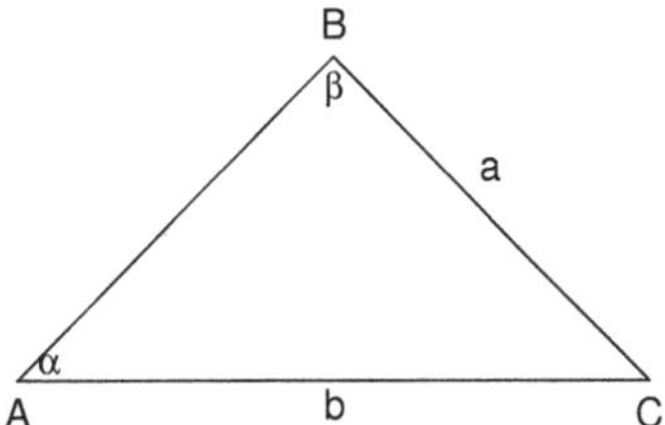

Figure 2.23: Basic inequality for triangles

Let us first prove that $a < b$ implies $\alpha < \beta$. With this purpose in mind we assume that $a < b$. Then we can choose a point D on $\overline{AC}$ such that $CD = a$ (Figure 2.24). By ITT, the two base angles of ΔCBD have the same measure γ. Applying EAT to ΔABD we get $\gamma > \alpha$. But $\beta > \gamma$ since D is an interior point of $\angle ABC$, consequently $\beta > \alpha$. Having proven the implication from the left to the right, it is time to prove

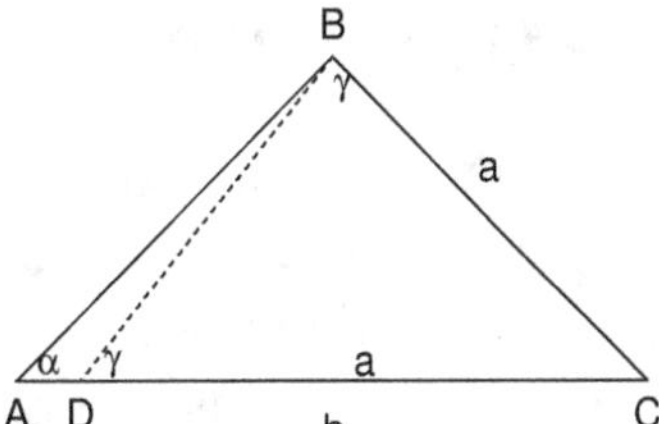

Figure 2.24: Proving the basic inequality for triangles

that $\alpha < \beta$ implies $a < b$. The logical contrapositive of the latter implication is simply

$$\text{If } b \leq a \text{ then } \beta \leq \alpha.$$

Let us assume that $b \leq a$. If $b = a$ we can conclude, using ITT, that $\alpha = \beta$, so $\beta \leq \alpha$. If $b < a$, thanks to the implication proven in the previous paragraph we will get that $\beta < \alpha$, so $\beta \leq \alpha$. QED

Remark

The reader can check that the same proof works for an obtuse triangle. The arguments used to prove BI do not depend on the fact that an acute triangle was chosen.

Triangle Inequality (TI)

In any triangle, the length of one side is less than the sum of the lengths of the two other sides. That is to say, given a triangle ABC, the following inequalities [9] are true : $a < b + c$, $b < a + c$, and $c < a + b$.

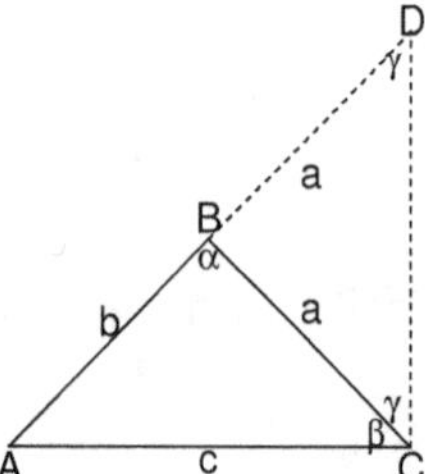

Figure 2.25: Proving the triangle inequality

Without loss of generality let us provide a proof of the inequality $c < a + b$. We start by extending $\overline{AB}$ so that $BD = a$ (Figure 2.25). Using ITT we get that $\angle BDC \cong \angle DCB$, with γ their common measure. Since $\beta + \gamma > \gamma$, the Basic Inequality (applied to ΔADC) implies that $c < a + b$. QED

The phenomenon of reflection of light

Let M be a plane mirror. A ray of light starts at A, hits the mirror at C and then passes through B (Figure 2.26). It has been experimentally determined that $\alpha = \beta$, where α is the measure of the angle of incidence and β is the measure of the angle of reflection[10]. We pose the following question: Has the ray of light chosen the path of least distance when going from A to B once it hits the mirror? Suppose that the ray of light were to choose the path from A to F and from F to B, where $F \neq C$. Without loss of generality we may suppose that $F > C$. Then prolong $\overline{BC}$ in such a way that it meets $\overrightarrow{AD}$, perpendicular to the mirror, at E. VAT implies that $\angle DCE \cong \angle BCF$, thus $\Delta ADC \cong \Delta EDC$ by ASA. Therefore

$$AD = ED \ \ and \ \ AC = EC. \tag{2.10}$$

Next build segment $\overline{EF}$. We realize that $\Delta ADF \cong \Delta EDF$ thanks to SAS. Hence

$$AF = EF. \tag{2.11}$$

[9] As usual, a denotes the length of the side opposite the vertex A, b denotes the length of the side opposite vertex B, and c denotes the length of the side opposite the vertex C.

[10] This is the well-known law of reflection of light.

The triangle inequality, applied to ΔEBF, implies that $EB < EF + FB$, that is, $EC + CB < EF + FB$. Using (2.10) and (2.11) we get $AC + CB < AF + FB$, thus proving that the ray of light has chosen the path of least distance.

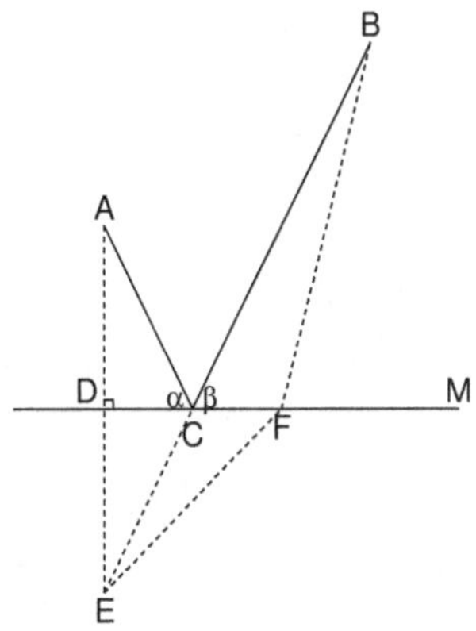

Figure 2.26: Reflection of light

Remark

Given a point P and a line L, P not on L, drop the perpendicular $\overline{PA}$ from P to L and prolong $\overline{PA}$ up to a point Q in such a way that $AQ = PA$ (Figure 2.27). The point Q is then known as the **reflection of** P **over** L.

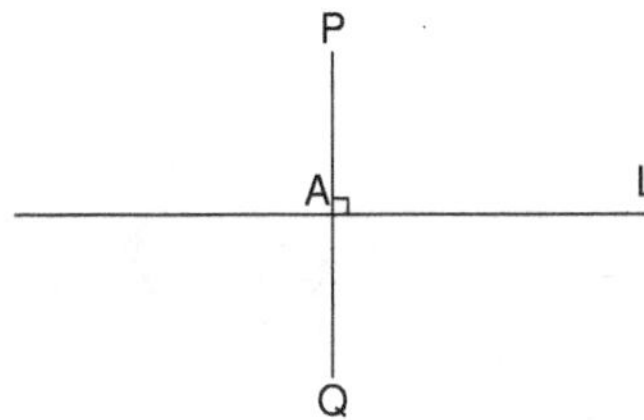

Figure 2.27: Reflecting a point through a line

2.8 AAS and HL congruence

It should not have come as a surprise that the Exterior Angle Theorem implies the triangle inequality; after all, EAT is a statement about two inequalities. However, we probably do not expect that EAT implies AAS congruence. The latter, like any congruence, is a statement about equalities.

Theorem (AAS congruence)

Given ΔABC and ΔDEF, assume that $\angle A \cong \angle D$, $\angle C \cong \angle F$ and $AB = DE$. Then $\Delta ABC \cong \Delta DEF$ (Figure 2.28).

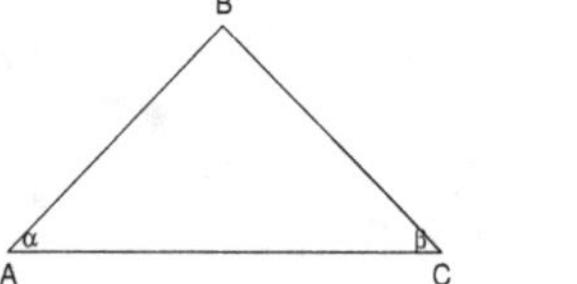

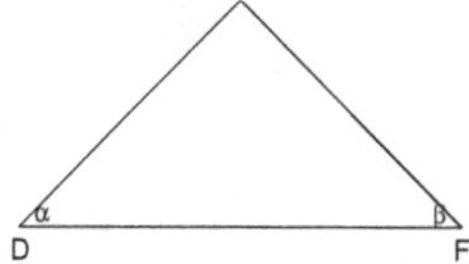

Figure 2.28: AAS congruence

Proof

If $AC = DF$ then both triangles are congruent by SAS. Suppose that $AC \neq DF$. Without loss of generality let us suppose that $DF < AC$, and choose G on $\overline{AC}$ such that $AG = DF$ (Figure 2.29). Then, thanks to SAS, we will have that $\Delta BAG \cong \Delta EDF$. In particular, $\angle AGB \cong \angle DFE$, hence $m\angle AGB = \beta$. Using EAT we get the inequality $\beta > \beta$, an impossibility. Thus, we must have $AC = DF$. As we mentioned at the beginning of the proof, under these circumstances $\Delta ABC \cong \Delta DEF$. QED

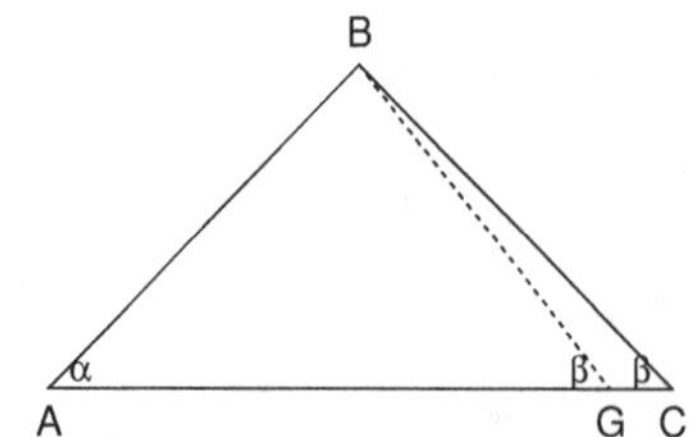

Figure 2.29: Proving AAS congruence

Corollary (HL congruence)

Let ΔABC and ΔDEF be right triangles with $AB = ED$ and $BC = EF$. Then $\Delta ABC \cong \Delta DEF$ (Figure 2.30). That is to say, two right triangles are congruent if their hypotenuses are congruent and a pair of legs are congruent too.

Figure 2.30: HL congruence

Proof

Prolong $\overline{AC}$ to the left so that $AG = DF$ and build the segment $\overline{GB}$ (Figure 2.31). Then SAS implies that

$$\Delta GAB \cong \Delta FDE, \tag{2.12}$$

consequently $GB = FE$. Since ΔGBC is isosceles, ITT implies that $\angle G \cong \angle C$; thus, using AAS we can assert that

$$\Delta GAB \cong \Delta CAB. \tag{2.13}$$

From (2.12) and (2.13) it follows that $\Delta FDE \cong \Delta CAB$, hence $\Delta ABC \cong \Delta DEF$. QED

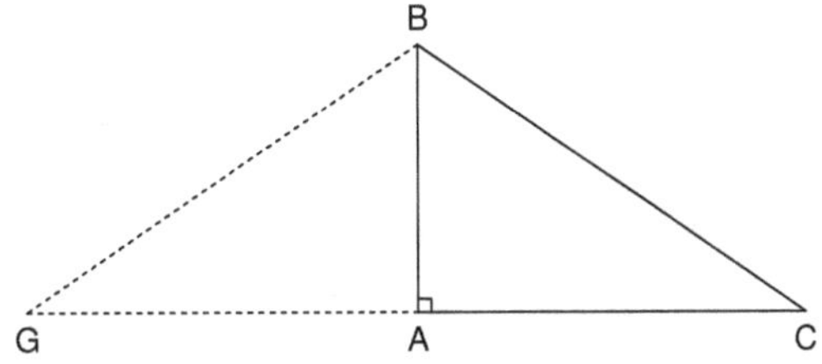

Figure 2.31: Proving HL congruence

2.9 Characterization of the angle bisector

Let us consider an arbitrary angle formed by the rays L_1 and L_2, which start at a point O, and let L be the angle bisector. Then

$$L = \{P : d(P, L_1) = d(P, L_2)\}.$$

That is, L is the set of all points whose distance to L_1 is equal to their distance to L_2.

Proof

As we did with the proof about the characterization of the perpendicular bisector of a segment, we have to divide the proof in two parts since it is necessary to show that one set is a subset of the other.

1. Assume that Q belongs to L and drop the perpendiculars $\overline{QM}$ and $\overline{QN}$ from Q to L_1 and L_2 respectively (Figure 2.32). By AAS we can assert that $\Delta MOQ \cong \Delta NOQ$, hence $MQ = NQ$ and therefore $d(Q, L_1) = d(Q, L_2)$. Thus, Q belongs to the set $\{P : d(P, L_1) = d(P, L_2)\}$.

2. Now assume that Q belongs to $\{P : d(P, L_1) = d(P, L_2)\}$, hence $d(Q, L_1) = d(Q, L_2)$. Next build the ray that starts at O and passes through Q (Figure 2.32). Using HL we have that $\Delta MOQ \cong \Delta NOQ$; thus, in particular, $\angle MOQ \cong \angle NOQ$. Then the above-mentioned ray is precisely the angle bisector$L$ of the given angle, so Q belongs to L. QED

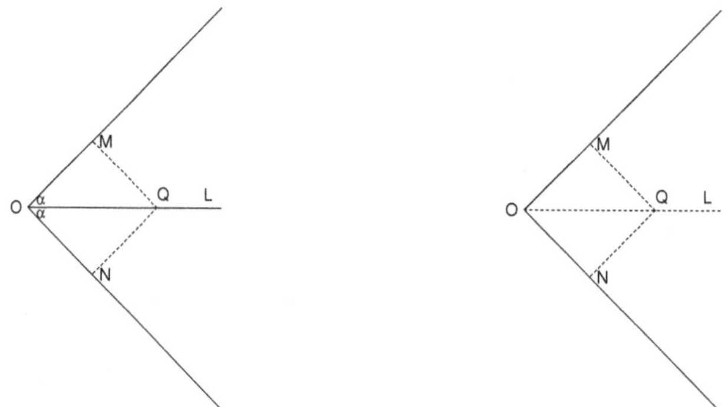

Figure 2.32: Proving the characterization theorem for angle bisectors

Proposition

The three angle bisectors of any triangle meet at a point.

Proof

Given an arbitrary triangle ABC let I be the point of intersection of the angle bisectors of $\angle B$ and $\angle C$. Then drop the perpendicular segments $\overline{IF}$, $\overline{IG}$, and $\overline{IH}$ to the sides $\overline{AB}$, $\overline{BC}$, and $\overline{AC}$, respectively. The characterization theorem for angle bisectors, applied twice, implies that $IF = IG$ and $IG = IH$; thus, $IF = IH$. A third application of the above-mentioned theorem implies that I belongs to the bisector of $\angle A$. We have been able to show that the bisector of $\angle A$ passes through I. This point of intersection of the three angle bisectors is called the **incenter** of the triangle. QED

2.10 SsA congruence

There is no SSA congruence property. For instance (Figure 2.33) ΔABC and ΔABD, with $BC = BD$, have two sides of the same length and share $\angle A$. However, they are obviously not congruent. Nonetheless, if we add an inequality then congruence is established.

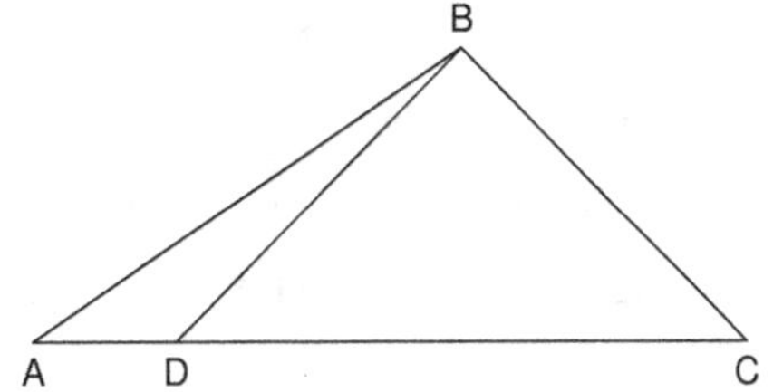

Figure 2.33: There is no SSA congruence property

Theorem (SsA congruence)

Let ΔABC and ΔDEF be given, with $BC = EF$, $AB = DE$, $\angle A \cong \angle D$, and $BC \geq AB$. Under these hypotheses it is true that $\Delta ABC \cong \Delta DEF$ (Figure 2.34).

Figure 2.34: The SsA property of congruence

Proof

If $AC = DF$ then $\Delta ABC \cong \Delta DEF$ by SAS and we are done. Assume that $AC \neq DF$, say $DF < AC$. Hence, we can choose a point G on $\overline{AC}$ such that $AG = DF$ (Figure 2.35). Then $\Delta AGB \cong \Delta DFE$ by SAS and, in particular, $GB = FE$. But $BC = FE$ by hypothesis, so $BG = BC$; therefore ΔCBG is isosceles. Thus ITT implies that

$$\alpha = m\angle CGB = m\angle GCB.$$

Defining $\beta = m\angle CAB$, and applying EAT to ΔGAB and exterior angle CGB, we can conclude that $\alpha > \beta$.

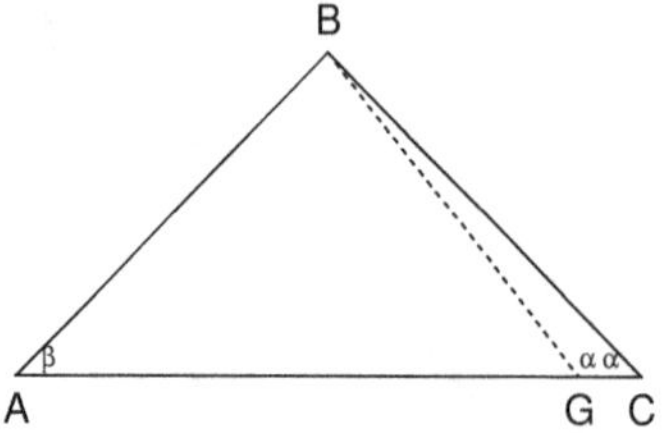

Figure 2.35: Proving the SsA congruence property

Next we pay attention to ΔABC. Since $\alpha > \beta$, the Basic Inequality (BI) implies that $AB > AC$, which is an impossibility since $BC \geq AB$ by hypothesis. Therefore we cannot have $AC \neq DF$ and consequently, as we mentioned at the beginning of the proof, it must be true that $\Delta ABC \cong \Delta DEF$. QED

A new proof of HL congruence

SsA congruence provides a path to a simple proof of HL congruence . Indeed, let ΔABC and ΔDEF be right triangles, with right angles at A and D respectively, and such that $BC = EF$, $AB = DE$ (Figure 2.36). Letting $\alpha = m\angle BCA$ we observe that

$\alpha + 90 < 180$ since the sum of the measures of any two interior angles of a triangle is less than 180, a result that was proved in the second remark after the proof of the Exterior Angle Theorem. Hence $\alpha < 90$, which in turn implies, thanks to the Basic Inequality, that $BC > AB$. Then SsA leads to $\Delta ABC \cong \Delta DEF$. QED

Figure 2.36: A new proof of HL congruence

2.11 The Hinge Theorem

Let us consider ΔABC and ΔDEF , with $AB = DE$ and $AC = DF$. Furthermore, assume that $m\angle A > m\angle D$. Then $BC > EF$ (Figure 2.37).

Figure 2.37: The Hinge Theorem

Proof

This theorem belongs to the category of those results in plane geometry that look so obvious that we might think that a proof is superfluous. Nonetheless, a proof needs to be given. Let us start by constructing $\angle PAC$ so that $m\angle PAC = m\angle D$ and choose Q on $\overrightarrow{AP}$ in such a way that $AQ = DE$ (Figure 2.38). Using SAS we can assert that $\Delta AQC \cong \Delta DEF$, which in turn leads to $QC = EF$.

Let $\overrightarrow{AR}$ be the angle bisector of $\angle BAQ$. According to the Crossbar Theorem, $\overrightarrow{AR}$ will intersect $\overline{BC}$ at a certain point S. Then SAS implies that $\Delta ASB \cong \Delta ASQ$, so $SB = SQ$. The Triangle Inequality, applied to ΔSQC, allows us to conclude that $QS + SC > QC$. But we proved that $QS = BS$ and $QC = EF$, consequently

$BS + SC > EF$. Finally, since $BS + SC = BC$ we reach the inequality that we want to prove, namely $BC > EF$. QED

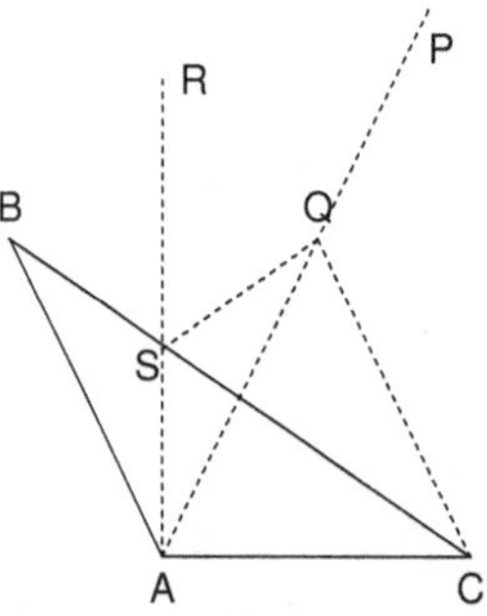

Figure 2.38: Proving the Hinge Theorem

Remarks

1. The reader must have noticed that we used the Crossbar Theorem in the previous proof. In section 2.2, when discussing ITT we preferred to avoid its use because there were alternatives like Euclid's original approach. Now we do not have to hesitate since the Crossbar Theorem is essential in the quest for a proof of the Hinge Theorem. Of course, one might think that we are using, without proof, an evident mathematical proposition in order to prove another evident mathematical proposition. However, in any logical ordering of the subject, the Crossbar Theorem comes before the Hinge Theorem.

2. A sort of converse of the Hinge Theorem is valid: Let ΔABC and ΔDEF be given, with $AB = DE$, $AC = DF$, and $BC > EF$. Then $m\angle A > m\angle D$. We will prove the logical contrapositive, namely, if $m\angle A \leq m\angle D$ then $BC \leq EF$. Assume that $m\angle A \leq m\angle D$. There are two possibilities: either $m\angle A = m\angle D$ or $m\angle A < m\angle D$. If $m\angle A = m\angle D$ then SAS congruence implies that $\Delta ABC \cong \Delta DEF$, hence $BC = EF$, which in turn implies that $BC \leq EF$. On the other hand, if $m\angle A < m\angle D$ then the Hinge Theorem implies that $BC < EF$, so $BC \leq EF$. QED

Exercises

1. Given triangles ABC and ADC with $AB = BC$ and $AD = DC$, show that $\overrightarrow{BD}$ is the angle bisector of $\angle ABC$ and $\angle ADC$.

2. Given a quadrilateral $ABCD$ build its diagonals AC and BD, which meet at a point O. Assuming that $AO = OC$, $BO = OD$, and $DA = DC$, show that $AB = BC$ and find the measure of $\angle AOB$.

3. Given a quadrilateral $ABCD$ let O be the point of intersection of the diagonals. Assuming that $AO = BO$, $CO = DO$ and $AD = 5$, find BC and show that $\angle DAB \cong \angle ABC$.

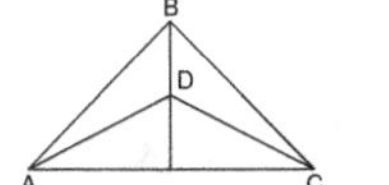

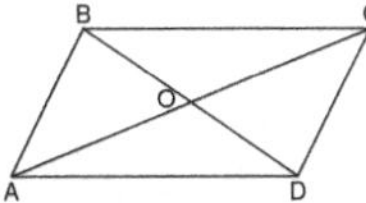

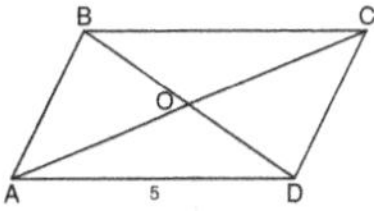

Figure 2.39: Figures related to exercises 1, 2, 3.

4. Assume that $BC = BD$ and $\angle ABC \cong \angle FBD$. Can we assert that ΔABF is isosceles?

5. It is given that a "kite" $ABCD$ is built in such a way that $\angle BAC \cong \angle BCA$ and $\angle DAC \cong \angle DCA$. Assuming that the diagonals meet at a point E, show that $\overleftrightarrow{ED}$ is the perpendicular bisector of $\overline{AC}$.

6. Triangle ABC is isosceles with $AB = BC$. Two of its medians, $\overline{CD}$ and $\overline{AE}$, meet at point F. Given that $\overrightarrow{BF}$ intersects $\overline{AC}$ at G, show that $\overleftrightarrow{BG}$ is the perpendicular bisector of $\overline{AC}$.

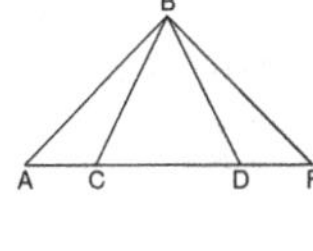

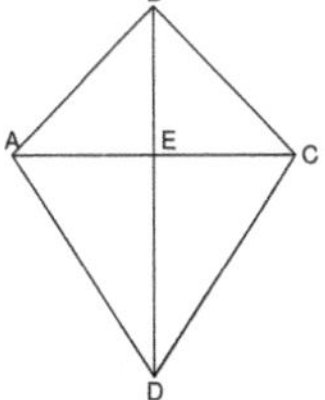

Figure 2.40: Figures related to exercises 4,5.

7. Let $\overline{AE}$ and $\overline{DC}$ be the bisectors of the base angles of the isosceles ΔABC, where $AB = BC$. Assuming that the angle bisectors meet at F and $\overrightarrow{BF}$ intersects $\overline{AC}$ at G, show that $\overleftrightarrow{BG}$ is the perpendicular bisector of $\overline{AC}$.

8. Triangle ABC is isosceles, with $AB = BC$. Assuming that $\alpha = \beta$ and $\gamma = \delta$, show that $\overline{BD} \perp \overline{AC}$.

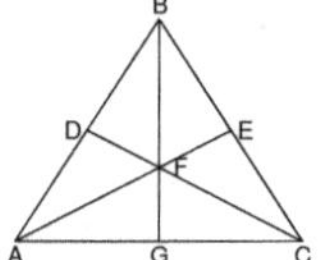

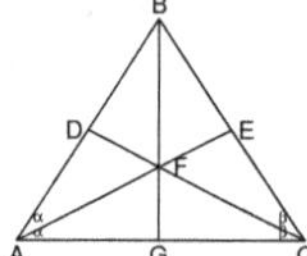

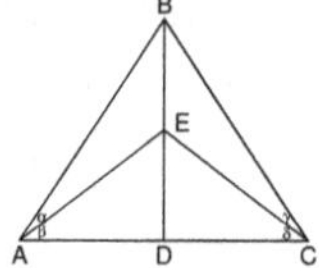

Figure 2.41: Figures related to exercises 6,7,8.

9. It is given that $GI = IH = FI$. Show that the perpendicular bisectors of $\overline{GF}$, $\overline{GH}$, and $\overline{FH}$ pass through I.

10. Show that $BD < BC$ (Figure 2.42)(Hint: Use the basic inequality (BI) on ΔBDC after applying EAT on ΔBAD.)

11. Let us consider an isosceles triangle, with $AB = BC$ (Figure 2.42). Show that $BD > AB$.

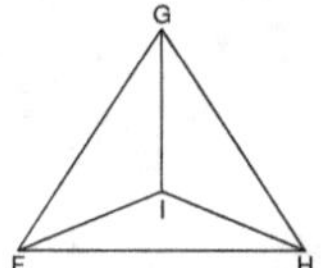

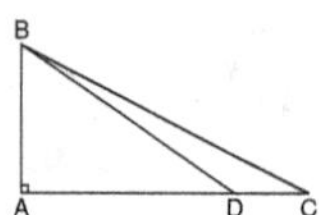

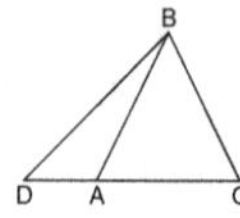

Figure 2.42: Figures related to exercises 9,10, 11.

12. Let us consider a simple quadrilateral (see Figure 3.12). Show that its perimeter is bigger than the sum of the lengths of its two diagonals, and that this sum is bigger than half the perimeter (Hint: Make use of the triangle inequality several times.)

13. Let us consider ΔABC and ΔDEF such that $AB = DE$, $BC > EF$, and $m\angle ABC = m\angle DEF \geq 90$. Show that $CA > FD$ (Hint: Extend $\overline{EF}$ up to a point G so that $EG = BC$).

14. We have $\angle ABC$ and its angle bisector $\overrightarrow{BD}$. Choose two arbitrary points P and Q that lie on $\overrightarrow{BA}$ and $\overrightarrow{BC}$, respectively (Figure 2.44). Let R be the intersection of $\overrightarrow{BD}$ and $\overrightarrow{PQ}$. Show that $QR < BQ$ and $PR < BP$ (Hint: EAT and BI, applied properly, will do the job).

Figure 2.43: Exercise 13

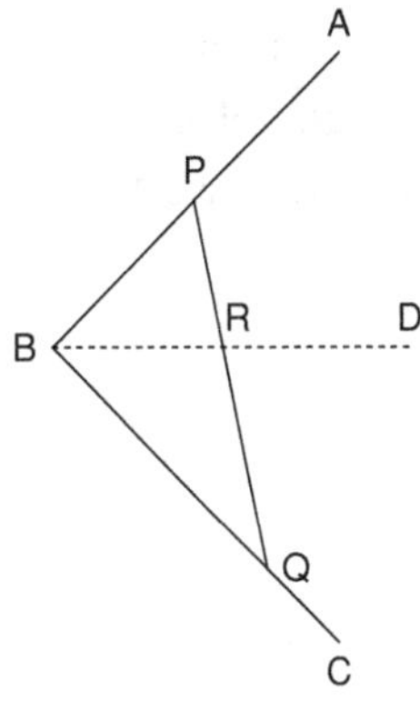

Figure 2.44: Exercise 14

15. Use the preceding exercise to develop a second proof, different from the one analyzed in section 2.7, of the triangle inequality.

16. Assume that $\overline{BD}$ is an angle bisector and a median of the obtuse triangle ABC (with $\angle ABC > 90$). Show that it must be isosceles[11], that is, $AB = BC$ (Hint: Let E be the midpoint of $\overline{AC}$. Then extend $\overline{BE}$ in such a way that $BE = ED$.)

17. Prove SSS congruence under the assumption that one of the triangles is a right triangle.

18. Assuming that the sum of the internal angles of any triangle is 180, show that the angles at the top of any Saccheri quadrilateral are right angles.

19. Assuming that the sum of the internal angles of any triangle is less than 180, show that the angles at the top of any Saccheri quadrilateral have a measure of less than 90.

[11]This very same problem was solved for acute triangles at the beginning of section 2.4.

20. Assuming that a ray of light travels from P to Q touching a plane mirror, and does so choosing the shortest path, show that the angle of incidence is congruent to the angle of reflection (Hint: Drop the perpendicular $\overline{PC}$, where C lies on the mirror. Then prolong it up to a point B so that $PC = CB$. That is, B is the reflection of P over the line that represents the plane mirror. Next build $\overline{BQ}$.)

Enrichment Note: Saccheri-Legendre Theorem

We learned in section 2.7 (second remark, right after the proof of EAT) that the sum of the measures of two angles of a triangle is less than 180. It is our purpose to prove that, in neutral geometry, the sum of the measures of the three angles of any triangle is less than or equal to 180. Firstly, we need a preliminary result.

Lemma

Given any triangle ABC , with angle measures α, β, γ, there exists a triangle ACE such that

$$\alpha_1 + \beta_1 + \gamma_1 = \alpha + \beta + \gamma \text{ and } \alpha_1 \leq \tfrac{\alpha}{2},$$

where $\alpha_1, \beta_1, \gamma_1$ are the measures of the internal angles of ΔACE.

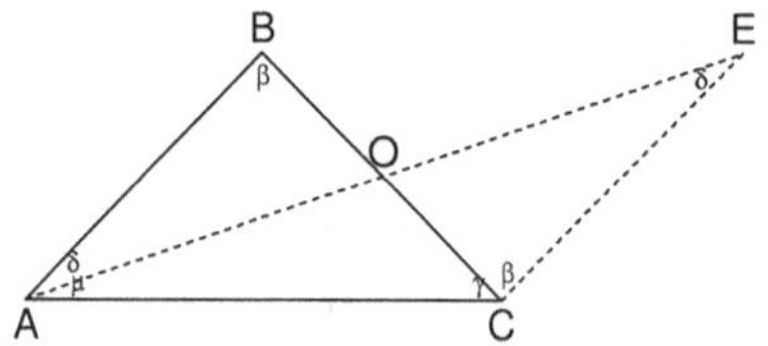

Figure 2.45: A lemma

Proof

Let us start by choosing the midpoint O of $\overline{BC}$ and build $\overline{AE}$ in such a way that $AO = OE$ (Figure 2.45). Defining $\delta = m\angle BAO$ and $\mu = m\angle CAO$ we get that $\alpha = \delta+\mu$. Due to SAS congruence it follows that $\Delta ABO \cong \Delta ECO$, thus $m\angle OEC = \delta$ and $m\angle ECO = \beta$. We observe that

$$\alpha + \beta + \gamma = \delta + \mu + \beta + \gamma.$$

It must be true that $\delta \leq \frac{\alpha}{2}$ or $\mu \leq \frac{\alpha}{2}$ because otherwise $\delta > \frac{\alpha}{2}$ and $\mu > \frac{\alpha}{2}$ implies $\delta + \mu > \alpha$, that is, $\alpha > \alpha$.

If $\delta \leq \frac{\alpha}{2}$ define $\alpha_1 = \delta$, $\beta_1 = \mu$, $\gamma_1 = \gamma + \beta$. On the other hand, if $\mu \leq \frac{\alpha}{2}$ define $\alpha_1 = \mu$, $\beta_1 = \delta$, $\gamma_1 = \gamma + \beta$. We have then proven that $\alpha + \beta + \gamma = \alpha_1 + \beta_1 + \gamma_1$ and $\alpha_1 \leq \frac{\alpha}{2}$. QED

Theorem (Saccheri-Legendre)

Given any triangle,

$$\alpha + \beta + \gamma \leq 180,$$

where α, β, γ are the measures of the corresponding internal angles.

Proof

It is enough to show that

$$\alpha + \beta + \gamma < 180 + \epsilon \text{ for every } \epsilon > 0.$$

Let $\epsilon > 0$ be given. Applying the preceding lemma we can assert that there exists a triangle Δ_1 such that

$$\alpha + \beta + \gamma = \alpha_1 + \beta_1 + \gamma, \; \alpha_1 \leq \tfrac{\alpha}{2}.$$

Applying the lemma again, there exists a triangle Δ_2 such that

$$\alpha_1 + \beta_1 + \gamma_1 = \alpha_2 + \beta_2 + \gamma_2, \; \alpha_2 \leq \tfrac{\alpha_1}{2}.$$

Hence $\alpha_2 \leq \frac{\alpha}{2^2}$. This process is repeated indefinitely; thus, for any positive integer n there exists a triangle Δ_n such that

$$\alpha + \beta + \gamma = \alpha_n + \beta_n + \gamma_n, \quad \alpha_n \leq \frac{\alpha}{2^n}. \tag{2.14}$$

Next choose n such that $\frac{\alpha}{2^n} < \epsilon$. Obviously $\alpha_n < \epsilon$, hence

$$\alpha_n + \beta_n + \gamma_n < \epsilon + \beta_n + \gamma_n.$$

But in neutral geometry the sum of the measures of any two internal angles is less than two right angles; thus, $\beta_n + \gamma_n < 180$. Consequently

$$\alpha_n + \beta_n + \gamma_n < 180 + \epsilon.$$

Then, from (2.14), it follows that

$$\alpha + \beta + \gamma < 180 + \epsilon \quad \text{for every } \epsilon > 0.$$

Therefore[12]

$$\alpha + \beta + \gamma \leq 180.$$

QED

In the next chapter we will prove that if we assume the Axiom of Parallels[13] then the sum of the measures of the internal angles of any triangle is 180. Moreover, Legendre's theorem (enrichment note at the end of chapter 3) asserts that, in neutral geometry, if the sum of the measures of the internal angles of any triangle is 180, then the Axiom of Parallels is true. Girolamo Saccheri (1667-1733) and Adrien Legendre (1752-1833) did, without having been their purpose, pioneering work in what nowadays is called neutral geometry.

[12] We are using a very simple fact, namely, if $a < b + \epsilon$ for every $\epsilon > 0$ then $a \leq b$.

[13] That is, if a line L is given and P is a point not on L, at most one line parallel to L passes through P.

Chapter 3

Parallelism

At the beginning of chapter 2 we defined the notion of parallelism between two lines: L_1 and L_2 are parallel, in symbols $L_1 \parallel L_2$, if and only if they have no points in common, that is, they never meet.

3.1 Basic results

Theorem

Let L_1, L_2, and L be lines such that $L_1 \perp L$ and $L_2 \perp L$. Then $L_1 \parallel L_2$.

Proof

Proceeding by contradiction assume that L_1 and L_2 meet at a point P. Then ΔAPB is born (Figure 3.1). But we have reached an impossibility since from the point P we cannot drop two different perpendicular segments to L (such a scenario contradicts EAT). QED

Figure 3.1: Perpendicularity and parallelism

Corollary

Let L_1 be a line and let P be a point that does not lie on it. Then there exists a line L_2 that passes through P and is parallel to L_1.

Proof

Let L be the perpendicular line to L_1 that passes through P (in section 2.5 we proved that such a line exists; moreover, we showed how to construct it). Then build[1] the line L_2 that passes through P and is perpendicular to L (Figure 3.2). The preceding theorem implies that $L_1 \parallel L_2$. QED

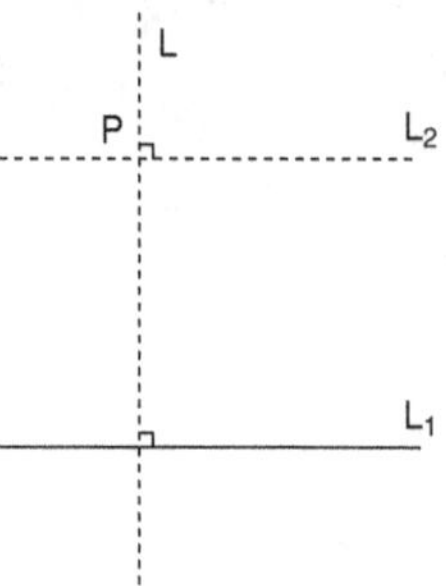

Figure 3.2: A corollary

Definition

Given two lines L_1 and L_2 we will say that a line L is a **transversal** of both lines if L meets L_1 and L_2 at different points.

Definition

Let us consider lines L_1 and L_2 and a transversal L (Figure 3.3). The pairs (γ, ρ) and (μ, δ) are called **alternate interior angles**[2] . The pairs (α, μ), (β, ρ), (γ, ξ), and (δ, ω) are called **corresponding angles** while the pairs (γ, μ) and (δ, ρ) are the **interior angles** of the system of lines L, L_1, L_2.

[1]The axiom of construction of angles justifies this step.

[2]Usually we use lower case letters from the Greek alphabet to denote the measure of an angle rather than the angle itself, but sometimes they are used to denote both the angle and its measure.

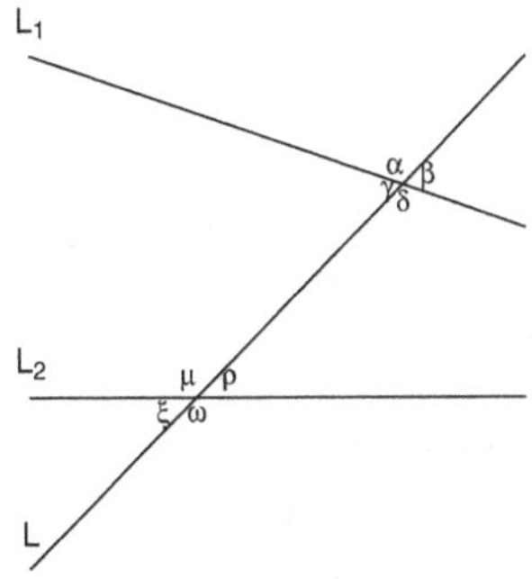

Figure 3.3: A transversal line

Theorem

If a pair of alternate interior angles are congruent then $L_1 \parallel L_2$.

Proof

We will prove the logical contrapositive: if L_1 is not parallel to L_2 then any pair of alternating interior angles are not congruent. Assume that L_1 is not parallel to L_2 and let α and β be the measures of a pair of alternating interior angles. Since L_1 is not parallel to L_2 we can assert that both lines intersect. Suppose that they do so to the right of the transversal (Figure 3.4). Then the Exterior Angle Theorem implies that $\alpha > \beta$, so $\alpha \neq \beta$. If L_1 and L_2 intersect to the left of the transversal, a similar argument leads to $\alpha \neq \beta$. QED

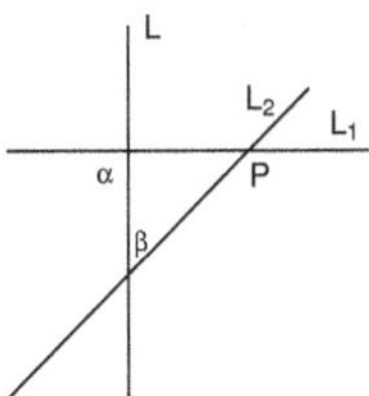

Figure 3.4: Two lines are parallel if a pair of alternating angles are congruent

Corollaries

1. Given lines L_1, L_2, and a transversal L, if two corresponding angles are congruent then $L_1 \parallel L_2$.

2. Given lines L_1, L_2, and a transversal L, suppose that the sum of the measures of two internal angles is 180. Then $L_1 \parallel L_2$.

The proofs of both corollaries is straightforward. To justify the first corollary we need to apply VAT and the preceding theorem, while for the second corollary it is a matter of applying the linear pair axiom and the preceding theorem.

Remark

Everything we have discussed so far in this chapter belongs to the realm of neutral geometry. However, to prove the converse of the last theorem, namely, if $L_1 \parallel L_2$ then both pairs of alternating angles are congruent, the axiom of parallels is needed.

Axiom of Parallels

Through a point P, not lying on a line L, at most one parallel to L can be drawn.

Comment

In the first corollary of the present section we proved that at least one parallel to L, passing through P, can be drawn. The Axiom of Parallels then implies that one, and only one, parallel to L and passing through P exists. Thus, if L_1 and L_2 pass through P and are parallel to L, it must be the case that $L_1 = L_2$. The version of the axiom that we are using goes back to John Playfair (1748-1819). It can be proven that it is equivalent to Euclid's 5th postulate[3] . **From now on we will accept the Axiom of Parallels**. Thus, we will immerse ourselves in the realm of Euclidean geometry.

Theorem

Assume that $L_1 \parallel L_2$ and L is a transversal. Then any pair (α, β) of alternating internal angles are congruent.

[3] See the third remark at the end of the present section.

Proof

We will prove the contrapositive, namely if $\alpha \neq \beta$ then L_1 is not parallel to L_2. Assume that $\alpha \neq \beta$, say $\alpha < \beta$. Then build the line L_3 that passes through P and makes an angle β with the transversal (Figure 3.5); evidently, $L_3 \neq L_1$ because otherwise we would have $\alpha = \beta$. Since L_3 makes an angle of measure β with the

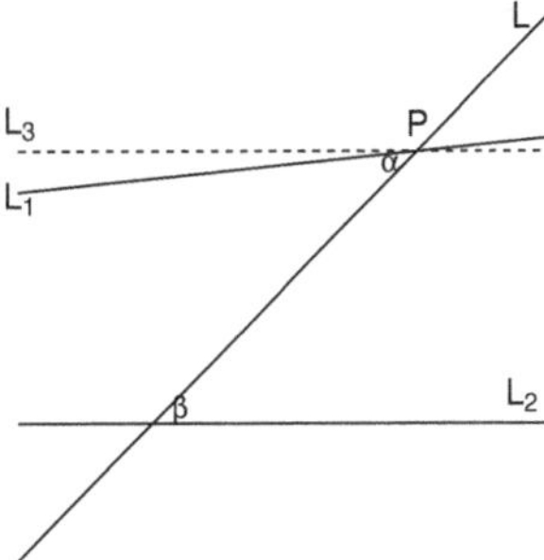

Figure 3.5: Proving that parallelism implies equality of alternating internal angles

transversal, which happens to be congruent with the angle that L_2 makes with the same transversal, and both are alternating interior angles, a well-known theorem from neutral geometry[4] implies that $L_3 \parallel L_2$. Then, using the Axiom of Parallels, we can conclude that L_1 is not parallel to L_2 because otherwise we would have two distinct lines (L_1 and L_3) that pass through P and are parallel to L_2. QED

Four Corollaries

1. Assume that $L_1 \parallel L_2$ and L is a transversal. Then $\alpha + \beta = 180$, where α and β are the measures of any pair of interior angles.

   The justification of this corollary is straightforward: Let γ be the measure of the angle that forms a linear pair with the angle whose measure is α; thus, γ and β are the measures of a pair of alternating interior angles. Since $L_1 \parallel L_2$, the preceding theorem implies that $\beta = \gamma$. But $\gamma + \alpha = 180$, consequently $\alpha + \beta = 180$. QED

2. Suppose that $L_1 \parallel L_2$ and L is a transversal. If $L \perp L_1$ then $L \perp L_2$.

[4]Proven in this section, before the Axiom of Parallels was established.

This corollary follows directly from the previous theorem. Indeed, any pair of alternating internal angles will have to have the same measure. Then VAT implies that this common measure is precisely 90. QED

3. Assume that L intersects L_1 at P, $L \neq L_1$, and $L_1 \parallel L_2$. Then L must intersect L_2.

   If L were not to intersect L_2 then $L \parallel L_2$. Thus, L and L_1 are parallel to L_2, pass through P, and are distinct. This is a flagrant violation of the Axiom of Parallels[5]. QED

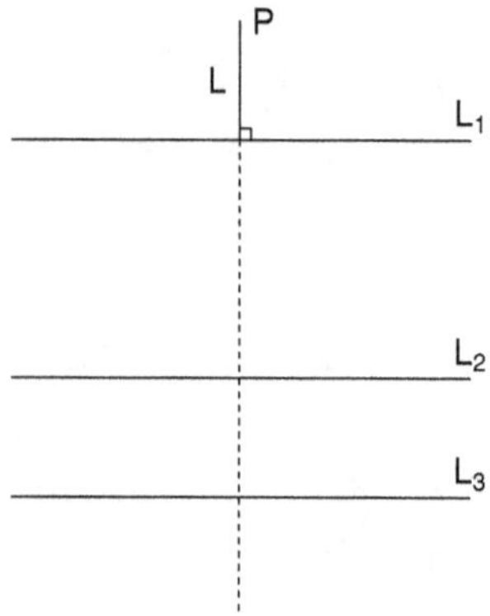

Figure 3.6: The transitivity of parallelism

4. Suppose that $L_1 \parallel L_2$ and $L_2 \parallel L_3$. Then $L_1 \parallel L_3$.

   This corollary asserts that parallelism is a transitive relation. To justify it let us choose a point P 'above' L_1 and let L be the perpendicular to L_1 that passes through P (Figure 3.6). Since $L_1 \parallel L_2$, the preceding corollary implies that L, once extended, will intersect L_2. In a similar fashion, since L intersects L_2 and $L_2 \parallel L_3$, it follows that L will intersect L_3 too. Thus, L is a transversal both of L_1 and L_2, and L_2 and L_3. The first corollary of the present list implies that $L \perp L_2$. Then the same corollary implies that $L \perp L_3$. Finally, since $L_1 \perp L$ and $L_3 \perp L$, the first theorem of this section leads to the conclusion that $L_1 \perp L_3$. QED

[5]The second corollary does not stem from the previous theorem but is a direct consequence from the Axiom of Parallels. It has been placed here because it will be needed to prove the next corollary.

Theorem (The sum of the internal angles)

A very important theorem asserts that the sum of the measures of the internal angles, of any triangle, is 180.

Most readers must have seen the proof of this proposition in High School: Let L be the line that goes through C and is parallel to $\overleftrightarrow{AB}$ (Figure 3.7). Since α and δ are alternating interior angles we can assert that $\delta = \alpha$. Similarly, μ and β are alternating interior angles, so $\mu = \beta$. But $\delta + \gamma + \mu = 180$, consequently $\alpha + \beta + \gamma = 180$. QED

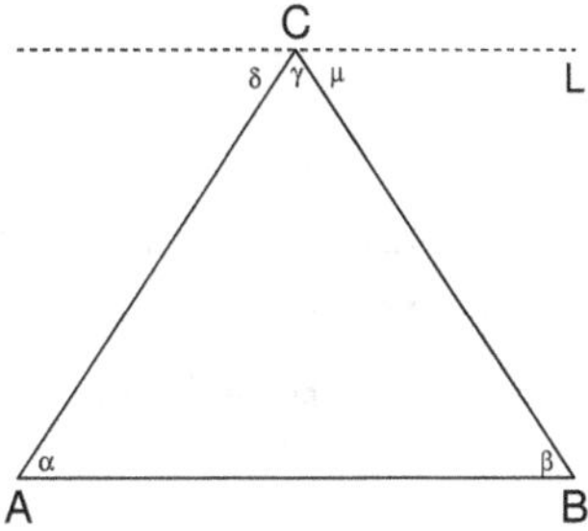

Figure 3.7: The sum of the internal angles

Remarks

1. Right away we realize that given an equilateral triangle, that is, a triangle with its three sides congruent, its three internal angles have to be congruent. If α is their common measure it follows that $3\alpha = 180$; thus, $\alpha = 60$.

2. Given an arbitrary triangle (Figure 3.8), the measure of any exterior angle (ρ) is equal to the sum of the measures of the two non-adjacent interior angles (β and γ), that is,

$$\rho = \beta + \gamma.$$

   The acronym for this property is **SEAT**. The justification of SEAT is very simple[6]: $\rho + \alpha = 180 = \alpha + \beta + \gamma$, consequently $\rho = \beta + \gamma$.

3. In *The Elements*, Euclid put forward the 5th postulate: *If a straight line falling on two straight lines makes the interior angles on the same side less than two right angles, then the two straight lines, if produced indefinitely meet on that side on which are the angles less than two right angles.* That is, given lines L_1

[6] In neutral geometry we have EAT but not SEAT.

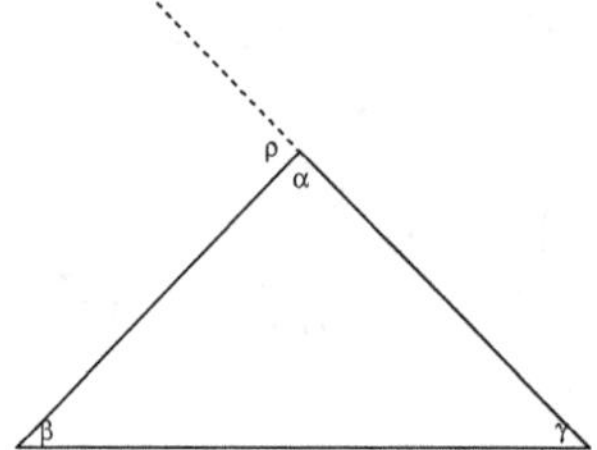

Figure 3.8: SEAT property

and L_2 and a transversal L, and a pair of interior angles with measure α, β, if $\alpha + \beta < 180$ then L_1 and L_2 will intersect on the side where the internal angles are given (Figure 3.9). We will prove that the Axiom of Parallels (Playfair's version) is equivalent to Euclid's 5th postulate.

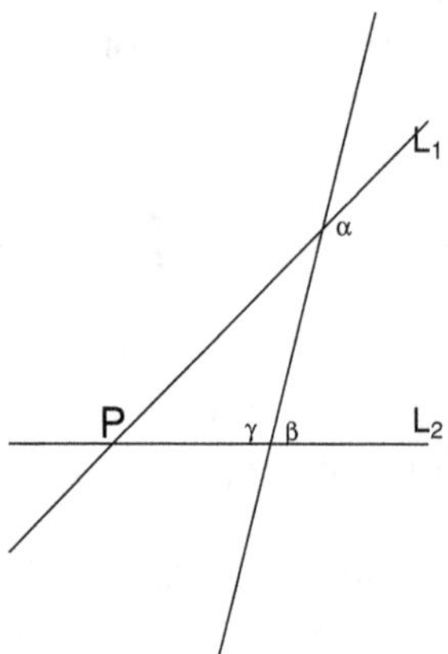

Figure 3.9: The Axiom of Parallels implies Euclid's $5th$

(a) Let us assume the Axiom of Parallels and the antecedent of Euclid's 5th, namely $\alpha + \beta < 180$. Then L_1 and L_2 have to intersect because otherwise they would be parallel, which in turn implies that $\alpha + \beta = 180$. It only remains to show that L_1 and L_2 intersect on the same side of the angles with measure α and β. Suppose that they were to intersect on the other side, at a point P (Figure 3.9). Since $\gamma + \beta = 180$ and $\alpha + \beta < 180$ we can conclude that $\alpha + \beta < \gamma + \beta$, so $\alpha < \gamma$. This is an impossibility due to the fact that EAT implies that $\alpha > \gamma$. Thus, L_1 and L_2 meet on the same side of the interior angles being considered.

(b) Let us prove that Euclid's 5th implies the Axiom of Parallels. With this purpose in mind, assume that L_1 is any line and P is a point exterior to it. Then construct a line, L_3, that passes through P and is perpendicular to L_1 (recall the third construction of section 2.5) and build a line L_2 that passes through P and is perpendicular to L_3; then $L_2 \parallel L_1$. Let L, $L \neq L_2$, be any line that passes through P. Since $L \neq L_2$, obviously $\alpha < 90$ where α is the measure of the acute angle made by L and the transversal L_3 (Figure 3.10 and Figure 3.11); thus, $\alpha + 90 < 180$. By Euclid's 5th it then follows that L and L_1 must meet, that is, they cannot be parallel. We have just succeeded in showing that there is no line, different from L_2, that passes through P and is parallel to L_1. QED

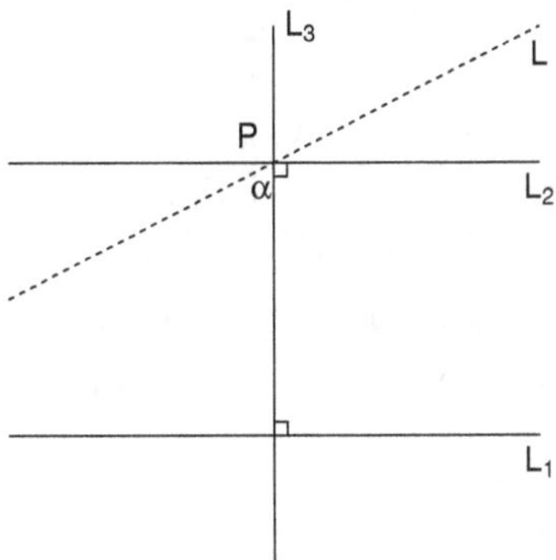

Figure 3.10: Euclid's 5th implies the Axiom of Parallels (first possibility)

4. Euclid's 5th postulate does not have the simplicity of the first four postulates, namely:

   (a) Between any two points a line can be drawn.

   (b) Any segment can be extended indefinitely, thus producing a line.

   (c) Given an arbitrary segment, a circle can be drawn having the segment as its radius and one endpoint of the segment as its center.

   (d) All right angles are congruent.

   This lack of simplicity of the fifth postulate, when compared to the other four postulates, led many mathematicians to wonder whether it could be proven to be a theorem. In the 19th century the answer was found, following the

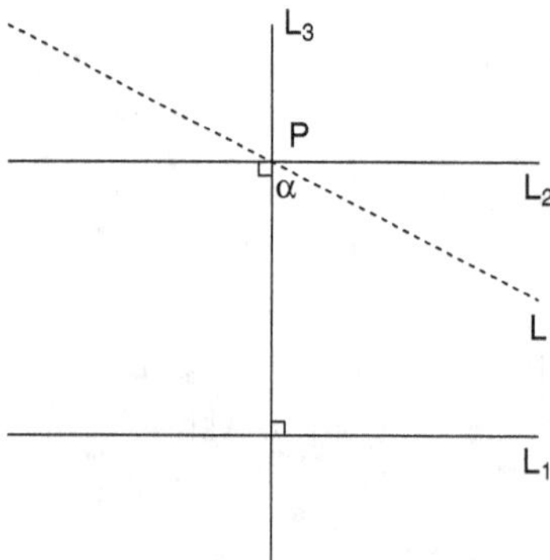

Figure 3.11: Euclid's 5th implies the Axiom of Parallels (second possibility)

pioneering works of Nikolai Lobachevsky (1792-1856) and Janos Bolyai (1802-1860): The Axiom of Parallels is independent of the other axioms of geometry and consequently it is needed to build the body of Euclidean geometry. If we add the negation of the Axiom of Parallels then hyperbolic geometry is born, which has the same 'right of existence' as Euclidean geometry.

3.2 Parallelograms

Given four points A, B, C, D, no three of which are collinear, a **quadrilateral** is the union of the segments $\overline{AB}, \overline{BC}, \overline{CD}$, and $\overline{DA}$, which are the **sides** of the geometric figure under consideration. The points A, B, C, D are called **vertices**, and $\overline{AB}, \overline{CD}$ and $\overline{BC}, \overline{AD}$ are the **opposite sides**. Moreover, $\overline{AC}$ and $\overline{BD}$ are the **diagonals** [7]. A **parallelogram** is a quadrilateral whose opposite sides are parallel.

Two basic properties of parallelograms

1. Given a parallelogram $ABCD$, its opposite sides are congruent and its opposite angles are congruent.

   The proof is quite simple. Start by drawing the diagonal $\overline{AC}$ (Figure 3.13). Since $\overline{BC} \parallel \overline{AD}$ we can conclude that $\angle BCA \cong \angle CAD$. Similarly, since $\overline{AB} \parallel \overline{CD}$ it follows that $\angle BAC \cong \angle ACD$. Then, by ASA congruence we will have

[7] We will deal only with 'simple quadrilaterals', in the sense that each side lies in one of the half planes determined by the opposite side (Figure 3.12). These quadrilaterals have the property that their diagonals intersect.

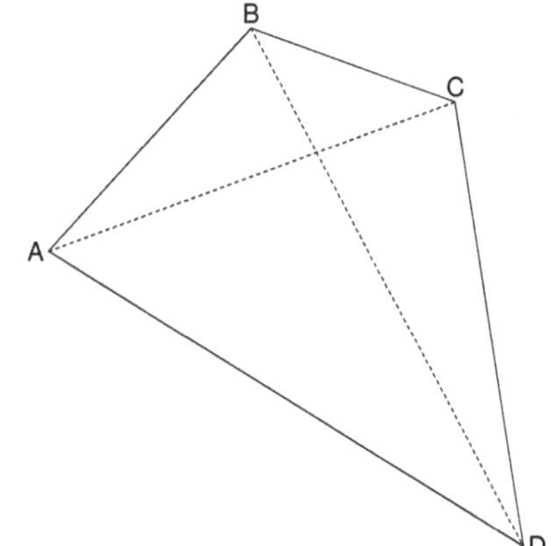

Figure 3.12: A simple quadrilateral

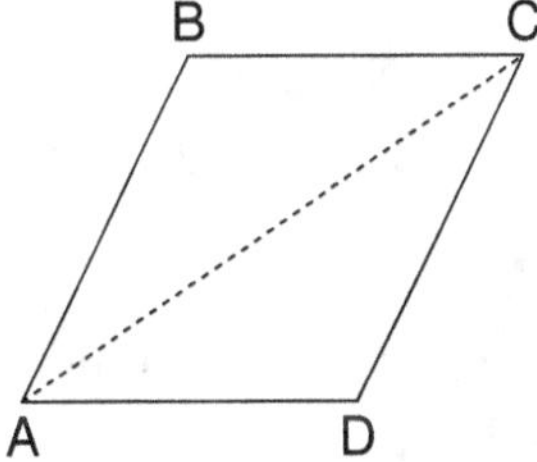

Figure 3.13: Opposite sides and opposite angles are congruent

that $\Delta ABC \cong \Delta CDA$. Thus $AB = CD$, $BC = DA$, $\angle B \cong \angle D$. Furthermore, obviously $\angle BAD \cong \angle BCD$. QED

2. The diagonals of any parallelogram, $ABCD$, bisect each other (Figure 3.14).

   Indeed, since $\overline{BC} \parallel \overline{AD}$ we get $\angle CBD \cong \angle ADB$ while VAT implies $\angle BOC \cong \angle AOD$. But the first property of parallelograms leads to $BC = AD$. Then AAS congruence implies that $\Delta CBO \cong \Delta ADO$. Therefore $BO = DO$ and $CO = AO$. QED

Four independent conditions under which a quadrilateral is a parallelogram

A quadrilateral will be a parallelogram if certain other, rather unexpected statements are true; statements different from the one used in the definition of a parallelogram.

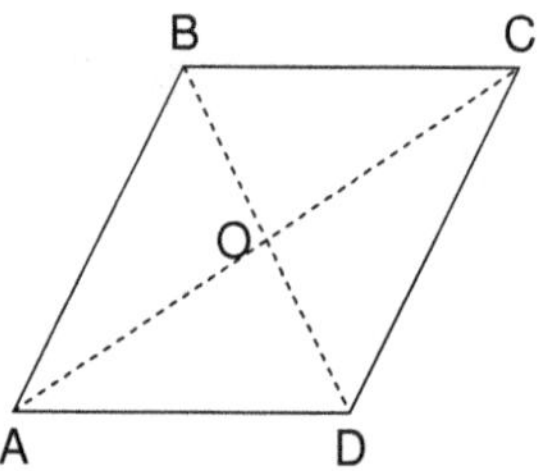

Figure 3.14: The diagonals of any parallelogram bisect each other

1. Suppose that opposite sides of a quadrilateral are congruent. Then it is a parallelogram.

   Why is this so? Draw the diagonal $\overline{AC}$ (Figure 3.13). Then, by SSS congruence, $\Delta ABC \cong \Delta CDA$. Hence $\angle BCA \cong \angle DAC$, consequently $\overline{BC} \parallel \overline{AD}$. Similarly, since $\Delta ABC \cong \Delta CDA$ we will also have $\angle BAC \cong \angle DCA$. Therefore $\overline{AB} \parallel \overline{CD}$. QED

2. Assume that a pair of opposite sides of a quadrilateral are congruent and parallel. Then the quadrilateral is a parallelogram.

   The proof is rather straightforward. Let $BC = AD$ and $\overline{BC} \parallel \overline{AD}$, and draw the diagonal $\overline{AC}$ (Figure 3.13). The parallelism between $\overline{BC}$ and $\overline{AD}$ implies that $\angle BCA \cong \angle DAC$. Then, using SAS congruence, we get $\Delta BCA \cong \Delta DAC$. In particular, $AB = CD$. Since both opposite sides are congruent, we can conclude that the quadrilateral under consideration is indeed a parallelogram. QED

3. Assume that the diagonals of a quadrilateral bisect each other. Then the quadrilateral is a parallelogram.

   Indeed, let $AO = OC$ and $BO = OD$ (Figure 3.14). Applying VAT we see that $\angle BOC \cong \angle AOD$, thus $\Delta BOC \cong \Delta DOA$ thanks to SAS congruence. In particular, $\angle BCO \cong \angle DAO$; hence $\overline{BC} \parallel \overline{AD}$. But $BC = AD$ due to the fact that $\Delta BOC \cong \Delta DOA$. We have shown that a pair of opposite sides are parallel and congruent, so the quadrilateral $ABCD$ is a parallelogram. QED

4. Assume that opposite angles of a quadrilateral are congruent. Then the quadrilateral is a parallelogram. Right away we have that $2\alpha + 2\beta = 360$ (Figure

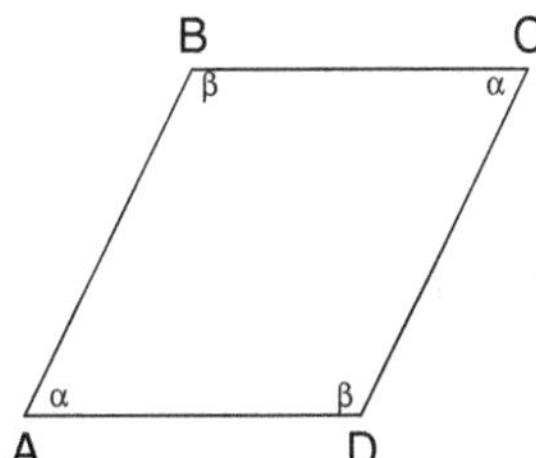

Figure 3.15: A quadrilateral is a parallelogram when opposite angles are congruent

3.15), hence $\alpha + \beta = 180$. Since $\angle B + \angle C = 180$ we get $\overline{AB} \parallel \overline{CD}$. Similarly, since $\angle A + \angle B = 180$ we can conclude that $\overline{BC} \parallel \overline{AD}$. Therefore the quadrilateral is a parallelogram. QED

Definition (distance between two parallel lines)

Given two parallel lines, L_1 and L_2, choose any point P on one of them and drop a perpendicular $\overline{PQ}$ to the other line. The distance between L_1 and L_2, $d(L_1, L_2)$, is defined as PQ.

Is this a 'good definition'? Suppose that R is another point on L_1 and we drop the perpendicular $\overline{RS}$ toward L_2 (Figure 3.16). We would have to show that $PQ = RS$. Indeed, since $L_1 \parallel L_2$ we can see that the four internal angles of the quadrilateral $PRSQ$ are right angles. In particular, the opposite angles of the quadrilateral are congruent, consequently the quadrilateral happens to be a parallelogram. Therefore $PQ = RS$.

The heights of any triangle are concurrent

This is an amazing result, which will follow from what we have studied so far. Let us recall that, given a triangle, a **height** is the segment that starts at a vertex and is perpendicular to the line that contains the opposite side. Obviously, any triangle has three heights. We will prove that the three lines, each containing a height, meet at a point. This point of concurrency, called the **orthocenter**, lies in the interior of the triangle if it is an acute triangle, and it lies outside the triangle if we are dealing with an obtuse triangle (Figure 3.17). Evidently, the orthocenter of a right triangle is the point where the two legs meet.

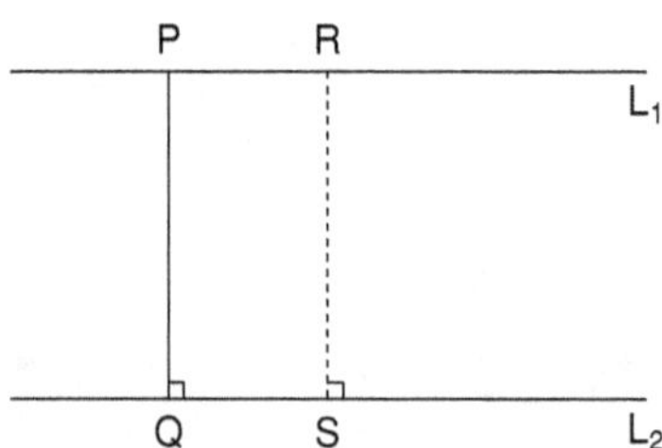

Figure 3.16: Distance between two parallel lines

Proof

Let ΔABC be an acute triangle and let $\overline{BD}$ be a height (Figure 3.18). Then draw lines parallel to each of the sides of the triangle. In this way we can build ΔEFG, each of whose lines is parallel to the corresponding side of ΔABC. Since $\overline{EF} \parallel \overline{AC}$ and $m\angle CDB = 90$ we can conclude that $m\angle DBF = 90$. Moreover, since $ABFC$ is a parallelogram it follows that $AC = BF$. Similarly, $AEBC$ is a parallelogram, so $AC = EB$ and consequently $EB = BF$. That is to say, B is the midpoint of $\overline{EF}$. We can then conclude that $\overline{BD}$ lies on the perpendicular bisector of $\overline{EF}$. In exactly the same way it can be proven that the height $\overline{CH}$ lies on the perpendicular bisector of $\overline{FG}$ and the height $\overline{AI}$ lies on the perpendicular bisector of $\overline{EG}$. But in the previous chapter we proved that the perpendicular bisectors of any triangle meet at a point. Hence the three lines that contain the heights, of any acute triangle, are concurrent. If the triangle were to be obtuse, we would follow the same steps. QED

3.3 Two theorems about triangles and parallel segments

In this section we will discuss two theorems, of considerable importance, that will allow us to prove many diverse propositions.

Theorem (TT1)

Let us consider ΔABC with $\overline{DE} \parallel \overline{AC}$ and $AD = DB$. Then $BE = EC$ and $DE = \frac{1}{2}AC$ (Figure 3.19).

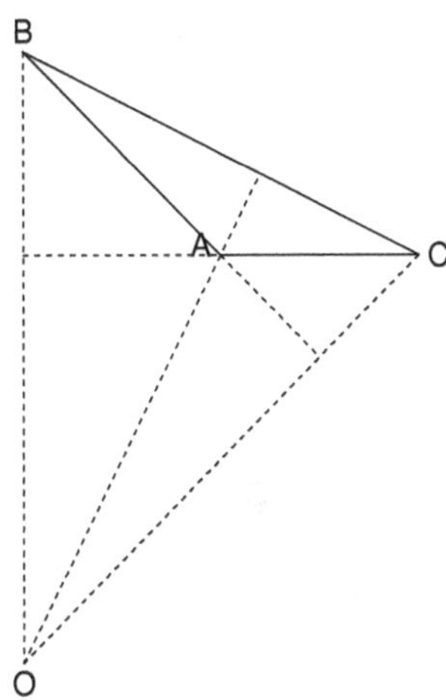

Figure 3.17: The lines that contain the heights meet at the orthocenter O

Proof

Build $\overline{EF}$ so that $\overline{EF} \parallel \overline{AD}$ (Figure 3.20). Since $L_3 \parallel L_4$ we get $\gamma = \delta$. Similarly, since $L_1 \parallel L_2$ and corresponding angles, linked to two parallel lines, are congruent, we can conclude that $\beta = \alpha$. But $ADEF$ is a parallelogram, thus $EF = AD$. Then, by AAS congruence, we will have $\Delta BDE \cong \Delta EFC$. In particular $DE = FC$ and $BE = EC$. We know that $DE = AF$ (recall that $DEFA$ is a parallelogram), consequently $DE = \frac{1}{2}AC$. QED

Theorem (TT2)

Let D and E be the midpoints of $\overline{AB}$ and $\overline{CB}$ respectively. Then $\overline{DE} \parallel \overline{AC}$ and $DE = \frac{1}{2}AC$ (Figure 3.21).

Proof

Prolong $\overline{DE}$ in such a way that $EF = DE$ (Figure 3.22). Using SAS we can conclude that $\Delta BED \cong \Delta CEF$, hence $\alpha = \beta$ and $FC = DB$. Since $\alpha = \beta$ it follows that $\overleftrightarrow{AB} \parallel \overleftrightarrow{CF}$ (note that $\overleftrightarrow{BC}$ is the transversal being considered), so $\overline{AD} \parallel \overline{FC}$. The quadrilateral $DFCA$ has two opposite sides, namely $\overline{AD}$ and $\overline{FC}$, parallel and congruent. Therefore it is a parallelogram and, in particular, $\overline{DE} \parallel \overline{AC}$. Moreover,

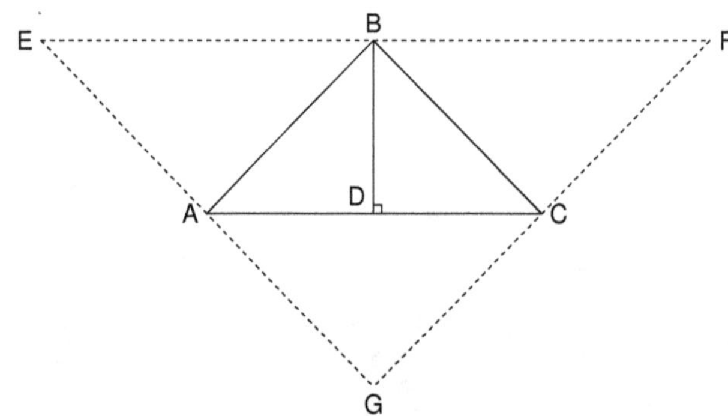

Figure 3.18: Proving that the three heights are concurrent

$DF = AC$ because opposite sides of a parallelogram are always congruent. But $DF = 2DE$, hence $AC = 2DE$, that is, $DE = \frac{1}{2}AC$. QED

The medians of any triangle are concurrent

Let ΔABC be an arbitrary triangle. Then the three medians of ΔABC meet at a point, which happens to be at the interior of the given triangle. It is called the **centroid**.

Proof

The medians $\overline{AE}$ and $\overline{CD}$ meet at a point P (Figure 3.23) and we prolong $\overrightarrow{BP}$ in such a way that $PF = BP$. Applying TT2 to ΔABF we get $\overline{DP} \parallel \overline{AF}$ and $DP = \frac{1}{2}AF$, thus $\overline{PC} \parallel \overline{AF}$. Again we apply TT2, this time to ΔFBC, and conclude that $\overline{PE} \parallel \overline{FC}$, $PE = \frac{1}{2}FC$; thus, $\overline{AP} \parallel \overline{FC}$.

We have just shown that $APCF$ is a parallelogram. Consequently, its diagonals bisect each other; so $AG = GC$. That is to say, $\overline{BG}$ is the third median. We have succeeded in proving that the three medians pass through a common point P. QED

Remarks

1. In Figure 3.23 let $DP = e_1$. We have shown that $AF = 2DP = 2e_1$, thus $PC = 2e_1$. In a similar fashion we can assert that if $PE = e_2$ then $AP = 2e_2$, while if $PG = e_3$ then $BP = 2e_3$. Hence the point of concurrency, P, divides each median in a fixed ratio: $\frac{1}{3}$ and $\frac{2}{3}$.

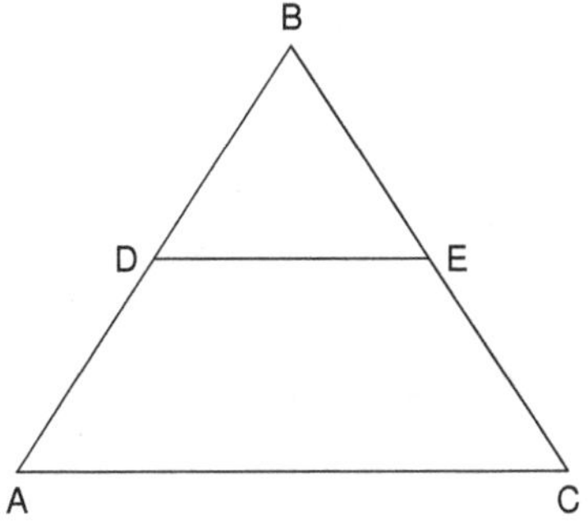

Figure 3.19: TT1 Theorem

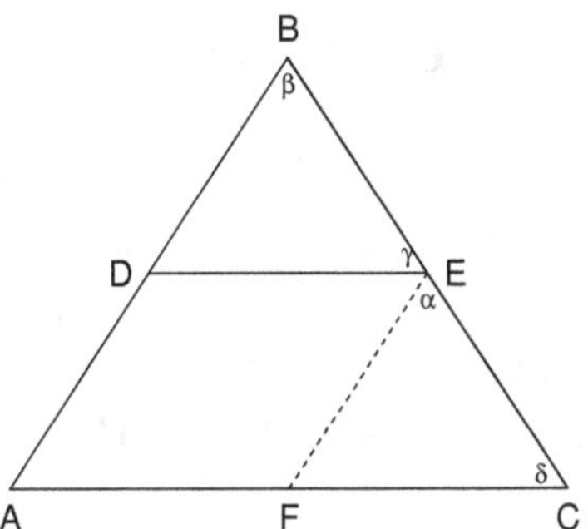

Figure 3.20: Proving the TT1 Theorem

2. There is a subtle fact that we have assumed implicitly, namely the inequality $PG < BP$. Our whole construction, which led to the proof of the concurrency of the medians, depended on the above-mentioned inequality. Can we prove it?

   Start by building $\overline{DE}$ (Figure 3.24). Applying TT2 to ΔABC we conclude that $\overline{DE} \parallel \overline{AC}$, so $\overline{DH} \parallel \overline{AG}$. Next we apply TT1 to ΔABG and conclude that $BH = HG$. Then

$$PG < HG = BH < BP,$$

   as we wished to prove.

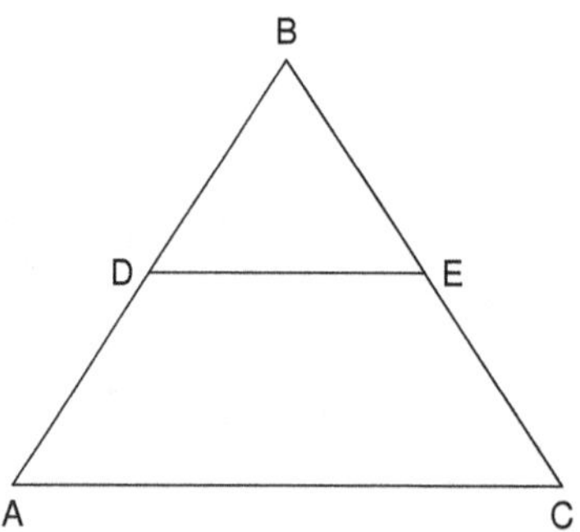

Figure 3.21: TT2 Theorem

3.4 Steiner-Lehmus Theorem

Suppose that ΔABC is isosceles, with $AB = BC$ (Figure 3.25), and let $\overline{AE}$ bisect $\angle BAC$ and $\overline{DC}$ bisect $\angle BCA$. Can we conclude that $DC = EA$? The answer is yes because ITT implies that $m\angle DAC = m\angle ECA = \alpha$, so $m\angle EAC = m\angle DCA = \frac{1}{2}\alpha$; thus, ASA congruence leads to $\Delta DAC \cong \Delta ECA$ and consequently $DC = EA$. We have just shown that if a triangle is isosceles then the segments that bisect the base angles are congruent.

Around 1840 Daniel Lehmus (1780-1863) asked whether the converse of the preceding proposition happened to be true. Jakob Steiner (1796-1863) provided a proof soon after Lehmus' question was raised. In the last 150 years or so many other proofs have been invented since Steiner-Lehmus Theorem is a topic that has fascinated mathematicians and amateurs alike. Part of the fascination probably stems from the fact that none of the proofs are straightforward despite the ease with which the converse of Steiner-Lehmus Theorem is proven (see previous paragraph).

Among the many different proofs of Steiner-Lehmus Theorem, we will discuss in detail one of them, which appeared in print fifty years ago (Fetisov, 1963). Suppose that $\overline{AE}$ and $\overline{CD}$ bisect the bottom angles of ΔABC (Figure 3.25) and assume that $AE = CD$. Then $AB = BC$.

Proof

Proceeding by contradiction assume that $AB \neq BC$. Without loss of generality suppose that $AB < BC$, so $2\beta < 2\alpha$ where $2\alpha = m\angle BAC$ and $2\beta = m\angle BCA$. Since $2\beta < 2\alpha$ and $AE = CD$, the Hinge Theorem (applied to triangles ACD and CAE)

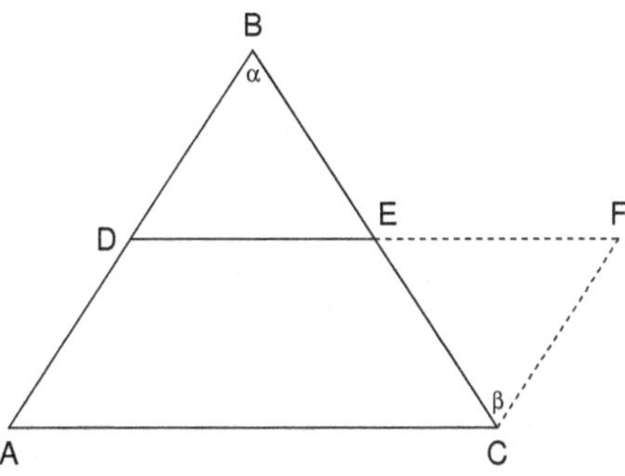

Figure 3.22: Proving the TT2 Theorem

leads to $AD < EC$.

Next build $\overline{DF}$ so that $\overline{DF} \parallel \overline{AE}$ and $DF = AE$ (Figure 3.25). Hence, the quadrilateral $ADFE$ happens to be a parallelogram, and consequently $EF = AD$; thus $EF < AC$. Applying BI (the basic inequality from the previous chapter) to ΔFEC we get $\delta = \gamma$, where $\delta = m\angle ECF$ and $\gamma = m\angle EFC$. On the other hand, $CD = AE = DF$ and therefore ΔCDF is isosceles. Since opposite angles of any parallelogram are congruent we can conclude that $m\angle DFE = \alpha$, hence ITT implies that

$$\alpha + \gamma = \beta + \delta. \tag{3.1}$$

But $\delta < \gamma$ and $\beta < \alpha$ lead to

$$\beta + \delta < \alpha + \gamma. \tag{3.2}$$

We note that (3.1) and (3.2) cannot be true at the same time. In other words, we have reached a contradiction. QED

3.5 Rectangles

A rectangle is a parallelogram that has an interior right angle.

Three properties shared by all rectangles

1. Given a rectangle $ABCD$ (Figure 3.26), with $m\angle D = 90$, we can assert that each of its internal angles is a right angle. This result follows from the fact that, by definition, a rectangle is a parallelogram; thus, $\angle B \cong \angle D$ and $\angle A \cong \angle C$. Let $\beta = m\angle A = m\angle C$. Since $2\beta + 180 = 360$ we get $\beta = 90$.

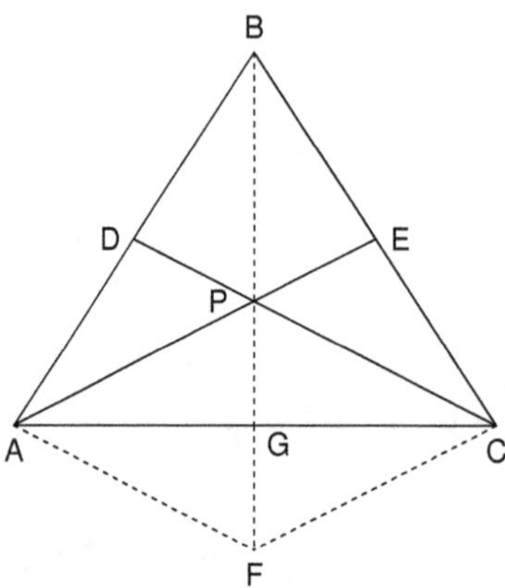

Figure 3.23: The medians are concurrent

2. The diagonals of any rectangle $ABCD$ are congruent (Figure 3.27). Why is this so? SAS congruence implies $\Delta BAD \cong \Delta CDA$, therefore $BD = AC$.

3. If the diagonals of a parallelogram $ABCD$ are congruent then it is a rectangle. The result follows from the observation that four triangles, namely ΔBAD, ΔABC, ΔBCD, and ΔCDA (Figure 3.27) are congruent thanks to the SSS property. Then

$$\angle BAD \cong \angle ABC \cong \angle BCD \cong \angle CDA.$$

Letting α be their common measure, since $4\alpha = 360$ it follows that $\alpha = 90$.

The median of a right triangle

Given a right triangle BAC, let $\overline{AO}$ be the median toward the hypotenuse (Figure 3.28). We will prove that

$$AO = \tfrac{1}{2}BC.$$

With this purpose in mind we extend $\overline{AO}$ in such a way that $OD = AO$ (Figure 3.28). Then we build $\overline{BD}$ and $\overline{DC}$. Since the diagonals of the quadrilateral $ABDC$ bisect each other, we can assert that $ABDC$ is a parallelogram. But $\angle A$ is a right angle, so $ABDC$ is a rectangle. Let us recall that the diagonals of a rectangle are congruent, hence $2AO = 2BO$. Consequently $AO = BO = OC$.

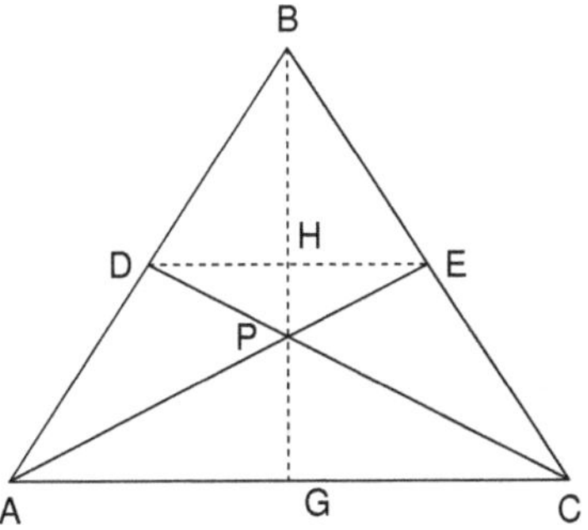

Figure 3.24: A subtle fact

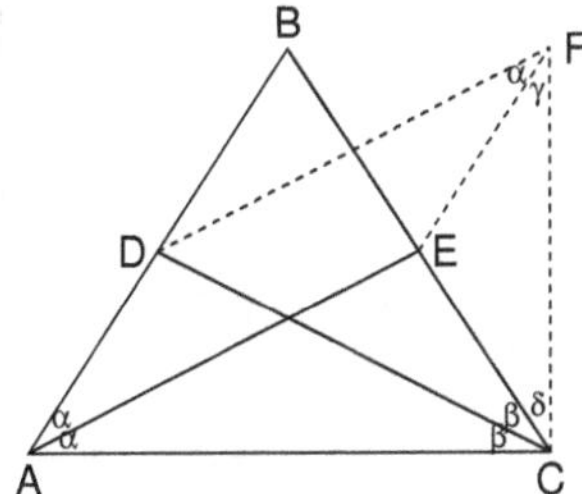

Figure 3.25: Fetisov's proof of the Steiner-Lehmus Theorem

3.6 Rhombuses

A rhombus is a parallelogram with all its sides congruent.

Three properties shared by all rhombuses

1. The diagonals of a rhombus are perpendicular to each other. Furthermore, they bisect the interior angles of the rhombus.

   Indeed, since a rhombus is a parallelogram we can assert that $AO = OC$ and $BO = OD$ (Figure 3.29). Then, through SSS congruence, we get $\Delta AOB \cong \Delta COB$; hence $\angle AOB \cong \angle COB$. Since these angles form a linear pair we can

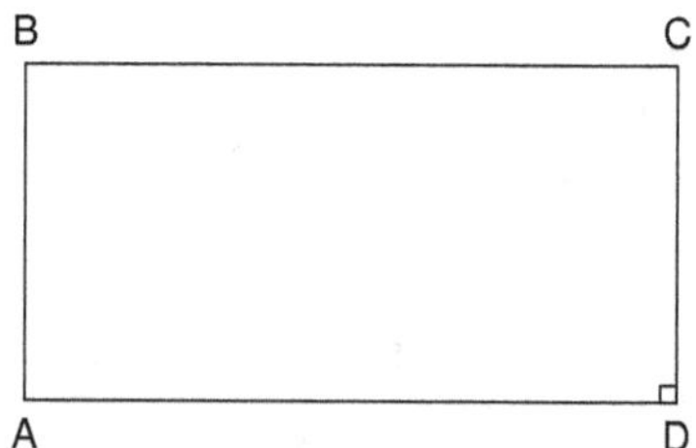

Figure 3.26: Definition of a rectangle

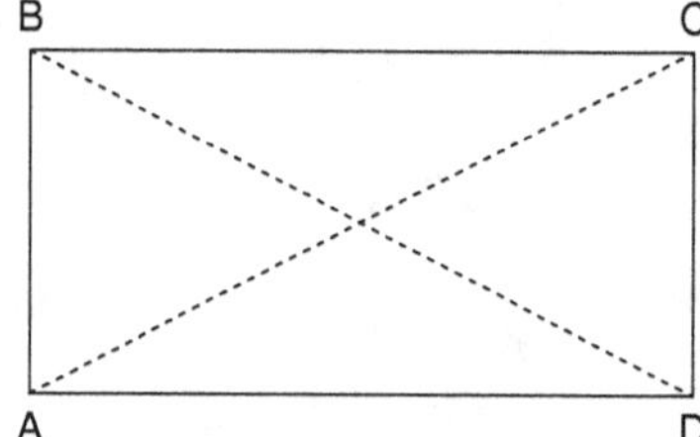

Figure 3.27: Diagonals of a rectangle

conclude that $m\angle AOB = 90$. Besides, SSS implies that $\Delta AOB \cong \Delta COB \cong \Delta AOD \cong \Delta COD$. Consequently $\angle OAB \cong \angle OAD$, $\angle ADO \cong \angle CDO$, $\angle BCO \cong \angle DCO$, and $\angle ABO \cong \angle CBO$. That is to say, the diagonals bisect the four interior angles of the rhombus.

2. If the diagonals of a parallelogram are perpendicular to each other then we have a rhombus.

   The proof is quite simple. Being a parallelogram, we know that $AO = OC$ (Figure 3.29). Then SAS congruence implies that $\Delta AOB \cong \Delta COB$, hence $AB = CB$. But opposite sides of any parallelogram are congruent, therefore $AB = CD$ and $BC = AD$. Then $AB = CB = CD = AD$. Thus, $ABCD$ is a rhombus.

3. If a diagonal of a parallelogram bisects an interior angle then the parallelogram is a rhombus.

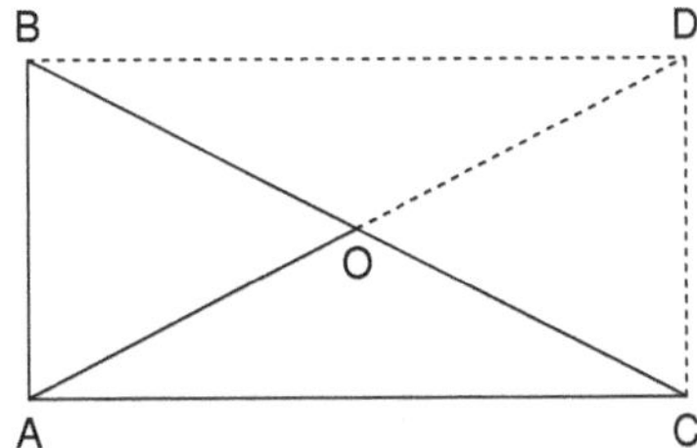

Figure 3.28: The median property of a right triangle

Let us provide a proof of this statement. Assume that $\overline{BD}$ bisects $\angle ADC$ (Figure 3.30). Since $\overline{AB} \parallel \overline{CD}$ and $m\angle BDC = \alpha$ we can conclude that $m\angle ABD = \alpha$. Then the converse of ITT implies that ΔBAD is isosceles. Consequently $AB = AD$, which in turn implies that $BC = AD = AB = CD$ due to the fact that opposite sides of a parallelogram are congruent.

3.7 Trapezia

A trapezoid is a quadrilateral with two parallel opposite sides, called the 'bases'. Moreover, the *median* of a trapezoid is the segment that joins the midpoints of the non-parallel sides while the *height* of a trapezoid is the distance between the parallel segments[8]. A trapezoid is said to be *isosceles* if the non-parallel sides are congruent.

Theorem (The length of the median)

The median of a trapezoid is parallel to the bases and its length is one-half the sum of the lengths of the bases, that is, $\overline{EF} \parallel \overline{AB}$ and $EF = \frac{1}{2}(AB+CD)$ (Figure 3.31).

Proof

Build $\overline{AF}$ and prolong it in such a way that it meets the ray $\overrightarrow{CD}$ at a point G (Figure 3.31). Since $\overline{AB} \parallel \overline{CD}$ we can assert that $m\angle ABF = m\angle GDF$. But VAT implies that $\angle BFA \cong \angle DFG$, therefore ASA allows us to conclude that $\Delta BFA \cong \Delta DFG$.

[8]We have in mind the notion of distance between the two parallel lines that contain the segments.

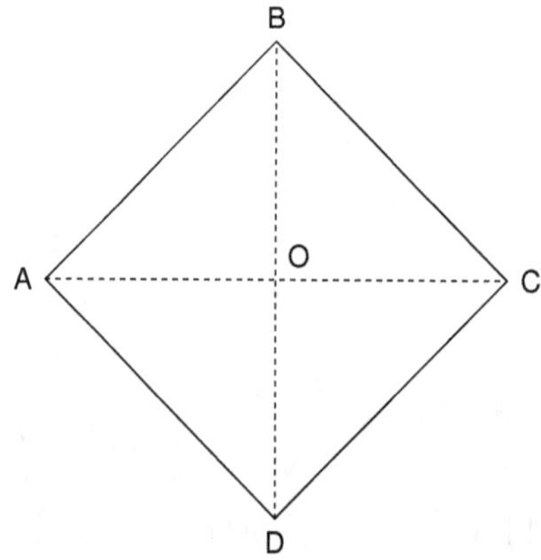

Figure 3.29: The diagonals of a rhombus

Hence $AB = DG$ and $AF = FG$. Then TT2, applied to ΔCAG, leads to $\overline{EF} \parallel \overline{CG}$ and $EF = \frac{1}{2}CG$. Consequently $EF = \frac{1}{2}(CD + DG)$. However, $DG = AB$, so $EF = \frac{1}{2}(CD + AB)$. On the other hand, $\overline{EF} \parallel \overline{AB}$ since $\overline{AB} \parallel \overline{CD}$. QED

Theorem (Two properties of isosceles trapezia)

If a trapezoid is isosceles then the 'base angles', as well as the diagonals, are congruent.

Proof

Start by dropping the perpendicular segments $\overline{AE}$ and $\overline{BF}$ (Figure 3.32). We have that $AC = BD$ and $AE = BF$. HL congruence implies that $\Delta ACE \cong \Delta BDF$. In particular, $\angle ACE \cong \angle BDF$. Next build the diagonals $\overline{AD}$ and $\overline{BC}$ (Figure 3.32). Since $AC = BD$ and the base angles, namely $\angle ACD$ and $\angle BDC$, were proven to be congruent, SAS congruence leads to $\Delta ACD \cong \Delta BDC$. Then $BC = AD$. QED

Two conditions under which a trapezoid is isosceles

1. Given a trapezoid, if the base angles are congruent then the trapezoid is isosceles.

   Indeed, build $\overline{AE}$ and $\overline{BF}$ (Figure 3.33). Since $AE = BF$, AAS congruence allows us to conclude that $\Delta AEC \cong \Delta BFD$. Then $AC = BD$.

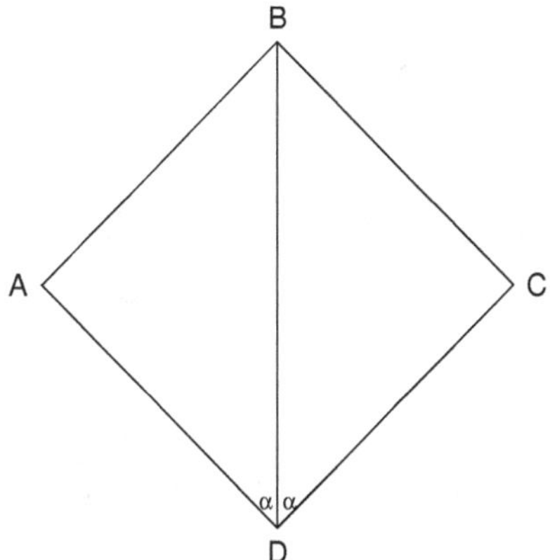

Figure 3.30: A condition under which a parallelogram is a rhombus

2. If the diagonals of a trapezoid are congruent then the trapezoid is isosceles.

   By hypothesis $AD = BC$. Then build $\overline{AE}$ and $\overline{BF}$, both segments perpendicular to $\overline{CD}$ (Figure 3.33). Then HL congruence implies $\Delta AED \cong \Delta BFC$, thus $CF = ED$ and consequently[9] $CE = FD$. Therefore $\Delta AEC \cong \Delta BFD$ due to SAS congruence, hence $AC = BD$.

3.8 Different strategies

By now the reader must have realized that, when solving a problem in geometry, it is often necessary to construct something, and these constructions are not usually unique. Let us illustrate these thoughts through a problem related to trapezoids: $ABCD$ is a trapezoid with diagonals $\overline{AC}$ and $\overline{BD}$. Assuming that E and F are the midpoints of the diagonals show that $\overline{EF} \parallel \overline{DC}$. Furthermore, find EF in terms of DC and AB (Figure 3.34)

First approach

Let H be the midpoint of $\overline{BC}$ (Figure 3.35). By TT2, applied to ΔDBC, we can conclude that $\overline{EH} \parallel \overline{DC}$. Let G be the point where $\overline{EH}$ meets $\overline{AC}$ and apply TT1

[9]Note that $CE + EF = EF + FD$ implies $CE = FD$.

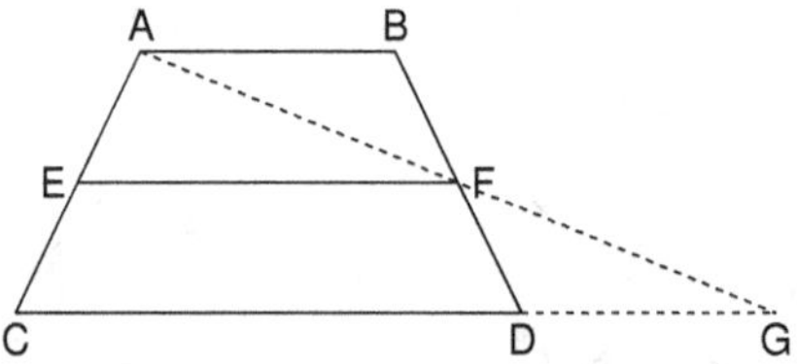

Figure 3.31: The median of a trapezoid

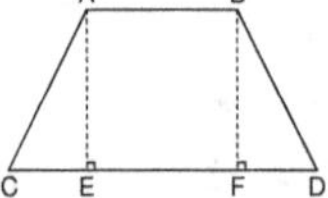

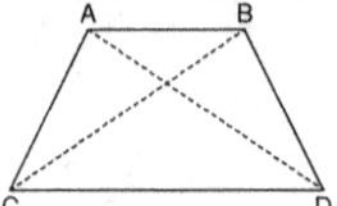

Figure 3.32: Two properties of isosceles trapezia

to ΔABC. Then $AG = GC$ and consequently[10] $G = F$. Thus, we have shown that $\overline{EF} \parallel \overline{DC}$. On the other hand, $EF + FH = \frac{1}{2}DC$ while $FH = GH = \frac{1}{2}AB$. Therefore $EF = \frac{1}{2}DC - \frac{1}{2}DC - \frac{1}{2}AB = \frac{1}{2}(DC - AB)$.

Second approach

Start by building $\overrightarrow{AE}$, which meets $\overline{DC}$ at G (Figure 3.36). We note that $\angle ABE \cong \angle EDG$ since $\overline{AB} \parallel \overline{DC}$. Then $\Delta AEB \cong \Delta GED$ thanks to ASA congruence[11]. Then $AB = GD$ and $AE = GE$. Next apply TT2 to ΔGAC and conclude that $\overline{EF} \parallel \overline{GC}$ and $EF = \frac{1}{2}GC$. Obviously $\overline{EF} \parallel \overline{DC}$ since both segments lie on the same line. Moreover, $GC = DC - DG$ and, since $DG = AB$, we will have

$$EF = \tfrac{1}{2}GC = \tfrac{1}{2}(DC - AB).$$

[10] A segment has a unique midpoint.

[11] Recall that $ED = EB$ by hypothesis.

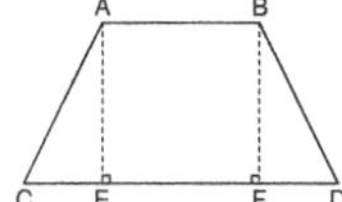

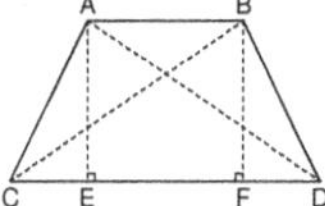

Figure 3.33: Two conditions under which a trapezoid is isosceles

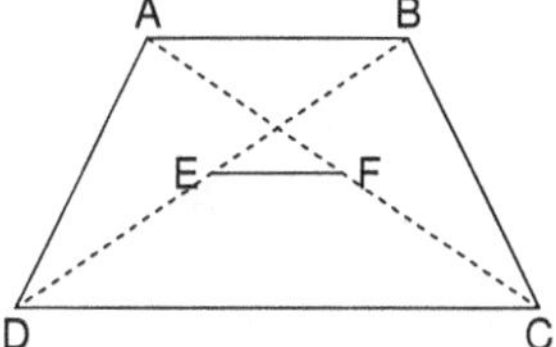

Figure 3.34: A challenging problem

Exercises

1. It is given that $OA = OB = BC$ (Figure 3.37). Find β in terms of ρ (Hint: Think about using ITT and SEAT.)

2. Axiom 10, the Axiom of Parallels, allowed us to prove that the relation 'paralelism between lines' is transitive (section 3.1). Show that if we were to accept that this relation is transitive then it would be possible to prove Axiom 10 on the basis of the first nine axioms.

3. We have an isosceles triangle ABC with $AB = BC$ and an inscribed equilateral triangle DEF (Figure 3.37). Find a relationship between α, β, and γ (Hint: As in exercise 1, choose where and when to apply ITT and SEAT.)

4. It is given that $\overrightarrow{BD}$ is the angle bisector of $\angle ABC$ and $\overrightarrow{CD}$ is the angle bisector of $\angle ACE$, while $m\angle BAC = 56$. Find $m\angle BDC$ (Figure 3.37) (Hint: Keep in mind that the sum of the internal angles of any triangle is 180. This fact and SEAT is all you need to use.)

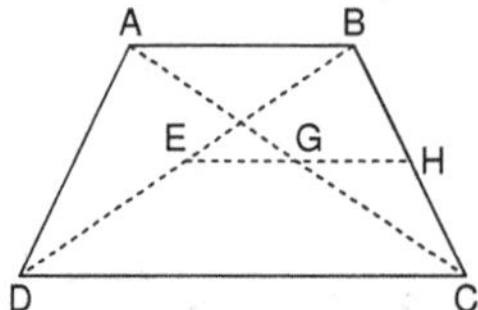

Figure 3.35: First approach

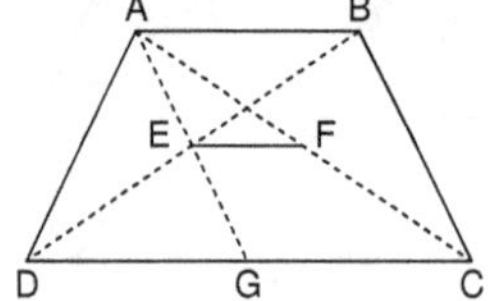

Figure 3.36: Second approach

5. Let us consider ΔABC, where the medians $\overline{AD}$ and $\overline{BE}$ are congruent. Show that ΔABC is isosceles (Hint: Build $\overline{ED}$ and apply the TT2 Theorem.)

6. $ABCD$ is an arbitrary quadrilateral and M, P, Q, R are corresponding midpoints. Show that $MPQR$ is a parallelogram.

7. It is given that $AB = BC$ and $CD = DE$ (Figure 3.38). Let F and H be the midpoints of $\overline{EG}$ and $\overline{AG}$, respectively. Assuming that $DF = 5$ cm calculate BH.

8. Let $MPQR$ be the midpoints of quadrilateral $ABCD$ (Figure 3.38). Assuming that $AC = 4$ cm and $BD = 5$ cm calculate the perimeter of the quadrilateral $MPQR$.

9. Show that a triangle is equilateral if and only if the circumcenter and the incenter coincide.

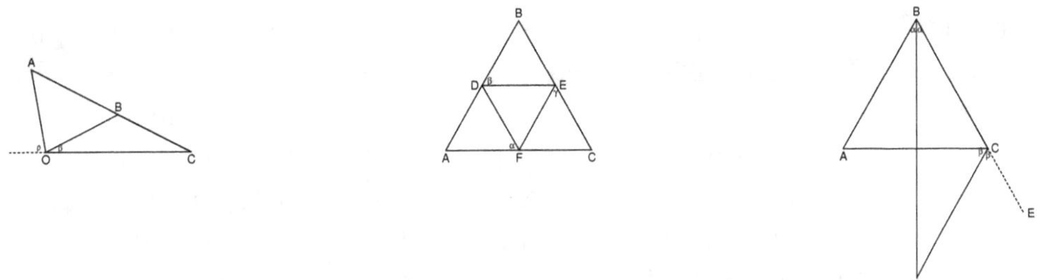

Figure 3.37: Figures related to exercises 1, 3, and 4.

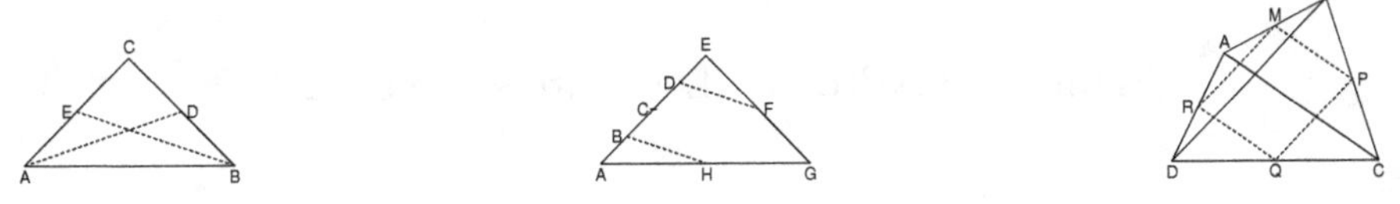

Figure 3.38: Figures related to exercises 5, 7, and 8.

10. Show that if a triangle is equilateral then the centroid coincides with the circumcenter and the incenter.

Enrichment Note: Legendre's Theorem

Adrien Legendre wrote an important book on geometry that went through twelve editions between 1794 and 1823. Therein he tried to prove, unsuccessfully, that the axiom of parallels is a theorem in what we nowadays call neutral geometry. Nonetheless, Legendre managed to prove that the axiom of parallels is a theorem if we were to assume that the sum of the internal angles of *any* triangle is 180. Since in section 3.1 we proved that the latter statement is true under the assumption of the axiom of parallels, both statements happen to be equivalent propositions in neutral geometry.

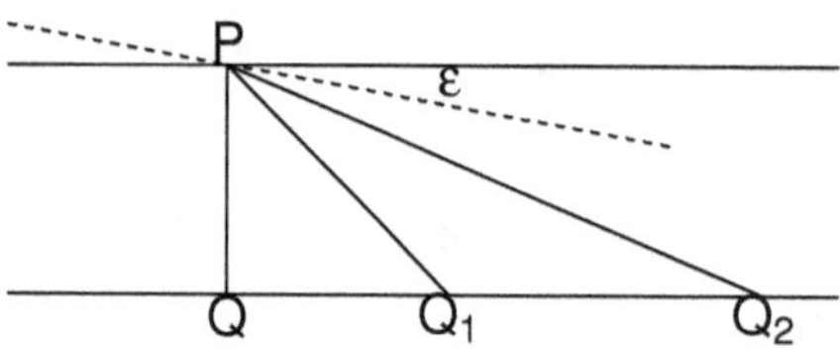

Figure 3.39: Proving Legendre's Theorem

Theorem

Assume, in neutral geometry, that the sum of the measures of the internal angles of any triangle is 180. Then the Axiom of Parallels must be true.

Proof

Let us consider a line L_1 and a point P not in L_1. Then drop a perpendicular segment $\overline{PQ}$ from P to the line L_1 and build the line L_2 that passes through P and is parallel to L_1. Let L be another line, different from L_2, that passes through P making an acute angle of measure ϵ between L and L_2 (Figure 3.39). We will show that L must meet L_1, that is, L cannot be parallel to L_1. With this purpose in mind, let us choose Q_1 on L_1 so that $QQ_1 = PQ$. Thereafter choose Q_2 on L_1 so that $Q_1Q_2 = PQ_1$, and so on. Due to the hypothesis we note that

$$m\angle QQ_1P = 45 = \tfrac{1}{2}90,$$

$$m\angle Q_1Q_2P = \tfrac{1}{2^2}90$$

and, in general,

$$m\angle Q_{n-1}Q_nP = \tfrac{1}{2^n}90.$$

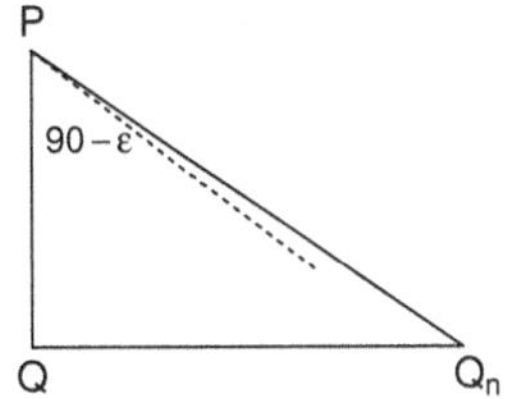

Figure 3.40: The triangle QPQ_n

Next choose n so that $\frac{1}{2^n}90 < \epsilon$. Hence

$$90 - \frac{1}{2^n}90 > 90 - \epsilon. \tag{3.3}$$

Let us pay close attention to ΔPQQ_n (Figure 3.40). We note that $m\angle QPQ_n = 90 - \frac{1}{2^n}90$ while $m\angle QPS = 90 - \epsilon$, where S belongs to L. From (3.3) we can conclude that S lies in the interior of ΔQPQ_n, so the Crossbar Theorem (a result from neutral geometry) implies that L must cross $\overline{QQ_n}$. Thus, L intercepts L_1. That is to say, L cannot be parallel to L_1. QED

If the negation of the axiom of parallels is adopted, that is, given a line L and a point P not on it, there exist at least two lines that pass through P and are parallel to L, then there exists a triangle such that the sum of the measures of its internal angles is less than 180. Under these circumstances, it can be proven that all triangles have an angle sum less than 180. Once in the realm of hyperbolic geometry (neutral geometry + negation of the axiom of parallels) a collection of surprising results come to life; for instance, if two triangles are similar then they are congruent!

Chapter 4

Area of Polygonal Regions

A **triangular region** is the union of a triangle and its interior. A **polygonal region** is a finite union of triangular regions, none of which intersect (except at borders or vertices). To every polygonal region R we assign a number $a(R)$, called the area of the region, such that

1. $a(R) > 0$.
2. If R_1 and R_2 are two congruent triangular regions, then $a(R_1) = a(R_2)$.
3. If R_1 and R_2 are two polygonal regions that are disjoint, or intersect only at borders or vertices, then $a(R_1 \cup R_2) = a(R_1) + a(R_2)$.
4. If R is a rectangle with base of length b and height h, then $a(R) = bh$.

The reader will surely agree that the preceding four properties seem natural and coincide with our intuition about the notion of area of a polygonal region. In reality, we are dealing with four axioms associated with the notion of area of a polygonal region.

4.1 Area of a right triangle

Let us consider a right triangle ABC, with right angle at B (Figure 4.1). Then build the segment $\overline{CD}$ in such a way that $\overline{CD}$ is perpendicular to $\overline{BC}$ and $CD = AB$. Since $\overline{CD} \parallel \overline{AB}$ and $CD = AB$ we can conclude that the quadrilateral $ABCD$ is a parallelogram. Moreover, $ABCD$ is a rectangle due to the fact that one interior angle of the parallelogram under consideration is a right angle. In particular, $m\angle D = 90$ and $AD = BC$. Using the SAS property of congruence we can assert that $\triangle ADC \cong \triangle CBA$, consequently $a(\triangle ADC) = a(\triangle CBA)$. But

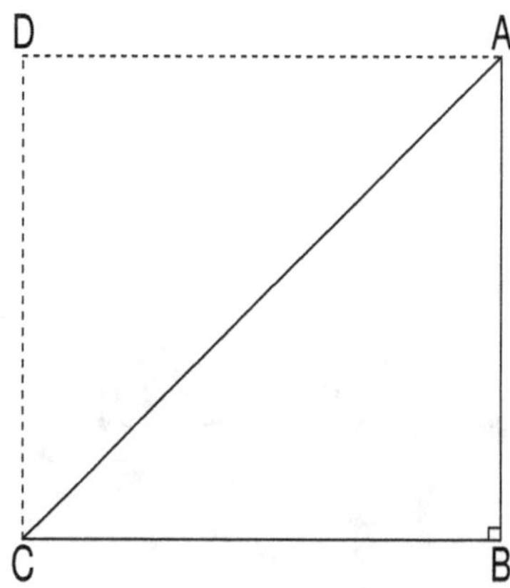

Figure 4.1: Finding the area of a right triangle

$$BC \times AB = \text{area of rectangle } ABCD = a(\triangle ADC) + a(\triangle CBA) = 2a(\triangle CBA)$$

Hence

$$a(\triangle ABC) = a(\triangle CBA) = \tfrac{1}{2}BC \times AB.$$

The reader must have noticed that we used the second, third, and fourth properties that characterize the area of a polygonal region.

4.2 Area of an arbitrary triangle

Firstly assume that the height $\overline{CD}$ falls inside the base $\overline{AB}$ (Figure 4.2). Then $a(\triangle ABC) = a(\triangle ADC) + a(\triangle BDC)$. But both $\triangle ADC$ and $\triangle BDC$ are right triangles. Therefore

$$a(\triangle ABC) = \tfrac{1}{2}AD \times DC + \tfrac{1}{2}DB \times DC = \tfrac{1}{2}(AD + DB)DC = \tfrac{1}{2}AB \times DC.$$

What can be done if the height $\overline{AD}$ falls outside the base $\overline{BC}$? Under these circumstances $a(\triangle ADC) = a(\triangle ADB) + a(\triangle ABC)$. Thus

$$\tfrac{1}{2}AD \times DC = \tfrac{1}{2}AD \times DB + a(\triangle ABC).$$

Then $\frac{1}{2}AD(DC - DB) = a(\triangle ABC)$, so

$$a(\triangle ABC) = \tfrac{1}{2}AD \times BC.$$

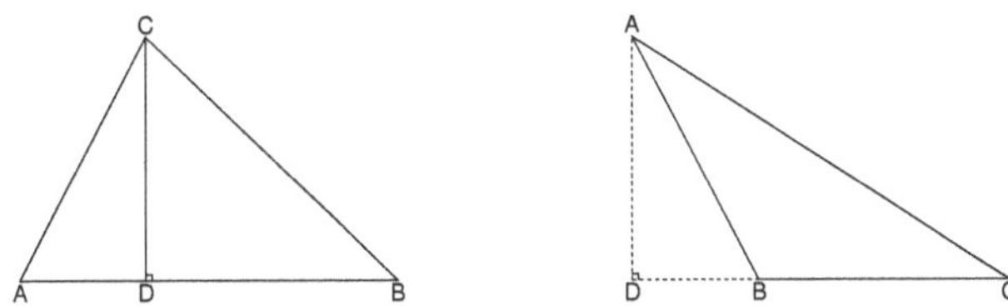

Figure 4.2: Finding the area of an arbitrary triangle

We have just shown that, given any triangle, it is 'legal' to choose an arbitrary vertex and drop the perpendicular towards the opposite side (called the corresponding base). Then the area of the triangle is one half the length of the base times the height.

Remark

Given an arbitrary triangle ABC, let X be any point on $\overline{BC}$ (Figure 4.3). Then

$$\frac{a(\Delta ABX)}{a(\Delta ACX)} = \frac{BX}{XC}.$$

Indeed, $a(\Delta ABX) = \frac{1}{2}BX \times h$ and $a(\Delta ACX) = \frac{1}{2}CX \times h$. Thus

$$\frac{a(\Delta ABX)}{a(\Delta ACX)} = \frac{BX}{XC}.$$

This simple result will be needed in section 4.10

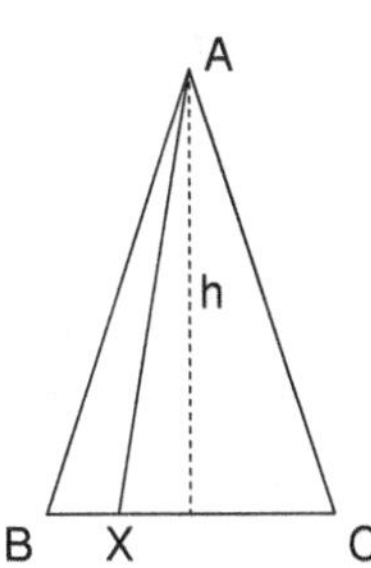

Figure 4.3: A simple property related to areas of triangles

4.3 Area of a trapezoid

Let $ABCD$ be a trapezoid (Figure 4.4). Let us build diagonal $\overline{AC}$ and drop the heights $\overline{CF}$ and $\overline{AE}$. Since $\overline{AB} \parallel \overline{CD}$ we can conclude that $AE = CF$. Hence

$$a(ABCD) = a(\triangle DAC) + a(\triangle CBA) = \tfrac{1}{2}DC \times AE + \tfrac{1}{2}AB \times FC = \tfrac{1}{2}DC + AB)h,$$

where $h = AE = CF$ is the height of the parallelogram.

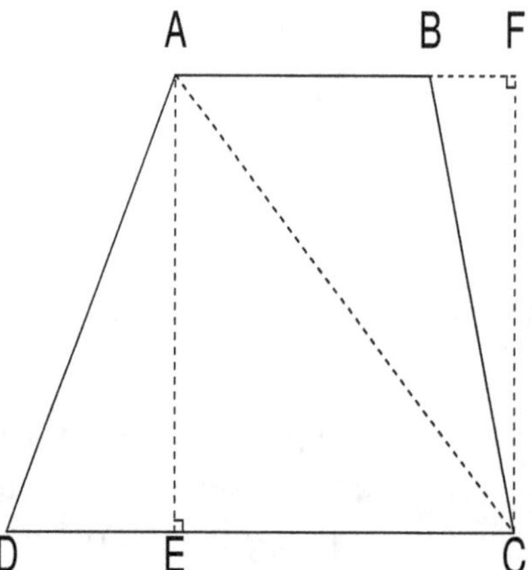

Figure 4.4: Finding the area of a trapezoid

4.4 Area of a parallelogram

A parallelogram is a particular case of a trapezoid. Therefore (Figure 4.5) $a(ABCD) = \frac{1}{2}(AB+DC) \times AE$. But opposite sides of a parallelogram are congruent, consequently

$$a(ABCD) = \tfrac{1}{2}(AB + AB) \times AE = AB \times AE.$$

Now that we have formulas for the area of a rectangle, triangle, trapezoid, and parallelogram, it is possible to solve multiple problems (see list of exercises at the end of the chapter). An unexpected bonus that stems from the notion of area is that it is possible to provide relatively straightforward proofs of Pythagoras' theorem and Thales' theorem, two of the most important theorems of plane geometry. The latter will play a pivotal role in the development of the theory of similarity among triangles. It goes without saying that the theory of similarity lies behind the foundations of trigonometry.

4.5 Pythagoras' Theorem

Given a right triangle ABC, with right angle at C (Figure 4.6), $c^2 = a^2 + b^2$. In words, the square of the hypotenuse is the sum of the squares of the legs.

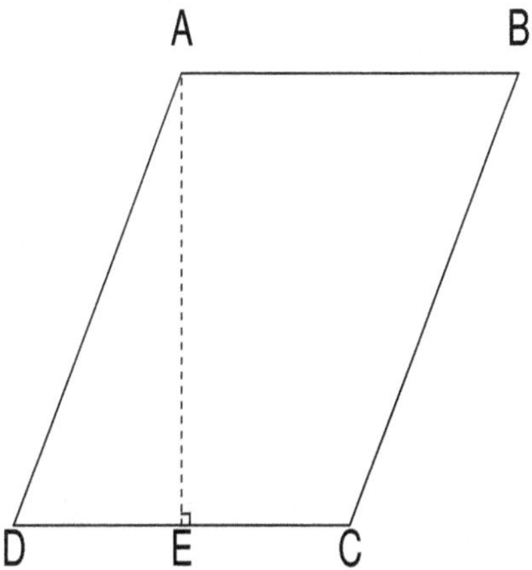

Figure 4.5: Finding the area of a parallelogram

Proof

We will analyze James Garfield's proof, one of the three hundred and sixty or so different existing proofs[1]. We start by prolonging $\overline{BC}$ in such a way that $BD = b$. Then we make a right turn and build $\overline{DE}$ so that $DE = a$. Next we build $\overline{BE}$. Using SAS congruency we can assert that

$$\triangle ACB \cong \triangle BDE. \tag{4.1}$$

In particular $BE = AB$, so $BE = c$. Also, from (4.1) we get $\angle CAB \cong \angle DBE$ and $\angle ABC \cong \angle BED$. Let us write $m\angle CAB = m\angle DBE = \alpha$, $m\angle ABC = m\angle BED = \beta$.

We note that $\overline{AC} \parallel \overline{ED}$, thus the quadrilateral $ACDE$ is a trapezoid with height $\overline{CD}$ and bases $\overline{AC}$ and $\overline{DE}$. Letting $\gamma = m\angle ABE$ we get the equality $\beta+\gamma+\alpha = 180$. But $\beta + \alpha = 90$, so $\gamma = 90$.

The area of the trapezoid is equal to the sum of the areas of the three right triangles. Therefore:

$$\tfrac{1}{2}(a+b)(a+b) = \tfrac{1}{2}ab + \tfrac{1}{2}c^2 + \tfrac{1}{2}ab.$$

Then $a^2 + 2ab + b^2 = c^2 + 2ab$, which in turn leads to $a^2 + b^2 = c^2$. QED

[1]James Garfield (1831-1881) was the 20th president of the U.S. Garfield developed a proof of Pythagoras' Theorem while he was a member of the House of Representatives.

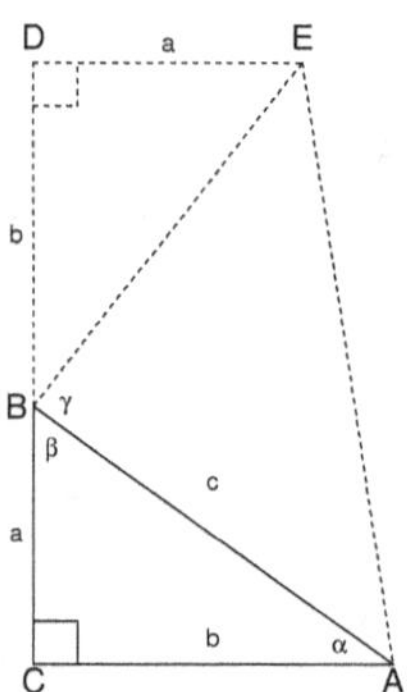

Figure 4.6: Garfield's proof of Pythagoras Theorem

4.6 Converse of Pythagoras' Theorem

Given $\triangle ABC$ with $AB = c$, $BC = a$, and $AC = b$ (Figure 4.7), assume that $c^2 = a^2 + b^2$. Then $m\angle C = 90$.

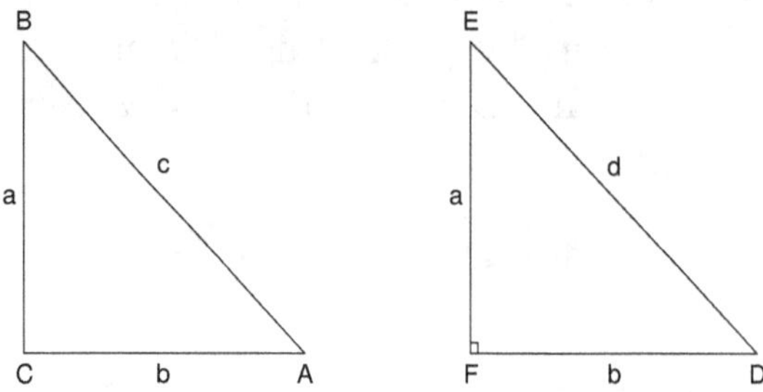

Figure 4.7: Proving the Converse of Pythagoras Theorem

Proof

Build $\triangle DEF$ with $DF = b$, $EF = a$ and $m\angle EFD = 90$. Let $d = ED$. Applying Pythagoras' Theorem we get $d^2 = a^2 + b^2$. Hence $c^2 = d^2$, and consequently $c = d$. By SSS congruency we get that

$$\triangle ABC \cong \triangle DEF.$$

In particular $\angle ACB \cong \angle EFD$. But $m\angle EFD = 90$, consequently $m\angle ACB = 90$. QED

An interesting application of Pythagoras' Theorem has to do with a formula to calculate the area of a triangle in terms of its sides. This result will play a crucial role in the solution of the isoperimetric problem for triangles.

4.7 Heron's Theorem

Heron of Alexandria (10-70 CE) was an engineer and mathematician. The formula that bears his name appeared in his book *Metrica*, published around 60 CE: Given a triangle ABC with sides of length a, b, c,

$$area(\triangle ABC) = \sqrt{s(s-a)(s-b)(s-c)},$$

where s is the semiperimeter, that is, $s = \frac{a+b+c}{2}$.

Proof

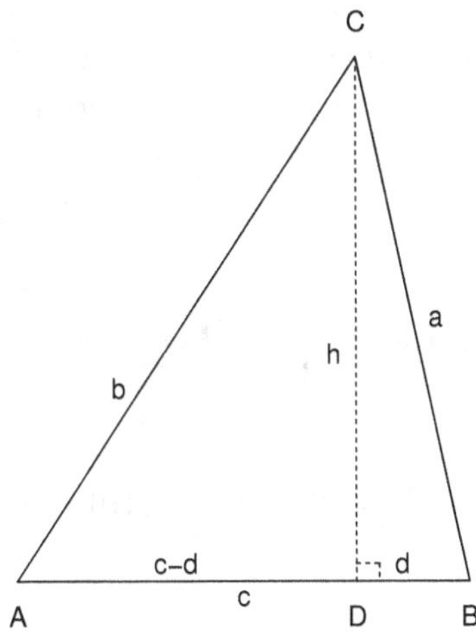

Figure 4.8: Diagram to accompany the proof of Heron's Theorem

Drop the perpendicular $\overline{CD}$ and let $h = CD$, $d = DB$ (Figure 4.8). Obviously $AD = c - d$. Applying Pythagoras' Theorem to $\triangle ACD$ and $\triangle BCD$ we get $a^2 - d^2 = h^2 = b^2 - (c-d)^2$. Thus

$$a^2 - d^2 = b^2 - c^2 + 2cd - d^2,$$

and consequently

$$d = \frac{a^2-b^2+c^2}{2c}.$$

Since $area(\triangle ABC) = \frac{ch}{2}$ we will have

$$4area^2 = c^2(a^2 - d^2) = c^2(a^2 - \frac{(a^2-b^2+c^2)^2}{4c^2}).$$

Therefore

$$16area^2 = 4a^2c^2 - (a^2 - b^2 + c^2)^2 = (2ac + a^2 - b^2 + c^2) \times (2ac - a^2 + b^2 - c^2) = ((a + c)^2 - b^2)(b^2 - (a - c)^2).$$

Consequently

$$area^2 = (\frac{a+c+b}{2})(\frac{a+c-b}{2}) \times (\frac{b+a-c}{2})(\frac{b-a+c}{2}) = s(s - b)(s - c)(s - a).$$

Finally $area = \sqrt{s(s - a)(s - b)(s - c)}$. QED

Remark

What happens if, when discussing a proof of Heron's Theorem, the triangle happens to be obtuse? We drop the height towards the side that lies opposite to the angle with measure higher than 90 and follow the same steps that led to a proof when the triangle is acute. It should be noted that the proof presented above uses symbolic algebra, a 16th century development. The proof that appeared in *Metrica* was geometric rather than algebraic, and is considerably more complicated (Oliver, 1993).

4.8 The isoperimetric problem for triangles

Among all triangles with fixed perimeter p, which one encloses the largest area? Let us consider any triangle with sides a, b, c and semiperimeter s, where $s = p/2$. Applying the AGM inequality ($n = 3$) we get

$$((s - a)(s - b)(s - c))^{1/3} \leq \frac{(s-a)+(s-b)+(s-c)}{3} = \frac{s}{3},$$

that is,

$$(s - a)(s - b)(s - c) \leq \frac{s^3}{3^3}.$$

Thus

$$s(s - a)(s - b)(s - c) \leq \frac{s^4}{3^3}.$$

Hence

$$area = \sqrt{s(s - a)(s - b)(s - c)} \leq \frac{s^2}{3\sqrt{3}}.$$

We have just shown that the area of any triangle with semiperimeter s is less than or equal $\frac{s^2}{3\sqrt{3}}$. Consequently, the best we can do is to have a triangle with area equal to $\frac{s^2}{3\sqrt{3}}$. Suppose then that

$$\sqrt{s(s-a)(s-b)(s-c)} = \frac{s^2}{3\sqrt{3}}.$$

Then $(s-a)(s-b)(s-c) = \frac{s^3}{3^3}$, so

$$((s-a)(s-b)(s-c))^{1/3} = \frac{s}{3}. \tag{4.2}$$

According to AGM we can conclude that $s-a = s-b = s-c$, therefore $a = b = c$. Obviously, if $a = b = c$ then AGM implies that (4.2) is true, thus $area = \frac{s^2}{3\sqrt{3}}$. In summary, the answer to the original question is to build an equilateral triangle with side $p/3$.

4.9 Thales' Theorem

This theorem is sometimes called The Basic Theorem of Proportionality[2]. Let us state it: Assume that $\overline{BD} \parallel \overline{AE}$ (Figure 4.9). Then

$$\frac{AC}{BC} = \frac{CE}{CD}.$$

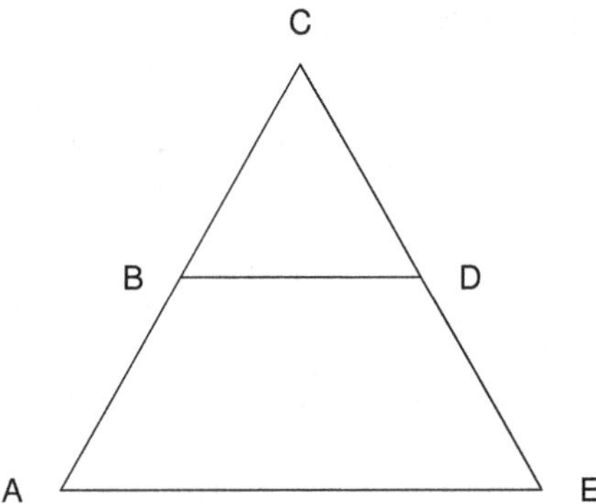

Figure 4.9: Thales' Theorem

[2]Thales (624 BCE-547 BCE) is considered to be the father of classic Greek mathematics.

Proof

We will divide the proof in three parts.

1. Build $\overline{AD}$ and the height p (Figure 4.10). Then $a(\Delta BCD) = \frac{pBC}{2}$, $a(\Delta ABD) = \frac{pAB}{2}$. Hence
$$\frac{a(\Delta BCD)}{a(\Delta ABD)} = \frac{BC}{AB}. \tag{4.3}$$

Figure 4.10: Proving Thales' Theorem

2. Build $\overline{BE}$ and height q (Figure 4.10). Then $a(\Delta BCD) = \frac{qCD}{2}$, $a(\Delta BDE) = \frac{qDE}{2}$. Hence
$$\frac{a(\Delta BCD)}{a(\Delta BDE)} = \frac{CD}{DE}. \tag{4.4}$$

3. Since $\overline{BD} \parallel \overline{AE}$ we can assert that $AF = GE$ (Figure 4.11). Denote this common length by the letter h. We have that
$$a(\Delta ABD) = \tfrac{hBD}{2} \text{ while } a(\Delta BDE) = \tfrac{hBD}{2}.$$
So
$$a(\Delta ABD) = a(\Delta BDE). \tag{4.5}$$
From (4.3), (4.4), and (4.5) we get $\frac{BC}{AB} = \frac{CD}{DE}$. Therefore
$$\frac{AB}{BC} = \frac{DE}{CD}. \tag{4.6}$$
Adding 1 to both sides of (4.6) we get $\frac{AB}{BC} + \frac{BC}{BC} = \frac{DE}{CD} + \frac{CD}{CD}$. Finally
$$\tfrac{AB+BC}{BC} = \tfrac{DE+CD}{CD},$$

that is,

$$\frac{AC}{BC} = \frac{CE}{CD}.$$

QED

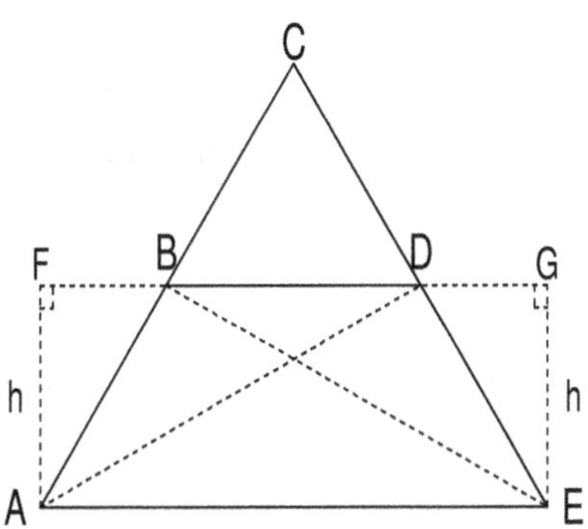

Figure 4.11: Third step in the proof of Thales' Theorem

The Angle Bisector Theorem

As an application of Thales Theorem we will prove that given ΔABC with $\angle ABD \cong \angle CBD$ (Figure 4.12), it must be true that

$$\frac{AB}{AD} = \frac{BC}{CD}.$$

Proof

Build the ray $\overrightarrow{CB}$ and draw the line that passes through A and is parallel to $\overline{BD}$. Let E be the point of intersection between the ray and the line (Figure 4.12). Using $\overleftrightarrow{AB}$ we can conclude that $m\angle EAB = \alpha$ where $\alpha = m\angle ABD = m\angle CBD$. Similarly, using $\overleftrightarrow{BC}$ it follows that $m\angle AEB = \alpha$. Hence ΔEBA is isosceles, so $EB = AB$. Then Thales Theorem, applied to ΔECA, implies that $\frac{EB}{AD} = \frac{BC}{CD}$. Therefore $\frac{AB}{AD} = \frac{BC}{CD}$. QED

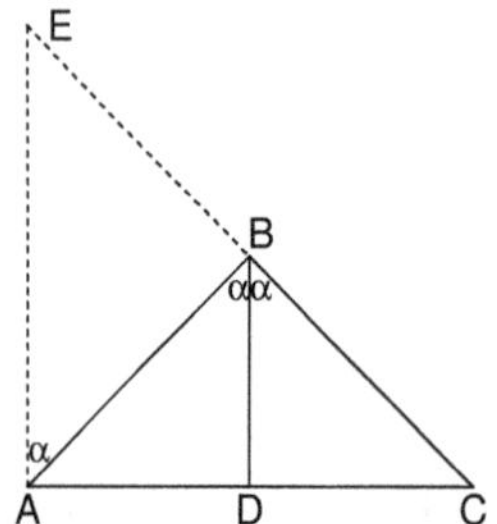

Figure 4.12: Proving the Angle Bisector Theorem

Remark

The Angle Bisector Theorem will be used in the next section in order to prove that the three angle bisectors of a triangle are concurrent.

4.10 Ceva's Theorem

Given an arbitrary triangle ABC, and segments $\overline{AY}$, $\overline{BX}$, and $\overline{CZ}$, which start at a vertex and end at the opposite sides[3], we can assert that the segments are concurrent if and only if

$$\frac{AX}{XC}\frac{CY}{YB}\frac{BZ}{ZA} = 1.$$

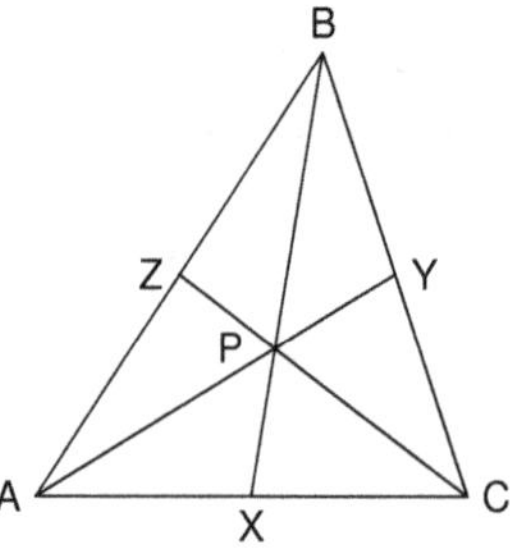

Figure 4.13: Proving Ceva's Theorem

[3]Segments of this type are called *cevians* in honor of Giovanni Ceva (1647-1734), who stated and proved the theorem that bears his name.

Proof

Suppose that the cevians meet at a point P (Figure 4.13). Using the remark from section 4.2 it follows that

$$\frac{AX}{XC} = \frac{a(\Delta BAX)}{a(\Delta BXC)} \text{ and } \frac{AX}{XC} = \frac{a(\Delta APX)}{a(\Delta PXC)}.$$

Then

$$AX \times a(\Delta BXC) = XC \times a(\Delta BAX), \tag{4.7}$$

$$AX \times a(\Delta PXC) = XC \times a(\Delta APX). \tag{4.8}$$

Subtracting (4.8) from (4.7) we get $AX \times a(\Delta BPC) = XC \times a(\Delta BAP)$. Thus

$$\frac{AX}{XC} = \frac{a(\Delta BAP)}{a(\Delta BPC)}. \tag{4.9}$$

In exactly the same fashion we can show that

$$\frac{CY}{YB} = \frac{a(\Delta APC)}{a(\Delta APB)}, \tag{4.10}$$

$$\frac{BZ}{ZA} = \frac{a(\Delta BPC)}{a(\Delta APC)}. \tag{4.11}$$

Multiplying together the last three expressions we get what we wanted, namely

$$\frac{AX}{XC}\frac{CY}{YB}\frac{BZ}{ZA} = 1.$$

Next we will prove that if $\frac{AX}{XC}\frac{CY}{YB}\frac{BZ}{ZA} = 1$ then the three cevians must meet at a point. Indeed, assume the antecedent and let P be the point of intersection between the cevians $\overline{AY}$ and $\overline{CZ}$. Furthermore, build the cevian $\overline{BW}$ that passes through P. Using the implication already proved, we have

$$\frac{AW}{WC}\frac{CY}{YB}\frac{BZ}{ZA} = 1.$$

Hence

$$\frac{AW}{WC}\frac{CY}{YB}\frac{BZ}{ZA} = \frac{AX}{XC}\frac{CY}{YB}\frac{BZ}{ZA}.$$

Therefore

$$\frac{AW}{WC} = \frac{AX}{XC}. \tag{4.12}$$

We claim that $W = X$ because $W \neq X$ contradicts (4.12). Indeed, suppose that W is to the left of X. Then $AW < AX$ and $CX < CW$, thus $AW \times CX < AX \times CW$, which in turn leads to $\frac{AW}{CW} < \frac{AX}{CX}$. In a similar fashion, if W is to the right of X we then have the inequality $\frac{AW}{CW} > \frac{AX}{CX}$. QED

Two important applications of Ceva's Theorem

The concurrency of the medians and the concurrency of the angle bisectors will follow almost immediately from Ceva's Theorem

1. Let the medians of ΔABC be $\overline{AY}$, $\overline{BX}$, and $\overline{CZ}$. Obviously

$$\frac{AX}{XC}\frac{CY}{YB}\frac{BZ}{ZA} = 1 \times 1 \times 1 = 1,$$

   consequently the medians are concurrent. As we know from section 3.3, wherein we used TT2 to prove the concurrency of the medians, their point of concurrency is called the **centroid**.

2. Let $\overline{AY}$, $\overline{BX}$, and $\overline{CZ}$ be the angle bisectors[4] of ΔABC. Applying the Angle Bisector Theorem, proven at the end of the preceding section, we have

$$\frac{AX}{XC} = \frac{AB}{BC}, \frac{CY}{YB} = \frac{AC}{AB}, \frac{BZ}{ZA} = \frac{BC}{AC}.$$

   Hence

$$\frac{AX}{XC}\frac{CY}{YB}\frac{BZ}{ZA} = \frac{AB}{BC}\frac{AC}{AB}\frac{BC}{AC} = 1.$$

   Therefore, the angle bisectors are concurrent, a result that was obtained in section 2.9. Recall that their point of concurrency is called the **incenter**.

Exercises

1. The square $ABCD$ has an area of 64 cm^2. Find the area of the equilateral triangle AED.

2. Given a square with side 5 cm, let P be the intersection of the diagonals and let Q be any point on $\overline{BC}$. Calculate the area of the non-convex quadrilateral $AQDPA$.

3. Let us consider the parallelogram $ABFE$ and the rectangle $ABCD$. Do they have equal areas?

4. You have a rectangular plank of wood, 2.10 m. wide and 4 m. long. Can you get it through a rectangular door that is 1 m. wide and 2 m. tall?

[4]Strictly speaking, an angle bisector is a ray but we are thinking about the cevians that lie on each angle bisector.

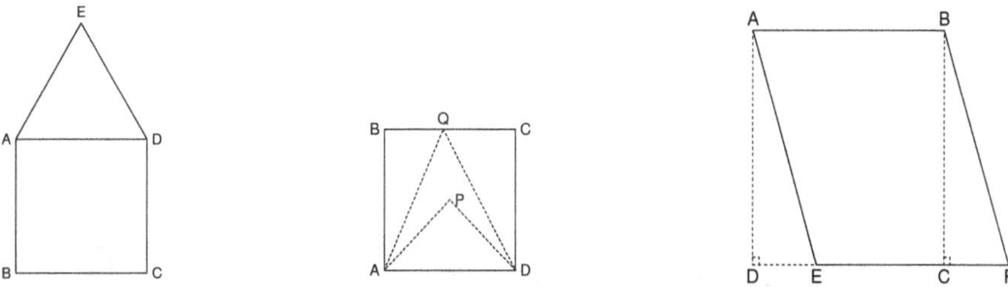

Figure 4.14: Figures corresponding to the first three exercises

5. Thanks to the converse of the Pythagorean theorem we can check that there exists a right triangle with sides 3, 4, 5. Are there infinitely many triplets (a, b, c) of positive integers such that $a^2 + b^2 = c^2$?

6. Given a right triangle ABC (Figure 4.15), with right angle at B, drop the height BD. Show that $BD^2 = AD \times DC$ (Hint: Apply Pythagoras' Theorem three times.)

7. You are given a parallelogram with equal sides and the length of both diagonals. Calculate the area of the parallelogram (Hint: Recall that the diagonals of a rhombus are perpendicular to each other.)

8. Build a square with side $1+2+3+4$ (Figure 4.15). Show that the area of each of the L-shaped regions is $2^3, 3^3, 4^3$ respectively. Then show that

$$(\tfrac{4(4+1)}{2})^2 = (1+2+3+4)^2 = 1^3 + 2^3 + 3^3 + 4^3.$$

9. Generalize the preceding exercise: Build a square with side $1 + 2 + 3 + ... + n$. Show that the area of each of the $n - 1$ L-shaped regions is $2^3, 3^3, ..., i^3, ..., n^3$ respectively. Then show that[5]

$$(\tfrac{n(n+1)}{2})^2 = 1^3 + 2^3 + 3^3 + ... + i^3 + ... + n^3 = \Sigma_{i=1}^{n} i^3.$$

10. Prove the converse of Thales' Theorem: If $\frac{AC}{BC} = \frac{CE}{CD}$ then $\overline{BD} \parallel \overline{AE}$ (Figure 4.9) (Hint: Proceed by contradiction assuming that $\overline{AE}$ is not parallel to $\overline{BD}$ and then draw $\overline{AF}$ so that $\overline{AF} \parallel \overline{BD}$.)

[5]This exercise provides an alternative justification to a formula found in section 1.2 .

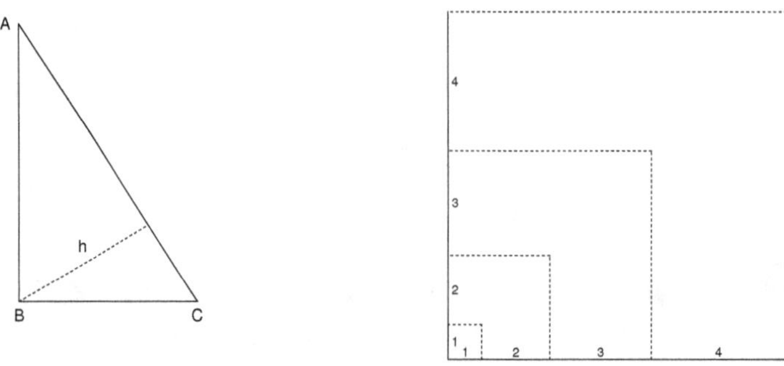

Figure 4.15: Figures corresponding to exercises 6 and 8

11. Provide a different proof of the Angle Bisector Theorem by choosing a point E on $\overrightarrow{CB}$ such that $EB = AB$ and proving that $\overline{AE} \parallel \overline{BD}$ (Hint: Choose E on $\overrightarrow{CB}$ so that $EB = AB$ and build $\overline{AE}$. ITT, SEAT and Thales' Theorem will lead to a proof.)

12. Given that ΔABC is isosceles, with $AB = AC$, $BD = 12$ ($\overline{BD}$ is a height), and $BC = 15$, calculate the area of ΔABC (Hint: Pythagoras' Theorem and a little bit of algebra is all you need to solve the problem at hand.)

Enrichment Note: Euclid's Proof of Pythagoras' Theorem

Proposition 47 from Book I of "The Elements" is none other than Pythagoras' Theorem. A modern rendition of Euclid's proof is as follows. Suppose that ABC is a right triangle with right angle at C (Figure 4.16). Then build squares on the three sides of the triangle and draw the segments $\overline{DB}$, $\overline{CH}$, as well as the perpendicular segment $\overline{CI}$. We observe that SAS congruence leads to $\Delta DAB \cong \Delta CAH$, hence

$$area(\Delta DAB) = area(\Delta CAH). \tag{4.13}$$

Moreover, $area(AKHI) = AH \times AK$ and $area(\Delta CAH) = \frac{1}{2}AH \times AK$. Therefore

$$area(AKHI) = 2area(\Delta CAH). \tag{4.14}$$

Analogously, $area(ADEC) = AD \times AC$ and $area(\Delta DAB) = \frac{1}{2}AD \times AC$, so

$$area(ADEC) = 2area(\Delta DAB). \tag{4.15}$$

Then from (4.13), (4.14), and (4.15) it follows that

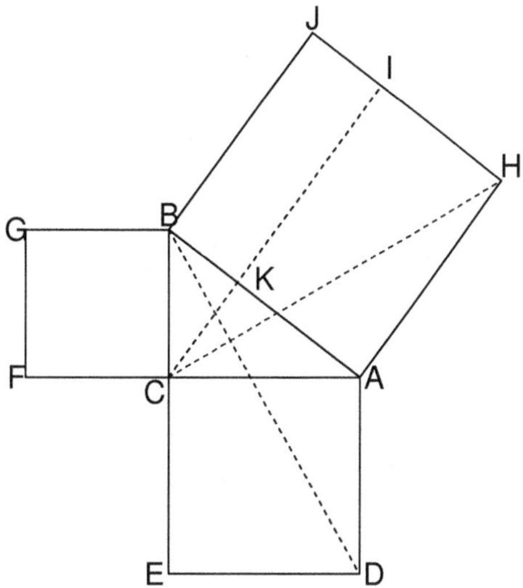

Figure 4.16: Euclid's proof of the Pythagorean proposition

$$area(AKIH) = area(ADEC).$$

In a similar fashion we can prove that

$$area(KBJI) = area(CFGB).$$

Then

$$area(ABJH) = area(AKIH) + area(KBJI) = area(ADEC) + area(CFGB).$$

Therefore, $c^2 = b^2 + a^2$ where c is the length of the hypotenuse and a, b are the lengths of the legs. QED

Chapter 5

Similarity of Triangles

5.1 AA similarity

Loosely speaking we might say that two triangles are similar if one is the photographic reduction, or enlargement, of the other. In a more precise way, given ΔABC and ΔDEF we will say that they are similar if their corresponding angles are congruent and their corresponding sides are proportional (Figure 5.1), that is,

1. $\angle A \cong \angle D$, $\angle B \cong \angle E$, and $\angle C \cong \angle F$.

2. $\frac{AB}{DE} = \frac{BC}{EF} = \frac{AC}{DF}$.

Under these circumstances we will write $\Delta ABC \sim \Delta DEF$.

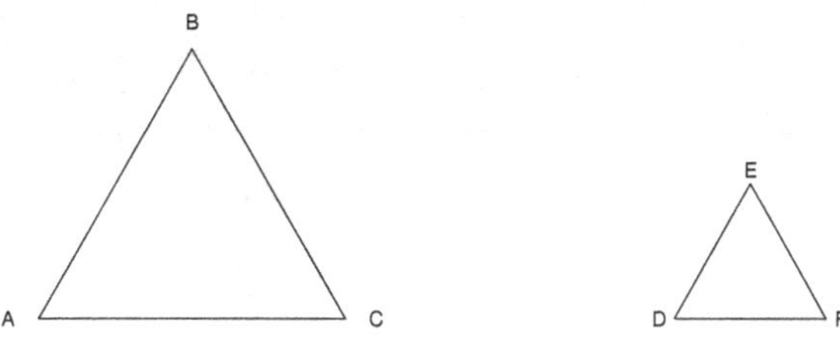

Figure 5.1: The concept of similarity

Since we are working in Euclidean geometry, it is only necessary to have two pairs of corresponding congruent angles; the third pair will be necessarily congruent. Moreover, congruence of triangles implies similarity since $\frac{AB}{A'B'} = \frac{BC}{B'C'} = \frac{AC}{A'C'} = 1$. Obviously, similarity does not imply congruence. Next we will prove a remarkable theorem that will save us a great deal of work because it states that two triangles are similar if and only if a pair of corresponding angles are congruent; nothing else is required since the proportionality of sides is implied by the theorem.

Theorem (AA similarity)

Given ΔABC and ΔDEF (Figure 5.2) assume that $\angle B \cong \angle E$ and $\angle A \cong \angle D$. Then $\Delta ABC \sim \Delta DEF$.

Proof

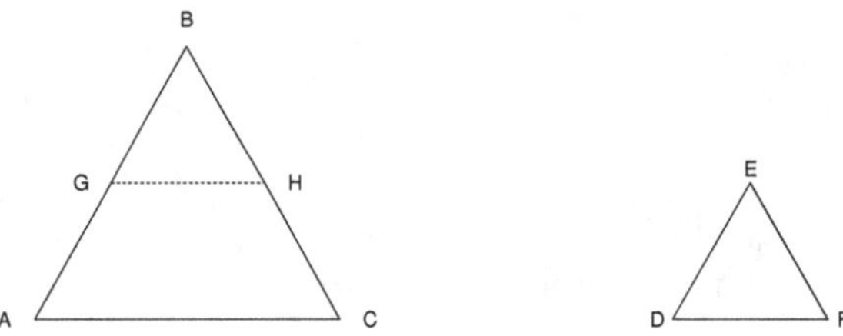

Figure 5.2: A proof of AA similarity

If $AB = DE$ then by ASA congruence it follows that $\Delta ABC \cong \Delta DEF$, so $\Delta ABC \sim \Delta DEF$. Suppose that $AB \neq DE$, say $DE < AB$. Choose G on $\overline{AB}$ in such a way that $BG = DE$ and, thereafter, build $\overline{GH}$ so that $\overline{GH} \parallel \overline{AC}$. Since the corresponding angles of two parallel lines are congruent, we can conclude that $\angle A \cong \angle G$. On the other hand, by ASA congruence we observe that $\Delta GBH \cong \Delta DEF$; thus, $BH = EF$. But Thales' theorem implies that $\frac{AB}{BG} = \frac{BC}{BH}$, consequently

$$\frac{AB}{DE} = \frac{BC}{EF}.$$

Proceeding in a similar fashion, with $\overline{JK} \parallel \overline{AB}$ and $JC = EF$ (see Figure 5.3), we can prove that

$$\frac{BC}{EF} = \frac{AC}{DF}.$$

QED

Corollary

Given ΔABC (Figure 5.4), assume that $\overline{DE} \parallel \overline{AC}$. Then $\Delta ABC \sim \Delta DBE$.

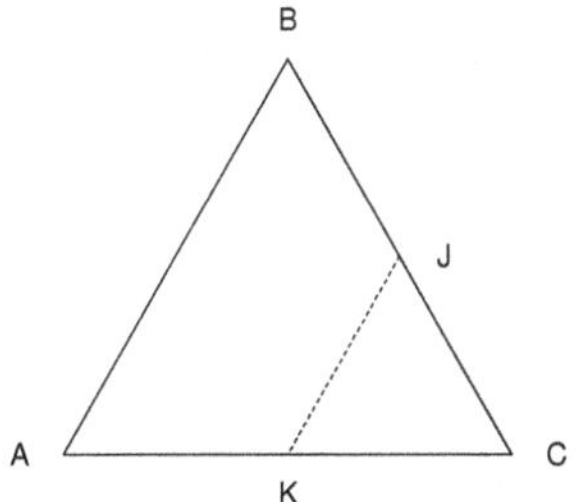

Figure 5.3: Completing a proof of AA similarity

Proof

Since $\overline{DE} \parallel \overline{AC}$ we can assert that $\angle BDE \cong \angle BAC$. But both triangles have angle B in common, so the AA similarity theorem implies that $\Delta ABC \sim \Delta DBE$.

QED

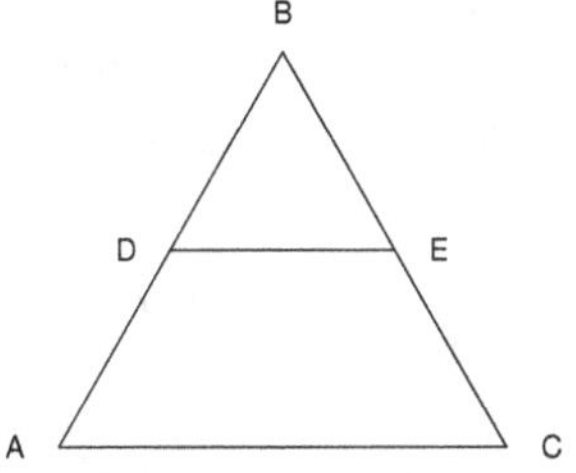

Figure 5.4: A corollary of the theorem on AA similarity

Examples

1. The theorem on AA similarity is extensively used in applications. Assume that two cottages, located at 40 m. and 50 m. from a river, on the same shore, get their water from a pump (Figure 5.5). Taking into consideration that $AB = 100$ m., find the point on $\overline{AB}$ where the pump has to be in order to minimize the amount of pipe used.

   We start by prolonging $\overline{AC}$ so that $AE = AC$. Then we build $\overline{ED}$, which

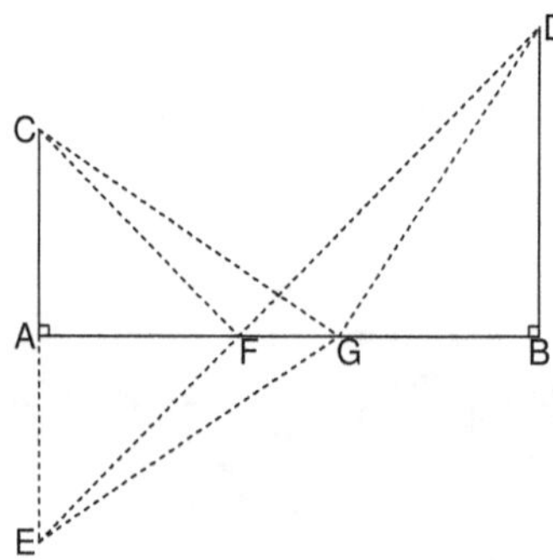

Figure 5.5: The two-cottages problem

intersects $\overline{AB}$ at F. We claim that F is the best choice. Indeed, let G be any other point[1] on $\overline{AB}$. By SAS congruency we can assert that $\Delta CAF \cong \Delta EAF$ and $\Delta CAG \cong \Delta EAG$. Thus, $CF = FE$ and $CG = GE$. Applying the triangle inequality to ΔEDG we get $ED < EG + GD$. Hence

$$CF + FD < CG + GD.$$

Therefore, our claim that F is the best choice has been substantiated.

A further question: What is the value of $x = AF$? Since $\Delta CAF \cong \Delta EAF$ we can assert that $\angle CFA \cong \angle EFA$. But $\angle AFE$ and $\angle BFD$ are opposite angles, so $\angle CFA \cong \angle DFB$. Then AA similarity implies that $\Delta CAF \sim \Delta DBF$. Therefore

$$\tfrac{x}{100-x} = \tfrac{40}{50}.$$

Consequently $x = 44.44$ m.

2. AA similarity provides a simple proof of Pythagoras' Theorem. Indeed, let us consider a right triangle ABC with right angle at C and drop the height $\overline{CD}$ (Figure 5.6). Let $a = BC, b = AC, c = AB, d = BD$. Using AA similarity we

[1]Without loss of generality we assume that G lies to the right of F.

can conclude that the three triangles are similar. Since $\Delta ABC \sim \Delta CBD$ we get $\frac{a}{c} = \frac{d}{a}$, thus

$$a^2 = cd. \tag{5.1}$$

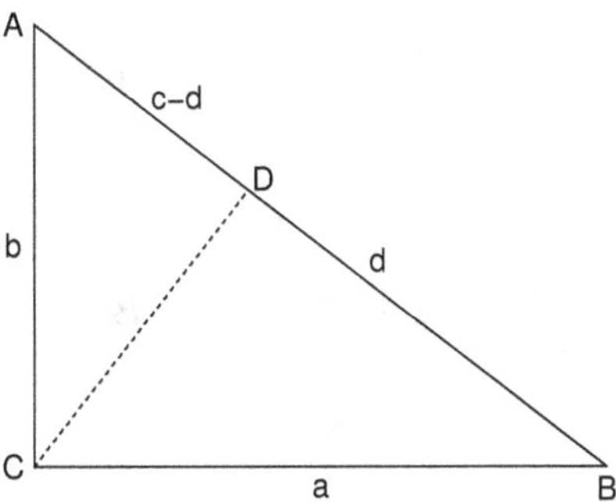

Figure 5.6: A proof of Pythagoras' Theorem through AA similarity

Besides, since $\Delta ABC \sim \Delta ACD$ we will have $\frac{b}{c} = \frac{c-d}{b}$. Hence

$$b^2 = c^2 - cd. \tag{5.2}$$

From (3.1) and (3.2) we get $a^2 + b^2 = c^2$.

QED

5.2 SAS similarity

In order to define the number π in the next chapter, as well as to study many other topics, the following result is needed.

Theorem

Given ΔABC and ΔDEF assume that $\angle B \cong \angle E$ and $\frac{AB}{DE} = \frac{BC}{EF}$. Then $\Delta ABC \sim \Delta DEF$.

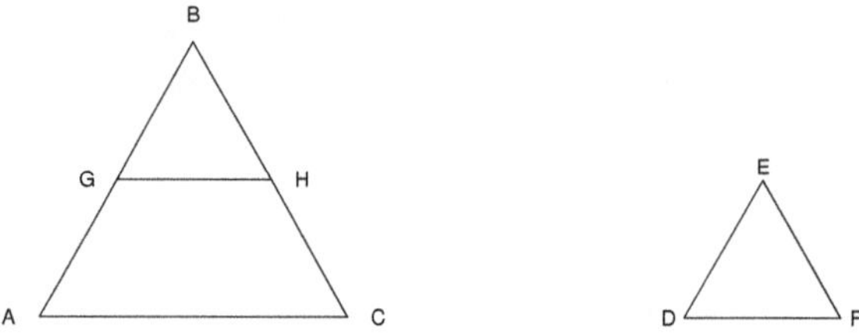

Figure 5.7: Proving SAS similarity

Proof

If $AB = DE$ and $BC = EF$ then $\Delta ABC \cong \Delta DEF$ by SAS congruence, which in turn implies that $\Delta ABC \sim \Delta DEF$ and we are done. Assume that $AB \neq DE$ or $BC \neq EF$. Without loss of generality let us assume that $DE < AB$. Then choose G on $\overline{AB}$ so that $BG = DE$ and build $\overline{GH}$, parallel to $\overline{AC}$ (Figure 5.7). Hence $\angle G \cong \angle A$ and by AA similarity we can conclude that

$$\Delta ABC \sim \Delta GBH. \tag{5.3}$$

Our strategy is to prove that $\Delta GBH \cong \Delta DEF$. With this goal in mind, all we have to do is show that $BH = EF$. Indeed, since $\frac{DE}{AB} = \frac{EF}{EC}$ by hypothesis and (5.3) holds, we will have

$$\frac{BH}{BC} = \frac{BG}{AB} = \frac{DE}{AB} = \frac{EF}{BC}.$$

Therefore $BH = EF$ and consequently

$$\Delta GBH \cong \Delta DEF. \tag{5.4}$$

From (5.3) and (5.4) it follows that $\Delta ABC \sim \Delta DEF$. QED

Alternative proof of a property of medians

In section 3.3 (first remark) we proved the 1/3, 2/3 property of medians of a triangle on the basis of the theory of parallelograms. Using SAS similarity it is possible to provide an alternative proof. Indeed, assume that $\overline{BE}$ and $\overline{DC}$ are two medians of ΔABC, intersecting at G (Figure 5.8). Then build $\overline{DE}$ and observe that

$$\frac{AB}{AD} = 2 = \frac{AC}{AE},$$

which in turn leads to $\Delta ABC \sim \Delta ADE$ due to SAS similarity[2]. In particular $\angle ADE \cong \angle ABC$, thus $\overline{DE} \parallel \overline{BC}$ and consequently $\angle EDG \cong \angle BCG$. Since $\angle DGE \cong \angle BCG$ we can conclude, using AA similarity, that $\Delta DGE \sim \Delta CGB$.

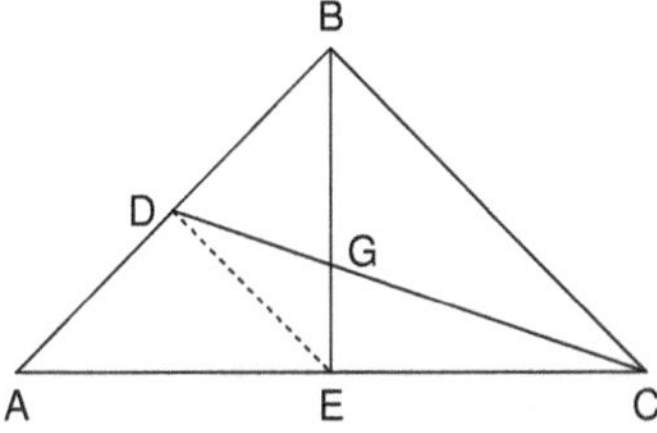

Figure 5.8: Proving the 2/3, 1/3 property of medians

Therefore

$$\frac{BG}{GE} = \frac{BC}{DE}.$$

But from the relationship $\Delta ABC \sim \Delta ADE$ we get $\frac{BC}{DE} = \frac{AB}{AD} = 2$, hence $\frac{BG}{GE} = 2$. Thus, $2GE = BG$ and consequently $3GE = GE + BG = BE$, so $GE = \frac{1}{3}BE$ and obviously $BG = \frac{2}{3}BE$. QED

Remark

The 1/3,2/3 property of medians provides a short proof of the concurrency of the medians. Let us consider an arbitrary triangle ABC and the medians $\overline{BE}$ and $\overline{CD}$, which intersect at a point G. Then draw the third median, namely $\overline{AF}$, which intercepts $\overline{BE}$ at G'. By the 1/3, 2/3 property of medians we have $EG = \frac{1}{3}BE$ and $EG' = \frac{1}{3}BE$, therefore $EG = EG'$. Since G and G' lie on $\overline{BE}$ it follows that $G = G'$.

5.3 SSS similarity

It is to be noted that the SSS property of similarity does hold. For instance, the triangle with sides $12, 9, 6$ is similar to the triangle with sides $4, 3, 2$.

[2]Both triangles share the same angle at the top.

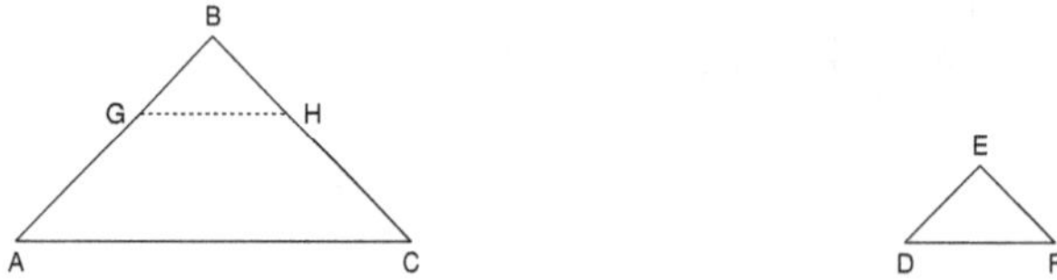

Figure 5.9: SSS similarity

Theorem

Given ΔABC and ΔDEF (Figure 5.9), assume that

$$\frac{AB}{DE} = \frac{BC}{EF} = \frac{AC}{DF}.$$

Then $\Delta ABC \sim \Delta DEF$.

Proof

If $AB = DE$, $BC = EF$, and $AC = DF$, both triangles are congruent, due to the SSS property of congruence, and the proof comes to an end because congruence implies similarity. Assume, without loss of generality, that $DE < AB$. Right away choose G on $\overline{AB}$ such that $GB = DE$ and build $\overline{GH}$ in such a way that $\overline{GH} \parallel \overline{AC}$. Then $\angle BGH \cong \angle BAC$, thus AA similarity implies that

$$\Delta ABC \sim \Delta GBH. \tag{5.5}$$

Next we will prove that $GH = DF$ and $BH = EF$. Indeed, using the fact that $GB = DE$, as well as the hypothesis and (5.5), we will have

$$\frac{GH}{AC} = \frac{GB}{AB} = \frac{ED}{AB} = \frac{DF}{AC}, \tag{5.6}$$

$$\frac{BH}{BC} = \frac{BG}{AB} = \frac{DE}{AB} = \frac{EF}{BC}. \tag{5.7}$$

From (5.6) we get $GH = DF$ while from (5.7) we get $BH = EF$, therefore $\Delta GBH \cong \Delta DEF$. Since $\Delta ABC \sim \Delta GBH$ (recall (5.5)) we finally reach the thesis, namely $\Delta ABC \sim \Delta DEF$. QED

Exercises

1. Assume that $\Delta ABC \sim \Delta DEF$. Let h_1 and h_2 be their respective heights and let r be the constant of proportionality between both triangles. Find a relationship between both heights and r.

2. Given that $\overline{AB} \parallel \overline{DE}$, build $\overline{AE}$ and $\overline{BD}$. The latter meet at a point C. Are both triangles similar?

3. Suppose that $\Delta BCD \sim \Delta FDE$. Is the quadrilateral $BDFA$ a parallelogram? (Figure 5.10)

4. Let us consider rectangles $ABCD$ and $EFGH$ (Figure 5.10). Show that $\Delta EBF \sim \Delta FCG$ (Hint: AA similarity is the key to the solution.)

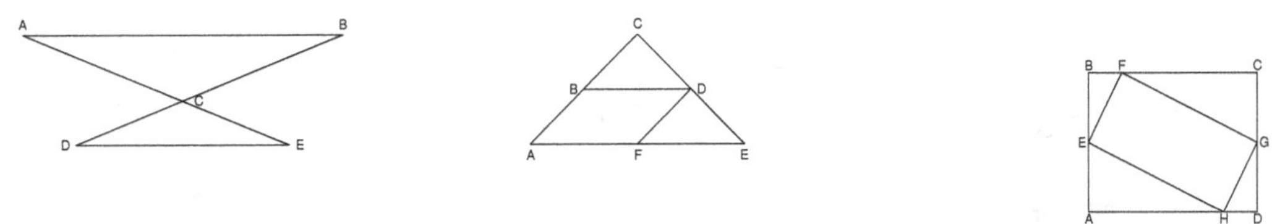

Figure 5.10: Figures related to exercises 2, 3, and 4.

5. Angles ABD and CBE are right angles and $CD = BD$ (Figure 5.11). Can you find any two similar triangles among all possible triangles? (Hint: Try to prove that $\Delta GCB \sim \Delta GBE$.)

6. Given that $BC = CD$ and $AB = DE$ (Figure 5.11), can we conclude that both triangles are similar?

7. Show that, in any right triangle, the product of the legs is equal to the product of the hypotenuse times the height that starts from the right vertex (the vertex corresponding to the right angle) and ends on the hypotenuse.

8. Solve exercise 6 (chapter 4) without using Pythagoras' Theorem (Hint: Just note that the height that starts at the right angle divides the triangle into two similar triangles).

9. Two cottages are located on the same side of a river. Their distances to the river are 100 m. and 175 m., respectively, and the distance between both parallel lines is 200 m. Where should we install a water pump in order to minimize the amount of pipe needed to carry water from the river?

10. Line $\overleftrightarrow{MN}$ meets line $\overleftrightarrow{AC}$ at a point P (Figure 5.11). Show that

$$\frac{BM}{MA}\frac{AP}{PC}\frac{CN}{NB} = 1.$$

(Hint: Draw a line that passes through B and is parallel to $\overleftrightarrow{AP}$.)

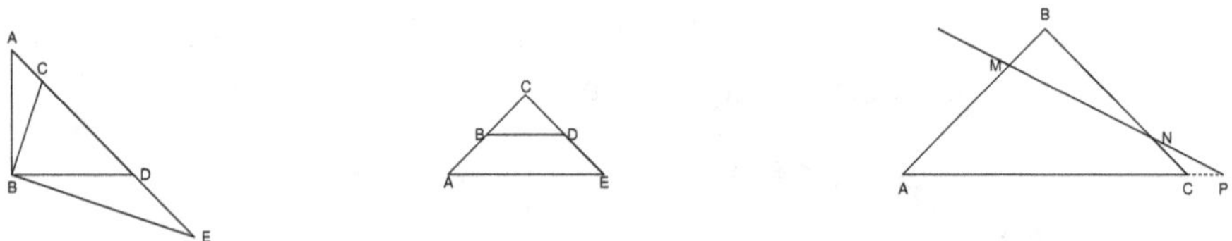

Figure 5.11: Figures related to exercises 5, 6, and 10.

Enrichment Note: The Euler Line

Given any non-equilateral triangle, its circumcenter, centroid, and orthocenter lie on a line. Leonhard Euler (1707-1783), a towering 18th century mathematician, stated and proved this result.

Proof

Let O be the circumcenter, that is, the point where the three perpendicular bisectors meet, and let G be the centroid or center of gravity of the given triangle (Figure 5.12). Start by drawing $\overleftrightarrow{GO}$ and define H on $\overleftrightarrow{GO}$ so that $HG = 2GO$ (H lies on the opposite side of G from O). Since $\overline{AM_1}$ is a median we can assert that $AG = 2GM_1$. Thus

$$\frac{AG}{GM_1} = 2 = \frac{HG}{GO}.$$

Then the SAS property of similarity implies that $\Delta AGH \sim \Delta M_1GO$ and, in particular, $\angle AHG \cong \angle M_1OG$; thus, $\overleftrightarrow{AH} \parallel \overleftrightarrow{OM_1}$. But, since O is the circumcenter, it follows that $\overline{OM_1} \perp \overline{BC}$ and consequently $\overline{AH_1} \perp \overline{BC}$. Therefore $\overline{AH_1}$ is a height.

In a similar fashion (Figure 5.13) $CG = 2GM_2$, so

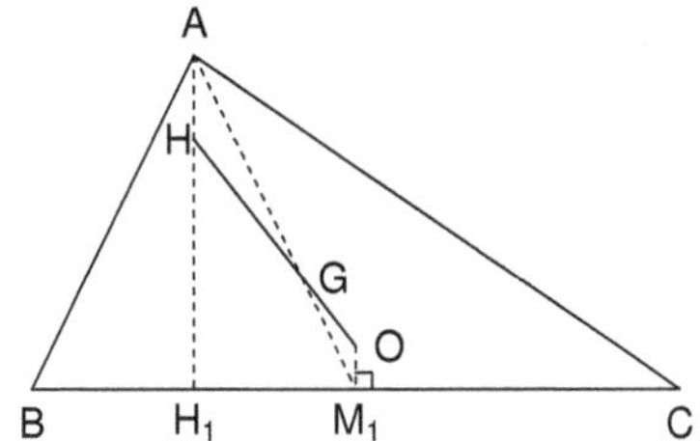

Figure 5.12: The Euler line (first step)

$$\frac{CG}{GM_2} = 2 = \frac{HG}{GO}.$$

Then the SAS property of similarity implies that $\Delta CGH \sim \Delta M_2GO$; thus, in particular, $\angle CHG \cong \angle M_2OG$, hence $\overline{CH_2} \parallel \overline{OM_2}$. But $OM_2 \perp \overline{AB}$, consequently $\overline{CH_2} \perp \overline{AB}$. That is to say, $\overline{CH_2}$ is a height. We have just shown that H is the point of intersection of two heights, hence H must be the orthocenter. In summary, we have been able to prove that H, G, and O lie on the same line and, moreover, $HG = 2OG$. QED

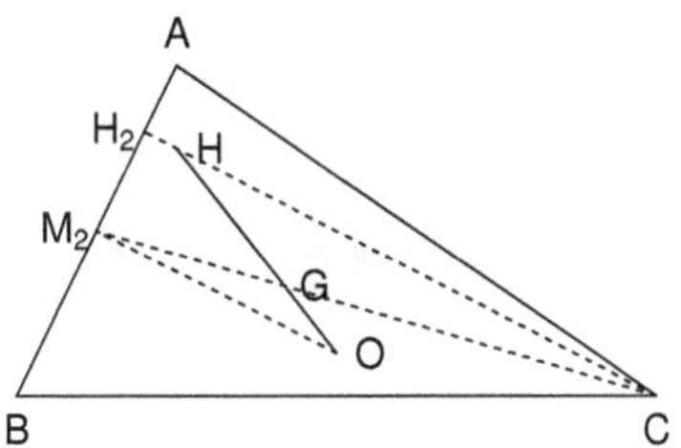

Figure 5.13: The Euler line (second step)

Chapter 6

Trigonometry of Acute Angles

6.1 Basic ideas from trigonometry

Given an arbitrary acute angle with measure α, made up from the rays L_1 and L_2 meeting at O (Figure 6.1), we choose an arbitrary point B on L_1 and drop the perpendicular $\overline{BC}$ towards L_2. Then we define the sine, cosine, and tangent of the angle with measure α:

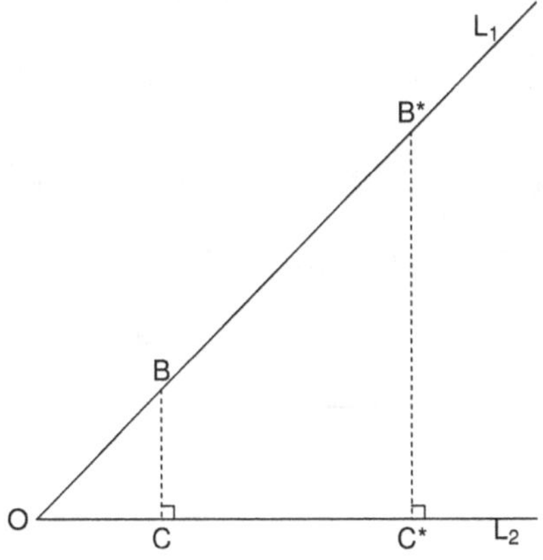

Figure 6.1: Definition of the sine, cosine, and tangent of an acute angle

$$\sin\alpha = \frac{BC}{OB},$$

$$\cos\alpha = \frac{OC}{OB},$$

$$\tan\alpha = \frac{BC}{OC}.$$

We are dealing with a 'good definition' since if we choose another point B^* on L_1 and drop the perpendicular $\overline{B^*C^*}$, both triangles are similar due to AA similarity. Therefore $\frac{BC}{OB} = \frac{B^*C^*}{OB^*}$, $\frac{OC}{OB} = \frac{OC^*}{OB^*}$, and $\frac{BC}{OC} = \frac{B^*C^*}{OC^*}$.

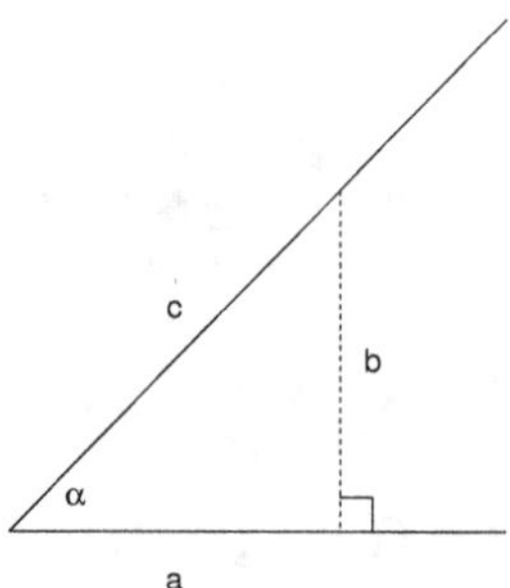

Figure 6.2: Justifying a fundamental trigonometric identity

Right away we notice that

$$\tan\alpha = \frac{\sin\alpha}{\cos\alpha}.$$

Moreover, from Figure 6.2 we get $\sin^2\alpha = \frac{b^2}{c^2}$ and $\cos^2\alpha = \frac{a^2}{c^2}$. Thus

$$\sin^2\alpha + \cos^2\alpha = \frac{b^2}{c^2} + \frac{a^2}{c^2} = \frac{a^2+b^2}{c^2}.$$

But Pythagoras' Theorem asserts that $a^2 + b^2 = c^2$. Therefore

$$\boxed{\cos^2\alpha + \sin^2\alpha = 1}. \tag{6.1}$$

This is a fundamental trigonometric identity.

6.2 Three particular angles

Let us consider an angle with measure 45° (Figure 6.3). Choose an arbitrary point A on L_1 and draw the perpendicular segment $\overline{AB}$, with B on L_2. Since $m\angle OAB = 45^o$, the converse of the Isosceles Triangle Theorem implies that $AB = OB = x$. Using Pythagoras Theorem we get $OA = x\sqrt{2}$. Therefore,

$$\sin 45^o = \frac{x}{x\sqrt{2}} = \frac{1}{\sqrt{2}} = \frac{\sqrt{2}}{2},$$

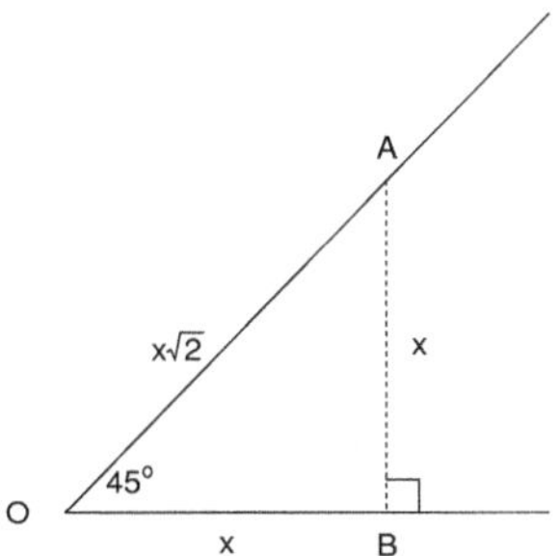

Figure 6.3: The trigonometry of an angle with measure 45°

$$\cos 45^o = \frac{x}{x\sqrt{2}} = \frac{1}{\sqrt{2}} = \frac{\sqrt{2}}{2},$$

$$\tan 45^o = \frac{x}{x} = 1.$$

Next let us consider an angle with a measure of 60° (Figure 6.4). Choose an arbitrary point A on L_1 and construct the perpendicular $\overline{AB}$, with B on L_2. Then build segment $\overline{AC}$ in such a way that $m\angle BAC = 30^o$. We can conclude that ΔDAC is equilateral since its three internal angles are congruent (their common measure is 60°), say $x = OA = AC = OC$. Using SAS, or ASA, it follows that $\Delta OAB \cong \Delta CAB$. In particular, $OB = BC = \frac{x}{2}$. Thus, $AB = \sqrt{x^2 - \frac{x^2}{4}} = \frac{\sqrt{3}}{2}x$. Hence

$$\sin 60^o = \frac{\frac{\sqrt{3}}{2}x}{x} = \frac{\sqrt{3}}{2},$$

$$\cos 60^o = \frac{\frac{x}{2}}{x} = \frac{1}{2},$$

$$\tan 60^o = \frac{\frac{\sqrt{3}}{2}}{\frac{1}{2}} = \sqrt{3}.$$

Analogously,

$$\sin 30^o = \frac{\frac{x}{2}}{x} = \frac{1}{2},$$

$$\cos 30^o = \frac{\frac{\sqrt{3}}{2}x}{x} = \frac{\sqrt{3}}{2},$$

$$\tan 30^o = \frac{\frac{1}{2}}{\frac{\sqrt{3}}{2}} = \frac{1}{\sqrt{3}} = \frac{\sqrt{3}}{3}.$$

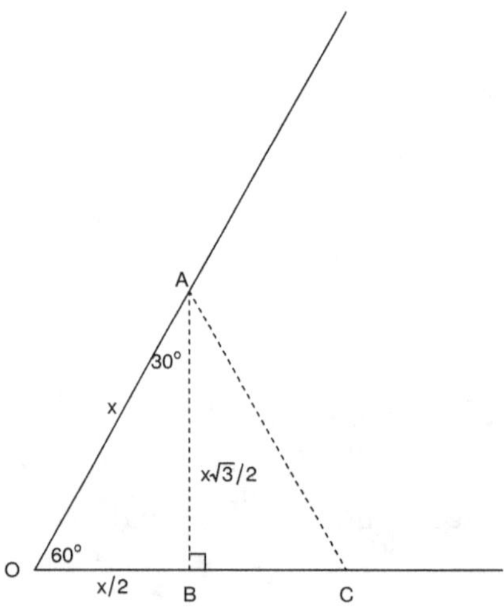

Figure 6.4: The trigonometry of a 30° and 60° angle

Later on we will be able to calculate the sine, cosine, and tangent of other important angles. In principle we could find approximate values for $\sin\alpha, \cos\alpha, \tan\alpha$ by measuring α with a protractor and using a good ruler to measure AB, BA, and DC. In practice, everyone uses nowadays an electronic calculator. These devices have software that do the needed calculations employing algorithms that stem from the theory of series. For instance, we carefully made measurements on a sheet of paper and found that $\sin 40^o = 0.65, \cos 40^o = 0.78$, and $\tan 40^o = 0.84$. An electronic calculator provides the following, very accurate, values: $\sin 40^o = 0.6427876, \cos 40^o = 0.76604443$, and $\tan 40^o = 0.8390996$.

6.3 A graphical representation

Let us consider a circle with radius 1. Assume that α is the measure of an acute angle (Figure 6.5). Then $\sin\alpha = \frac{AB}{1} = AB$, $\cos\alpha = \frac{OA}{1} = OA$, where $\overline{BA}$ is perpendicular to $\overline{OC}$. Moreover, we make a right turn at C and build a ray that goes upwards, which will meet at D the ray that starts at O and contains B. Using AA similarity we have $\frac{DC}{BA} = \frac{1}{OA}$, thus

$$DC = \frac{AB}{OA} = \frac{\sin\alpha}{\cos\alpha}.$$

Consequently $DC = \tan\alpha$.

From these graphical representations we can see that the sine of an acute angle increases as α increases. It is very close to zero insofar as the angle has a very

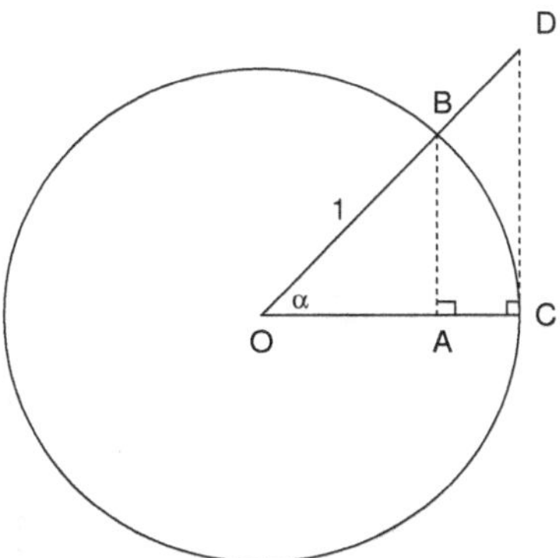

Figure 6.5: Graphical representation of the sine, cosine, and tangent

small measure, while it is very close to 1 when α gets very close to 90°. The opposite happens to $\cos\alpha$, from being very close to 1, when the angle has a very small measure, it decreases as α increases. Close to 90° its value is almost zero. Finally, $\tan\alpha$ increases, from being almost zero when the measure of α is close to zero, to a huge value as we get close to 90°. Obviously, the sine, cosine, or tangent of an acute angle is a positive number.

6.4 Another formula for the area of a triangle

Let us consider an arbitrary acute triangle ABC (Figure 6.6), with $m\angle C = \alpha$. We will show that

$$\text{area}(\Delta ABC) = \tfrac{1}{2}ab\sin\alpha.$$

where $a = BC$ and $b = AC$. Indeed, area$(\Delta ABC) = \frac{1}{2}bh$, where h is the height that starts from the vertex B. But $sin\alpha = \frac{h}{a}$, thus area$(\Delta ABC) = \frac{b}{2}a\sin\alpha$. The reader can check that if angle C is obtuse then area$(\Delta ABC) = \frac{1}{2}ab\sin(180 - \alpha)$.

The new formula for the area of a triangle is pretty useful. For instance, suppose that we wish to build a triangular tent of maximum volume (Figure 6.7). What angle at the top should we choose? Since the length is fixed, we have to choose θ so that the area of the triangle is maximized. Since area $\Delta = \frac{1}{2}a^2\sin\theta$ it is evident that we should build the tent in such a way that θ is as close to 90° as possible (recall that

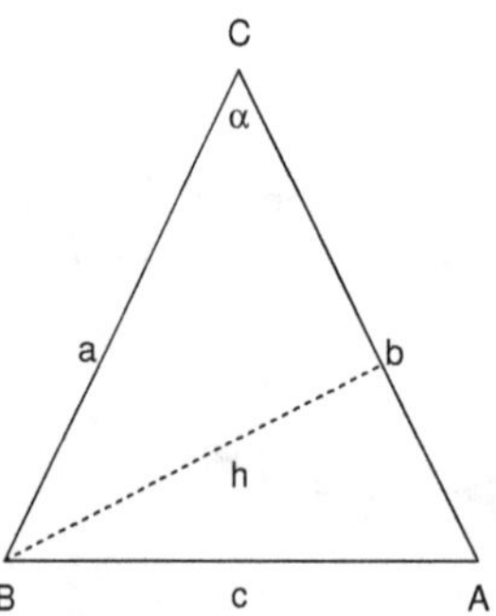

Figure 6.6: A useful formula for area

$\sin\theta$ increases its value as θ increases, $0 < \theta < 90$). In chapter 12 we will discuss this problem again, from a different perspective.

6.5 Law of sines

Given any acute triangle ABC (Figure 6.8) we have

$$\frac{\sin\alpha}{a} = \frac{\sin\beta}{b} = \frac{\sin\gamma}{c},$$

where $a = BC$, $b = AC$, and $c = AB$. If the triangle is obtuse (with obtuse angle at A, see Figure 6.8) then

$$\frac{\sin(180-\alpha)}{a} = \frac{\sin\beta}{b} = \frac{\sin\gamma}{c}.$$

Proof

Let us first consider the case when the triangle is acute. We start by dropping the perpendicular $\overline{BD}$ (Figure 6.9). Then $\sin\alpha = \frac{BD}{AB}$ and $\sin\gamma = \frac{BD}{BC}$. Hence $(\sin\alpha)AB = (\sin\gamma)BC$, thus

$$\frac{\sin\alpha}{BC} = \frac{\sin\gamma}{AB}. \tag{6.2}$$

Next build the perpendicular $\overline{AF}$ (Figure 6.9). Then $\sin\beta = \frac{AF}{AB}$, $\sin\gamma = \frac{AF}{AC}$. Thus, $(\sin\beta)AB = (\sin\gamma)AC$. Consequently

$$\frac{\sin\beta}{AC} = \frac{\sin\gamma}{AB}. \tag{6.3}$$

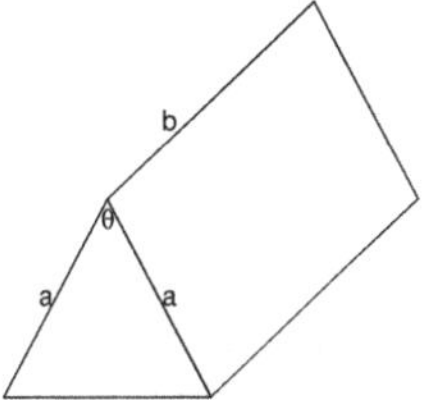

Figure 6.7: The tent problem

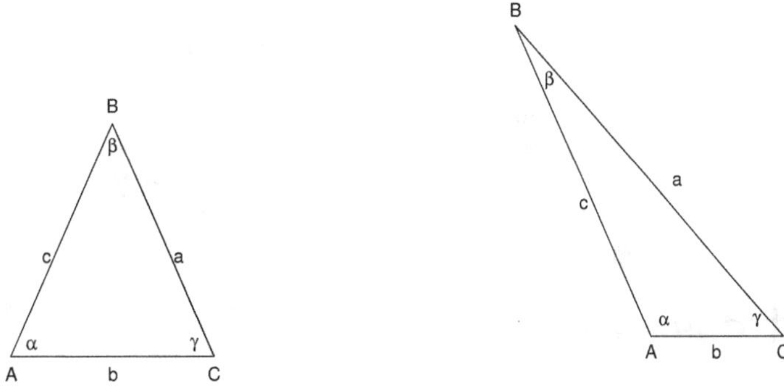

Figure 6.8: Law of sines

From (6.2) and (6.3) we get

$$\frac{\sin\alpha}{BC} = \frac{\sin\beta}{AC} = \frac{\sin\gamma}{AB}.$$

Let us now deal with an obtuse triangle. Start by building the perpendicular $\overline{BD}$ (Figure 6.9). We can see that $\sin(180-\alpha) = \frac{BD}{c}$, $\sin\gamma = \frac{BD}{a}$. Hence $c\sin(180-\alpha) = a\sin\gamma$, so

$$\frac{\sin(180-\alpha)}{a} = \frac{\sin\gamma}{c}. \tag{6.4}$$

Next build the perpendicular $\overline{AE}$. Therefore $\sin\beta = \frac{AE}{AB}$ and $\sin\gamma = \frac{AE}{b}$. Thus, $c\sin\beta = b\sin\gamma$, which in turn leads to

$$\frac{\sin\beta}{b} = \frac{\sin\gamma}{c}. \tag{6.5}$$

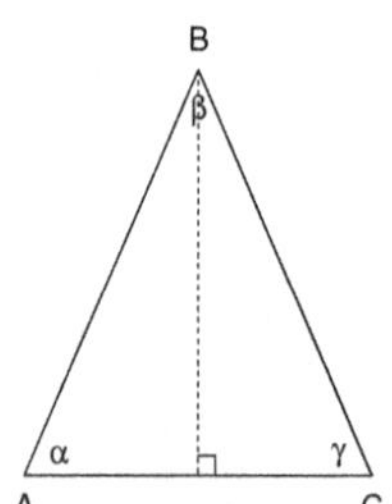

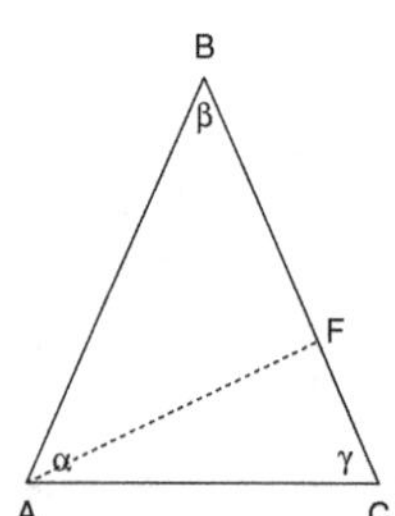

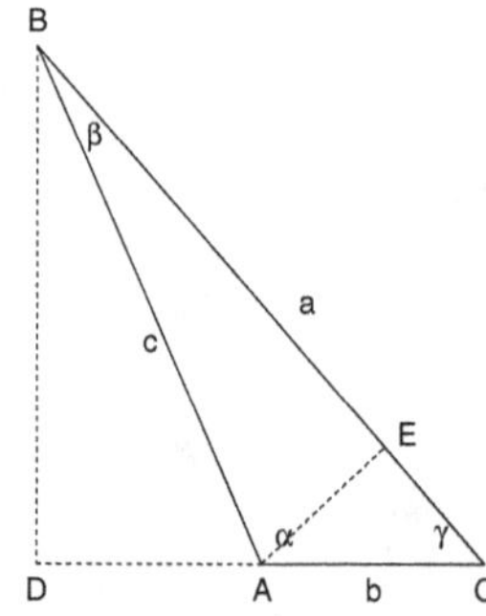

Figure 6.9: Proving the law of sines

From (6.4) and (6.5) we get

$$\frac{\sin(180-\alpha)}{a} = \frac{\sin\beta}{b} = \frac{\sin\gamma}{c}.$$

QED

6.6 Law of cosines

Given an acute triangle ABC (Figure 6.10) we have

$$a^2 = b^2 + c^2 - 2bc\cos\alpha,$$

where $\alpha = m\angle A$. If ΔABC is obtuse, with obtuse angle at A (Figure 6.10), the law

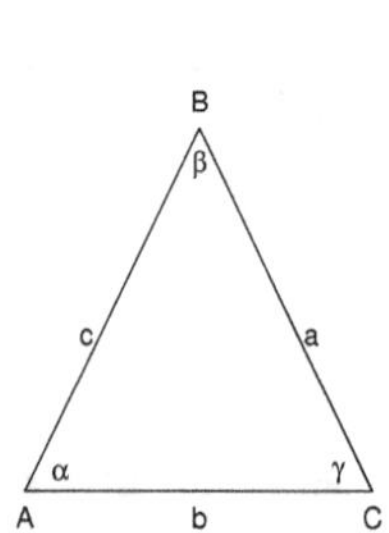

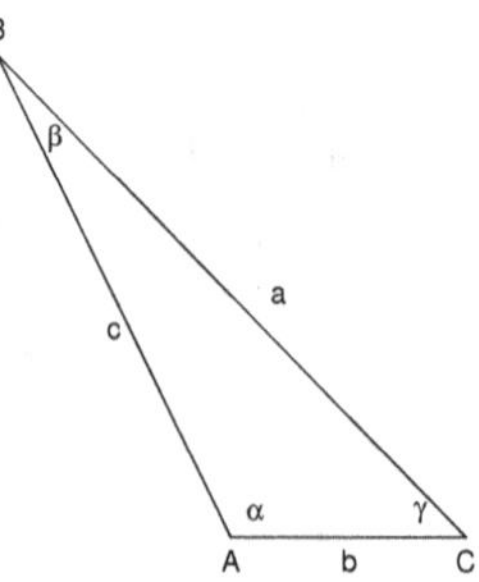

Figure 6.10: The law of cosines

of cosines becomes $a^2 = b^2 + c^2 + 2bc\cos(180 - \alpha)$.

Proof

Let us first provide a proof when the triangle is acute. Build the height $\overline{BD}$, and let $h = BD$. Then $\cos\alpha = \frac{x}{c}$, where $x = AD$. Applying Pythagoras' Theorem to triangles ABD and CBD we get

$$c^2 - x^2 = h^2 = a^2 - (b - x)^2.$$

Thus

$$c^2 - x^2 = a^2 - b^2 + 2bx - x^2,$$

that is,

$$a^2 = b^2 + c^2 - 2bx.$$

But $x = c\cos\alpha$, consequently $a^2 = b^2 + c^2 - 2bc\cos\alpha$. Let us now provide a proof

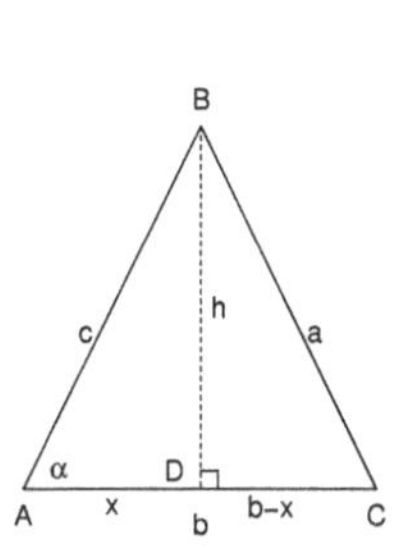

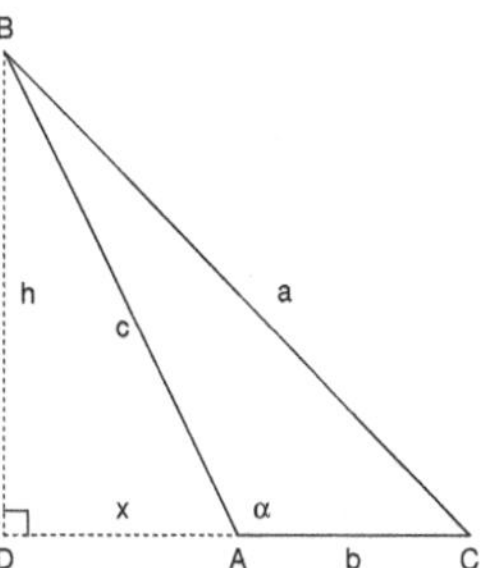

Figure 6.11: Proving the law of cosines

when the angle under consideration is obtuse. We start by building the perpendicular $\overline{BD}$ and writing $h = BD$ (Figure 6.11). Then $\cos(180 - \alpha) = \frac{x}{c}$, where $x = AD$. Applying Pythagoras Theorem twice we get

$$c^2 - x^2 = h^2 = a^2 - (b + x)^2.$$

Then

$$c^2 - x^2 = a^2 - b^2 - 2bx - x^2,$$

that is,

$$a^2 = b^2 + c^2 + 2bx.$$

But $x = c\cos(180 - \alpha)$, consequently $a^2 = b^2 + c^2 + 2bc\cos(180 - \alpha)$. QED

Remarks

1. If angle A is obtuse then the two other angles are acute and the following equalities hold:

$$b^2 = a^2 + c^2 - 2ac\cos\beta,$$

$$c^2 = a^2 + b^2 - 2ab\cos\gamma.$$

2. If angle A is obtuse, with $m\angle A = \alpha$, we can define[1] $\sin\alpha = \sin(180 - \alpha)$ and $\cos\alpha = -\cos(180 - \alpha)$. With this definition the law of cosines becomes

$$a^2 = b^2 + c^2 - 2bc\cos\alpha,$$

no matter whether $\angle A$ is acute or obtuse. Similarly, the law of sines happens to become

$$\frac{\sin\alpha}{BC} = \frac{\sin\beta}{AC} = \frac{\sin\gamma}{AB}$$

under all circumstances.

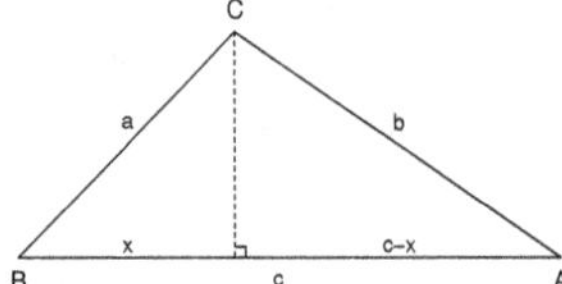

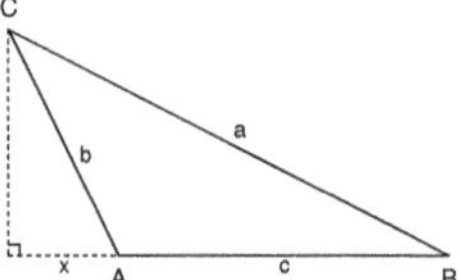

Figure 6.12: Projection property for triangles

3. The **projection property for triangles** asserts that given any ΔABC, with corresponding sides of length a, b, c,

$$c = a\cos B + b\cos A.$$

The proof is quite simple. Assume first that the triangle under consideration is acute (Figure 6.12). We observe that $\cos B = \frac{x}{a}$ and $\cos A = \frac{c-x}{b}$. Hence

$$a\cos B + b\cos A = x + c - x = c.$$

[1] In chapter 7 we will see that these definitions are compatible with the extension of the sine and cosine function to the whole real line.

If the triangle is obtuse (Figure 6.12), with obtuse angle at A, we proceed in an analogous way: $\cos B = \frac{c+x}{a}$, $\cos(180 - A) = \frac{x}{b}$. Hence

$$a\cos B - b\cos(180 - A) = c + x - x = c.$$

But $\cos(180 - A) = -\cos A$, consequently $a\cos B + b\cos A = c$. QED

6.7 Solving triangles

It seems that knowing two angles and a side, where the side lies between the angles or next to them, we can calculate the other sides and angles. Similarly, if we know two sides and the angle in between, or the three sides, we might be able to calculate the other angles and sides. Indeed, recalling the ASA, AAS, SAS, and SSS properties of congruency, such a task can be completed either using the law of sines or the law of cosines. Bowing to tradition, we will use the notation ρ^o instead of simply ρ when a specific value is given or found for the measure of an angle.

1. Given $AB = 5$, $m\angle CAB = 30^o$, and $m\angle ABC = 70^o$, calculate BC and AC (Figure 6.13). We notice that $m\angle C = 80^o$. Applying the law of sines we get

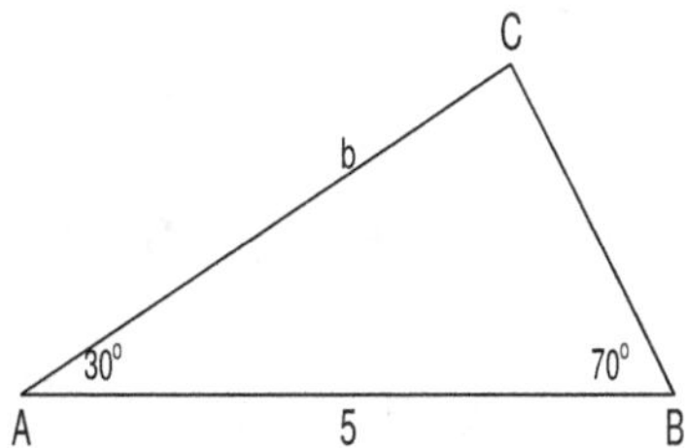

Figure 6.13: Example 1

$$\frac{\sin 30^o}{a} = \frac{\sin 80^o}{5}.$$

Hence $a = \frac{2.5}{\sin 80^o} \approx 2.54$. Similarly,

$$\frac{\sin 70^o}{b} = \frac{\sin 80^o}{5}.$$

Thus $b = \frac{5\sin 70^o}{\sin 80^o} \approx 4.77$.

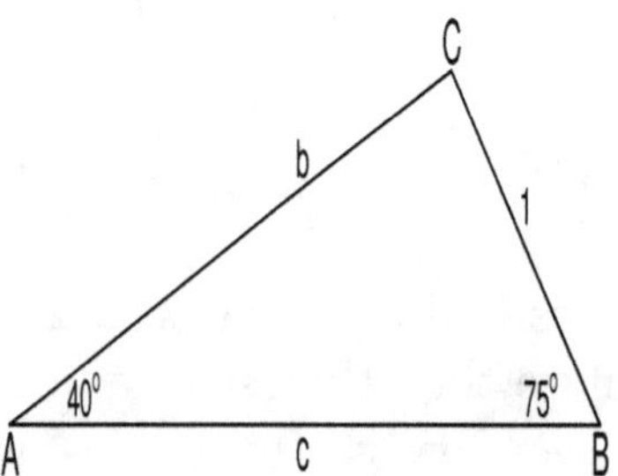

Figure 6.14: Example 2

2. Given that $BC = 1$, $m\angle CAB = 40$, and $m\angle CBA = 75^o$, calculate b and c (Figure 6.14).

   The law of sines implies

   $$\frac{\sin 75^o}{b} = \frac{\sin 40^o}{1}.$$

   Hence $b = \frac{\sin 75^o}{\sin 40^o} \approx 1.503$. We note that $m\angle ACB = 65^o$. Then

   $$\frac{c}{\sin 65^o} = \frac{1}{\sin 40^o},$$

   which in turn leads to $c = \frac{\sin 65^o}{\sin 40^o} \approx 1.41$.

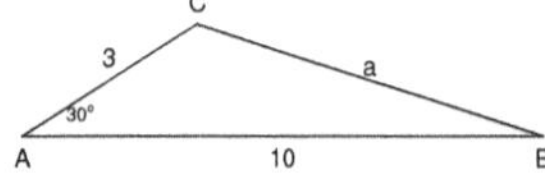

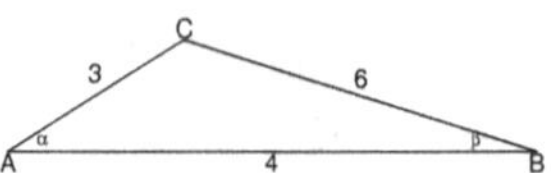

Figure 6.15: Diagrams related to examples 3 and 4

3. Given that $AC = 3$, $AB = 10$ and $m\angle CAB = 30^o$, calculate BC (Figure 6.15).

Let $BC = a$. The law of cosines implies that $a^2 = 3^2 + 10^2 - 2(3)(10)\cos 30^o$, then $a \approx 7.55$.

4. Given a triangle with sides 3, 4, 6, calculate the internal angles[2](Figure 6.15).

   Let α=m$\angle CAB$. The law of cosines implies that $6^2 = 3^2 + 4^2 - (2)(12)\cos\alpha$, thus $\cos\alpha = -11/24$. What angle has a cosine with the value of $-11/24$? Every graphing calculator has the $\cos^{-1}$ option[3], which provides the unique angle, between 0 and 180, such that its cosine is precisely $-11/24$: $\alpha = \cos^{-1}(-11/24) = 117.28^o$. Next we can use the law of sines:

   $$\frac{\sin\beta}{3} = \frac{\sin 117.28^o}{6}.$$

   Hence $\sin\beta = \frac{1}{2}\sin 117.28^o = 0.4444$, which in turn leads to $\beta = \sin^{-1}(0.4444) \approx 26.385^o$.

A special case

We have to pay special attention to the case SSA because there is no SSA congruence property. However, the reader might recall from chapter 2 that SsA congruence is valid in the following sense (Figure 6.16):

Assume that $BC = B^*C^*$, $AB = A^*B^*$, $\angle A \cong \angle A^*$, and $BC \geq AB$. Then $\Delta ABC \cong \Delta A^*B^*C^*$.

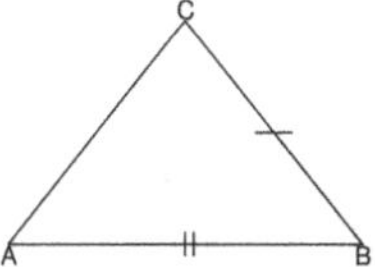

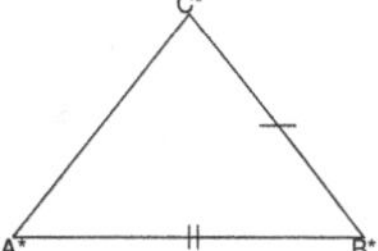

Figure 6.16: The SsA property of congruence

Keeping in mind SsA congruence, let us discuss three examples.

5. Given that $AC = 3$, $BC = 8$, and m$\angle A = 60^o$, calculate m$\angle ACB$ and AB (Figure 6.17).

Due to SsA congruence we should expect only one solution to the problem at hand. Indeed,

---

[2]We cannot assign arbitrary values a, b, c to the sides. The triangle inequality has to hold for such a triangle to exist; that is, $a < b + c, b < a + c, c < a + b$.

[3]The inverse trigonometric functions will be analyzed in more detail in chapter 8.

$$\frac{\sin \beta}{3} = \frac{\sin 60^o}{8},$$

where $\beta =$m$\angle ABC$. Hence $\sin \beta = 0.3248$, and consequently $\beta = \sin^{-1}(0.3248) \approx 18.951^o$. Then m$\angle ACB \approx 101.049^o$. Applying, once more, the law of sines we get

$$\frac{\sin 101.049^o}{AB} = \frac{\sin \beta}{3}.$$

Therefore $AB = \frac{3 \sin 101.049^o}{\sin 18.951^o} \approx 9.066$.

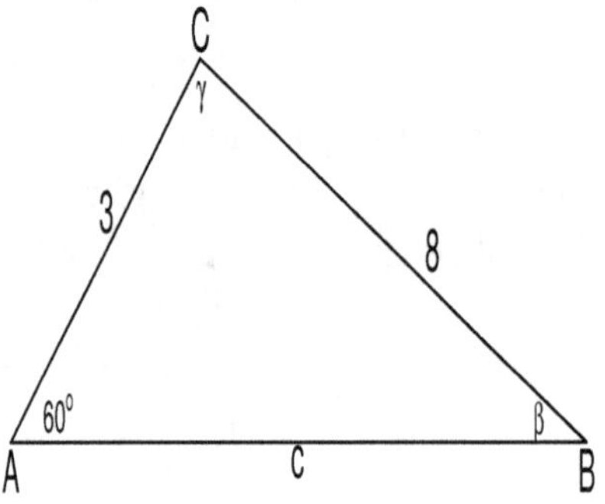

Figure 6.17: Example 5

6. Let $BC = 4, AC = 5$, and m$\angle A = 40^o$ (Figure 6.18). Calculate the third side and m$\angle B$.

   By the law of sines

$$\frac{\sin \beta}{5} = \frac{\sin 40^o}{4},$$

   where $\beta =$m$\angle ABC$. Hence $\sin \beta = \frac{5}{4} \sin 40^o \approx 0.8035$. Thus $\beta = \sin^{-1}(0.8035) \approx 53.47^o$. On the other hand, since m$\angle C = 180^o - (40^o + 53.47^o) = 86.53^o$, the law of sines leads to

$$\frac{\sin 86.53^o}{AB} = \frac{\sin 40^o}{4}.$$

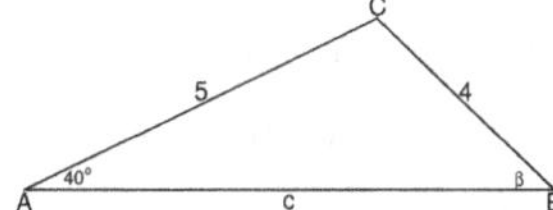

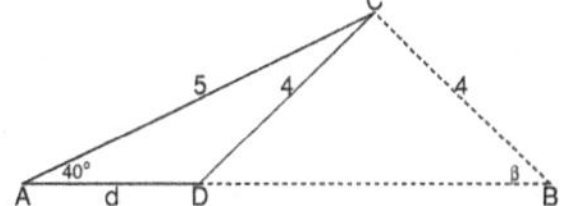

Figure 6.18: Example 6

Hence $AB = \frac{4 \sin 86.53^o}{\sin 40^o} \approx 6.21$. But there is another possibility (see Figure 6.18): Since ΔDCB is isosceles we have that $m\angle CDB = \beta = 53.47^o$. Thus, $m\angle ADC = 126.53^o$. Then $m\angle ACD = 13.47^o$. The law of sines implies that

$$\frac{\sin 13.47^o}{AD} = \frac{\sin 40^o}{4},$$

so $AD = 1.45$.

An interesting alternative procedure to solve problem 6 makes use of the law of cosines: $4^2 = c^2 + 5^2 - 2 \times 5c \cos 40^o$, which is equivalent to the quadratic equation $c^2 - 10(\cos 40^o)c + 9 = 0$, that is, $c^2 - 7.66044c + 9 = 0$. The two roots of this equation are, approximately, 6.21 and 1.45; precisely the solutions found before.

7. Let us consider ΔABC with $BC = 4$, $AC = 5$ and $m\angle A = 60^o$. Calculate the third side and the two other interior angles.

   Applying the law of sines we get

   $$\frac{\sin \alpha}{5} = \frac{\sin 60^o}{4},$$

   where $\alpha = m\angle B$. Hence $\sin \alpha = \frac{5}{4} \sin 60^o = 1.083$. But this is an impossibility since the sine of an angle has to be in the interval $[-1, 1]$. Thus, no triangle with the given data exists! As in the previous problem, a different approach can be used: If a triangle were to exist, satisfying the given data, the law of cosines would imply that $4^2 = c^2 + 5^2 - 2 \times 5c \cos 60^o$, which is equivalent to the quadratic equation $c^2 - 5c + 9 = 0$. But this equation has no real solutions since its discriminant, namely $5^2 - 4 \times 9$, is a negative number.

Remark

Observe that in the case SSA there might be a unique solution, two solutions, or no solution.

6.8 Applications

The law of sines and the law of cosines are widely used. We will present four applications, the first three dealing with finding the distance between two inaccessible points and the fourth with calculating the length of a median. The range of applicability of both laws is really astonishing.

Height of a hill

Suppose that we wish to calculate the height, h, of an inaccessible hill (Figure 6.19). With a transit we measure β and thereafter walk a distance d and measure α. Letting $x = BC$ we get $\tan\alpha = \frac{h}{x}$, so $x = \frac{h}{\tan\alpha}$. Moreover,

$$\tan\beta = \frac{h}{d+x} = \frac{h}{d+\frac{h}{\tan\alpha}}.$$

Consequently

$$d\tan\beta + \frac{h\tan\beta}{\tan\alpha} = h,$$

which in turn leads to

$$h = \frac{d\tan\alpha\tan\beta}{\tan\alpha - \tan\beta}.$$

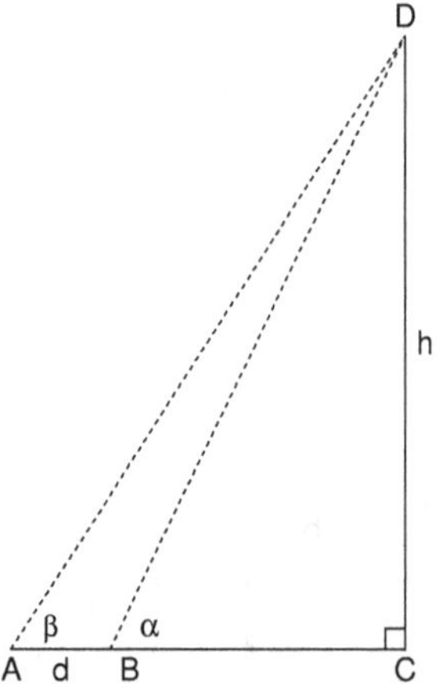

Figure 6.19: Height of a hill

The antenna problem

Assume that an antenna is on the top of a hill and we wish to calculate the height of the antenna (Figure 6.20).Using a transit we make the following measurements: $m\angle CAB = 15^o$, $m\angle BAE = 30^o$, and $m\angle CBD = 50^o$. Besides, we know that $AB =$ 200 m. It is our task to calculate CD.

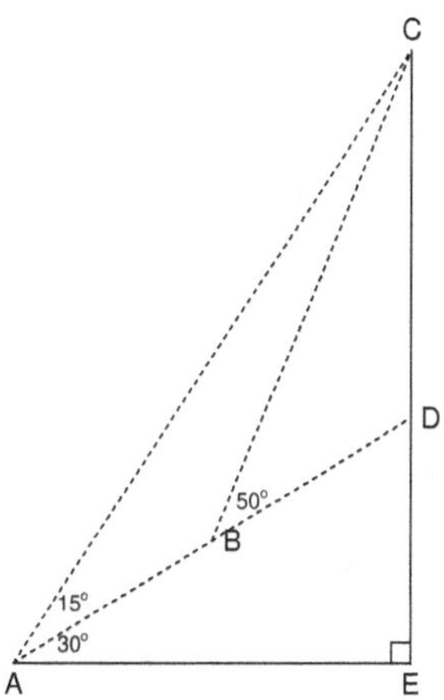

Figure 6.20: height of an antenna

We start by noticing that $m\angle ABC = 180^o - 50^o = 130^o$. Therefore $m\angle ACB = 180^o - (130^o + 15^o) = 35^o$. Applying the law of sines to ΔABC we get

$$\frac{BC}{\sin 15^o} = \frac{200}{\sin 35^o},$$

thus

$$BC = 200\frac{\sin 15^o}{\sin 35^o}.$$

On the other hand, $m\angle ADE = 90^o - 30^o = 60^o$, which in turn implies that $m\angle ADC = 120^o$. Applying the law of sines (this time to ΔCBD) we obtain the proportion

$$\frac{DC}{\sin 50^o} = \frac{BC}{\sin(180^o - 120^o)} = \frac{BC}{\sin 60^o}.$$

Therefore

$$DC = \frac{\sin 50^o}{\sin 60^o}BC = \frac{\sin 50^o}{\sin 60^o} \times 200\frac{\sin 15^o}{\sin 35^o}.$$

Finally, using a calculator we can assert that $DC \approx 79.83$ m.

The two-island problem

We need to find the distance AB between two islands (Figure 6.21), given the distance $CD = 300$ m. on the shore. Using a transit we get $m\angle ACB = 75^o, m\angle BCD = 45^o, m\angle ADC = 40^o, m\angle ADB = 60^o$. Right away we can calculate the measure of

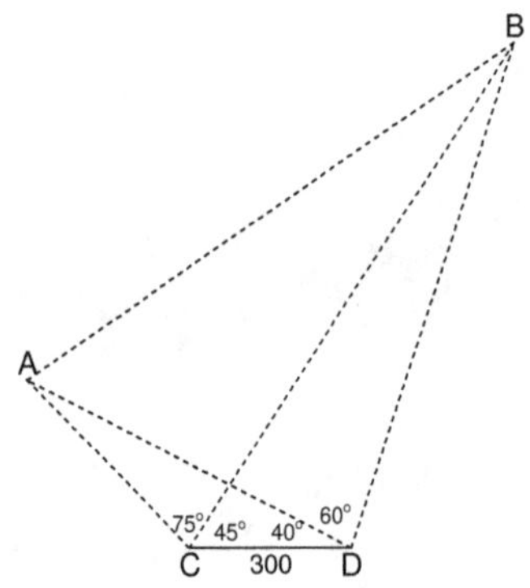

Figure 6.21: The distance between two islands

$\angle CAD$. Indeed, $m\angle CAD = 20^o$. The law of sines implies that

$$\frac{\sin 20^o}{300} = \frac{\sin 40^o}{AC},$$

thus

$$AC = \frac{300 \sin 40^o}{\sin 20^o} = 563.81557.$$

Applying the law of sines a second time, after realizing that $m\angle CBD = 35^o$, we get

$$\frac{BC}{\sin(180^o - 100^o)} = \frac{300}{\sin 35^o}.$$

Hence

$$BC = \frac{300 \sin 80^o}{\sin 35^o} \approx 515.08798.$$

Finally, the law of cosines, applied to ΔABC, implies

$$AB^2 = 563.81557^2 + 515.08798^2 - 2(563.81557)(515.08798)\cos 75^o = 432,873.9533.$$

Therefore $AB \approx 657.03$ m.

Calculating the median

Let us consider an acute triangle ABC (Figure 6.22) and let $\overline{CD}$ be a median, with $CD = m$. Without loss of generality we may assume that $\angle BDC$ is acute and $\angle ADC$ is obtuse. Let $\alpha = m\angle BDC$ and $\beta = m\angle ADC$. Applying the law of cosines twice (first to ΔCBD and thereafter to ΔCDA) we get

$$a^2 = m^2 + \frac{c^2}{4} - 2m\frac{c}{2}\cos\alpha, \tag{6.6}$$

$$b^2 = m^2 + \frac{c^2}{4} + 2m\frac{c}{2}\cos(180 - \beta) = m^2 + \frac{c^2}{4} + 2m\frac{c}{2}\cos\alpha. \tag{6.7}$$

Adding (6.6) and (6.7) we arrive at the equality $a^2 + b^2 = 2m^2 + \frac{c^2}{2}$. Then

$$m^2 = \tfrac{2a^2+2b^2-c^2}{4},$$

which in turn leads to $m = \frac{1}{2}\sqrt{2a^2 + 2b^2 - c^2}$.

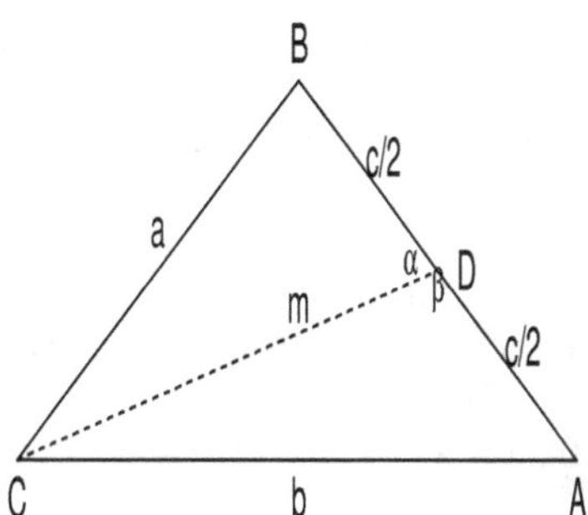

Figure 6.22: Calculating the length of a median

6.9 Addition formulas

There are two well-known addition formulas in trigonometry:

$$\sin(\alpha + \beta) = \sin\alpha\cos\beta + \cos\alpha\sin\beta,$$

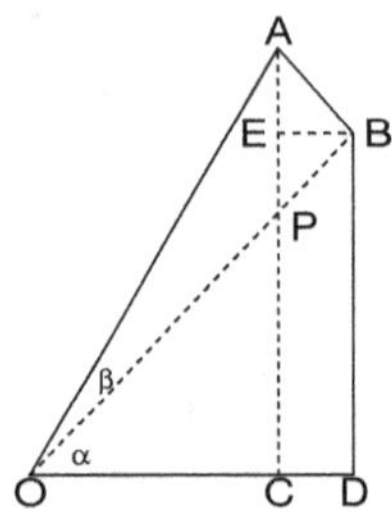

Figure 6.23: Diagram for addition formulas

$$\cos(\alpha+\beta)=\cos\alpha\cos\beta-\sin\alpha\sin\beta.$$

We will provide a proof assuming that both α and β, as well as $\alpha+\beta$, are less than 90. With this purpose in mind we build the right triangle BDO with $m\angle BOD=\alpha$ and the right triangle ABO with $m\angle AOB=\beta$ (Figure 6.23). Thereafter we drop the perpendicular segments $\overline{AC}$ and $\overline{BE}$. We observe that

$$\sin(\alpha+\beta)=\frac{AC}{OA}=\frac{AE+BD}{OA}=\frac{AE}{OA}+\frac{BD}{OA}=\frac{AE}{AB}\frac{AB}{OA}+\frac{BD}{OB}\frac{OB}{OA}. \tag{6.8}$$

But $\angle EAB\cong\angle BOD$ (their common measure is α) since $\Delta OPC\sim\Delta APB$ due to the fact that $\angle OPC\cong\angle APB$. Therefore

$$\sin\alpha=\frac{BD}{OB},\quad \sin\beta=\frac{AB}{AO},\quad \cos\alpha=\frac{AE}{AB},\quad \cos\beta=\frac{OB}{OA}. \tag{6.9}$$

Replacing in (6.8) the values shown in (6.9) we get

$$\sin(\alpha+\beta)=\sin\alpha\cos\beta+\cos\alpha\sin\beta.$$

Having been successful in providing a proof for the first formula, let us try to develop a proof for the second addition formula. Using the same diagram (Figure 6.23) we have

$$\cos(\alpha+\beta)=\frac{OC}{OA}=\frac{OD-CD}{OA}=\frac{OD}{OA}-\frac{CD}{OA}=\frac{OD}{OB}\frac{OB}{OA}-\frac{CD}{AB}\frac{AB}{OA}. \tag{6.10}$$

But $CD=EB$ and $\sin\alpha=\frac{EB}{AB}$, while $\cos\alpha=\frac{OD}{OB}$, $\cos\beta=\frac{OB}{OA}$, and $\sin\beta=\frac{AB}{OA}$. Replacing these values in (6.10) we get

$$\cos(\alpha+\beta)=\cos\alpha\cos\beta-\sin\alpha\sin\beta.$$

Remarks

1. From the addition formula for the sine we get $\sin(2\alpha) = 2\sin\alpha\cos\alpha$, assuming that $0 < 2\alpha < 90$. This is known as the *double angle formula.*

2. If α is the measure of an acute angle, then from the addition formula for the cosine we get

$$\cos\alpha = \cos(\tfrac{\alpha}{2}+\tfrac{\alpha}{2}) = \cos^2(\tfrac{\alpha}{2})-\sin^2(\tfrac{\alpha}{2}) = \cos^2(\tfrac{\alpha}{2})-(1-\cos^2(\tfrac{\alpha}{2})) = 2\cos^2(\tfrac{\alpha}{2})-1.$$

Therefore

$$\cos(\tfrac{\alpha}{2}) = \sqrt{\tfrac{1+\cos\alpha}{2}}.$$

This is the famous *half-angle formula.*

6.10 Subtraction formulas

From our earliest exposure to trigonometry we know the subtraction formulas

$$\sin(\alpha-\beta) = \sin\alpha\cos\beta - \cos\alpha\sin\beta,$$

$$\cos(\alpha-\beta) = \cos\alpha\cos\beta + \sin\alpha\sin\beta.$$

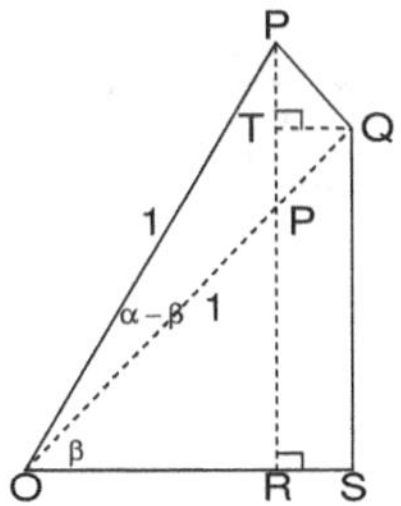

Figure 6.24: Diagram for subtraction formulas

Let us prove the latter and leave a proof of the former to chapter 8, where we will take advantage of Ptolemy's Theorem to provide a rather direct proof. Let us start by building a diagram[4] (Figure 6.24), where $OP = OQ = 1$ and $m\angle SOQ = \beta$,

[4]The proof that we will present assumes that α is the measure of an acute angle, with $\alpha > \beta$.

$m\angle SOP = \alpha$. Observe that ΔOPQ is not, necessarily, a right triangle. Right away we have several equalities: $\sin\alpha = PR$, $\cos\alpha = OR$, $\sin\beta = QS$, and $\cos\beta = OS$. Thus $PT = PR - QS = \sin\alpha - \sin\beta$ and $QT = OS - OR = \cos\beta - \cos\alpha$. Pythagoras' Theorem, applied to ΔPTQ, implies

$$PQ^2 = (\sin\alpha - \sin\beta)^2 + (\cos\beta - \cos\alpha)^2.$$

But the law of cosines, applied to ΔPOQ, leads to

$$PQ^2 = 1^2 + 1^2 - 2\cos(\alpha - \beta).$$

Therefore

$$\sin^2\alpha - 2\sin\alpha\sin\beta + \sin^2\beta + \cos^2\beta + \cos^2\alpha - 2\cos\alpha\cos\beta = 2 - 2\cos(\alpha - \beta).$$

However, $\sin^2\alpha + \cos^2\alpha = \sin^2\beta + \cos^2\beta = 1$. Consequently

$$1 - \sin\alpha\sin\beta - \cos\alpha\cos\beta = 1 - \cos(\alpha - \beta),$$

that is, $\cos(\alpha - \beta) = \cos\alpha\cos\beta + \sin\alpha\sin\beta$. QED

Exercises

1. Find the area of the quadrilateral $ABCD$ (Figure 6.25) where $AE = AD = ED = 2$ and $EB = BC = EC = 1$.

2. Let us consider the parallelogram $ABCD$ (Figure 6.25) where $AC = 10$, $BD = 20$ and $\alpha = 30^o$. Calculate the area of the parallelogram.

3. Let $ABCD$ be a square of side 3 m. On top of it we build an equilateral triangle DCH (Figure 6.25). Calculate the height DM of triangle ADH (Hint: Notice that $\sin 15^o = DM/3$.)

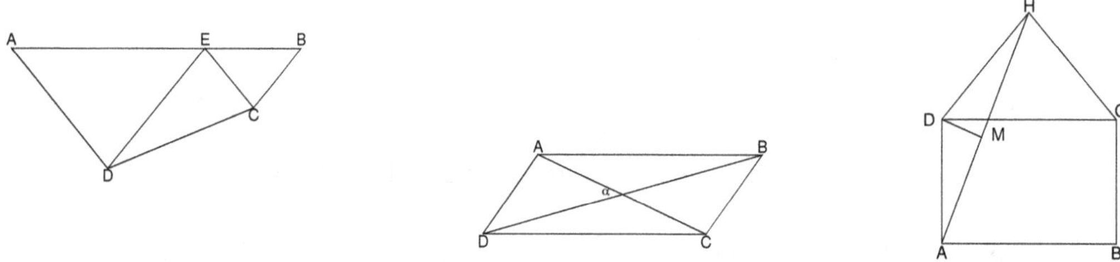

Figure 6.25: Figures corresponding to exercises 1, 2, and 3

4. You wish to build a tunnel between the points A and B on a hill (Figure 6.26). A surveyor measures AC, BC and $\angle ACB$: $AC = 1000$ m., $BC = 1100$ m. and $\text{m}\angle ACB = 60^o$. Calculate the length of the tunnel.

5. From the top of a 30-meter lighthouse an observer sees a boat at C and finds that $\text{m}\angle ABC = 40^o$ (Figure 6.26). What is the distance between the boat and the lighthouse?

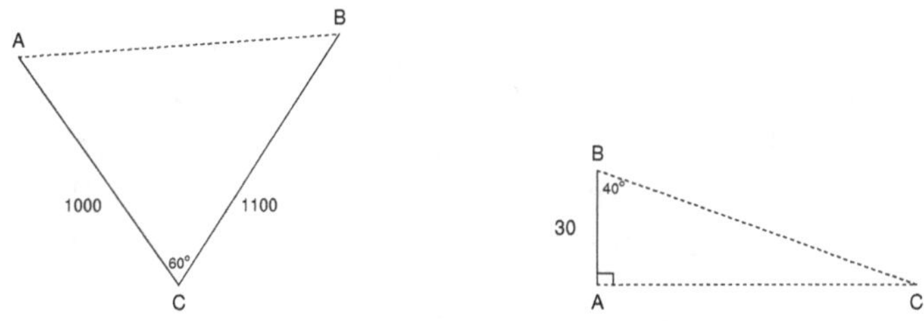

Figure 6.26: Figures corresponding to exercises 4 and 5

6. Solve the two-island problem depicted in Figure 6.27.

7. Solve the antenna problem depicted in Figure 6.27, given that $AB = 300m$.

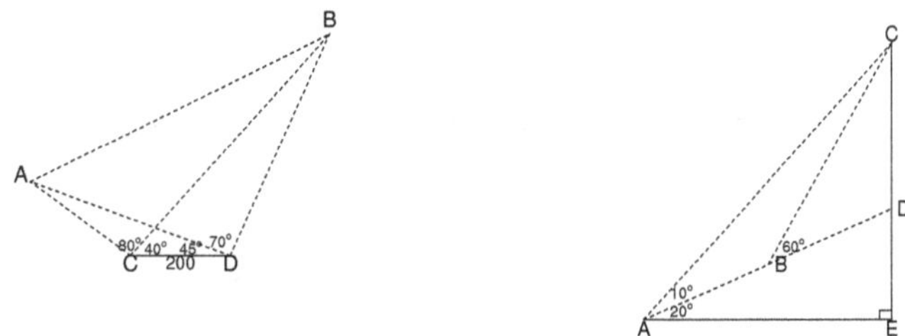

Figure 6.27: Figures corresponding to exercises 6 and 7

8. A pilot, flying h miles above ground, measures angles COB and COA simultaneously[5], where A and B are two lights on the ground (Figure 6.28). It is known that $AB = 5$ miles and the two above-mentioned measurements are: $\text{m}\angle COB = 20^o$, $\text{m}\angle COA = 31^o$. Calculate h.

9. Assume that α, β and $\alpha + \beta$ are acute angles. Then build ΔABC (Figure 6.28) in such a way that $\overline{AD}$ is perpendicular to $\overline{BC}$. Provide an alternative proof of

[5]These are called 'angles of depression'.

the addition formula for the sine using the formula for the area of a triangle in terms of two sides and the sine of the angle between them[6].

10. Find the interior angles of the triangle with sides 2,3,4.

11. Given that $AB = 4$, $BC = 3$, $m\angle BAC = 30^o$, calculate the third side and the measure of the other two angles.

12. Given ΔABC and ΔACD with $m\angle ADC = 45^o$, $AB = AC$, and $m\angle BAC = m\angle DAC$, calculate AB/AD (Figure 6.29) (Hint: Think about the Angle Bisector Theorem. Besides, ITT, SEAT, and the law of sines will be quite useful.)

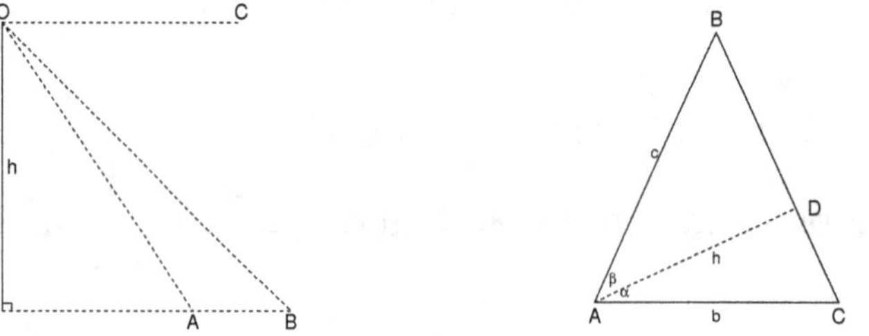

Figure 6.28: Figures corresponding to exercises 8 and 9

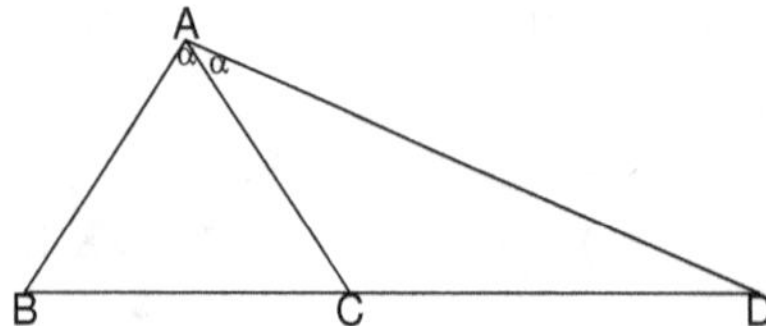

Figure 6.29: Diagram corresponding to exercise 12

[6]This formula was developed in section 6.4.

Enrichment Note: Napoleon's Theorem

Given an arbitrary triangle ΔABC, on each side of it build an equilateral triangle. Let G, H, and I be the centroids of the corresponding equilateral triangles. Then ΔGHI is equilateral (Figure 6.30).

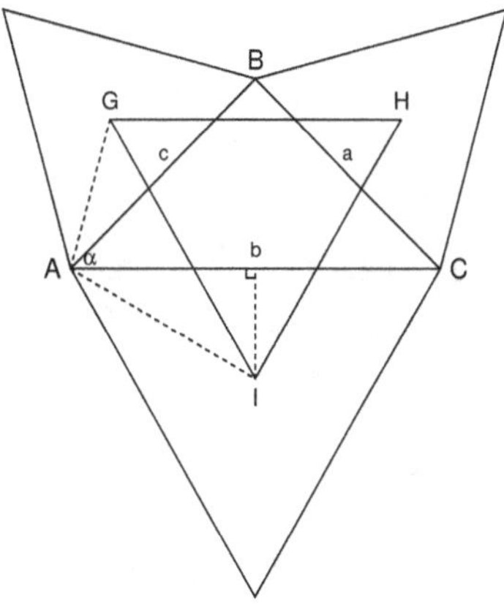

Figure 6.30: A proof of Napoleon's Theorem using trigonometry

Proof

We will follow an approach that uses the law of cosines to great advantage. Let us recall that, for any equilateral triangle, the centroid (the point of intersection of the medians) coincides with the incenter (the point of intersection of the angle bisectors) because any median is an angle bisector and vice versa. Thus, I is the incenter of the equilateral triangle built on the side $\overline{AC}$. Hence

$$m\angle IAC = m\angle ICA = 30^o.$$

Then drop a perpendicular from I to $\overline{AC}$. We have that $\sin 60^o = \frac{b/2}{AI}$, hence

$$AI = \frac{\frac{b}{2}}{\frac{\sqrt{3}}{2}} = \frac{b}{\sqrt{3}}.$$

Analogously, $AG = \frac{c}{\sqrt{3}}$. Next we apply the law of cosines to ΔAGI and we get, letting $\alpha = m\angle BAC$,

$$GI^2 = AG^2 + AI^2 - 2AG \times AI \cos(\alpha + 60^o) = \frac{c^2}{3} + \frac{b^2}{3} - 2\frac{c}{\sqrt{3}}\frac{b}{\sqrt{3}}(\tfrac{1}{2}\cos\alpha - \tfrac{\sqrt{3}}{2}\sin\alpha).$$

Thus

$$GI^2 = \frac{b^2+c^2}{3} - \frac{1}{3}bc\cos\alpha + \frac{bc}{\sqrt{3}}\sin\alpha. \tag{6.11}$$

On the other hand, the law of cosines (applied to the original triangle ABC) implies $a^2 = b^2 + c^2 - 2bc\cos\alpha$. Therefore

$$-bc\cos\alpha = \frac{1}{2}(a^2 - b^2 - c^2). \tag{6.12}$$

But

$$area(\Delta ABC) = \frac{1}{2}bc\sin\alpha. \tag{6.13}$$

From (6.11), (6.12) and (6.13) we get

$$GI^2 = \tfrac{b^2+c^2}{3} + \tfrac{1}{6}(a^2 - b^2 - c^2) + \tfrac{2}{\sqrt{3}}area(\Delta ABC).$$

That is,

$$GI^2 = \tfrac{1}{6}(2b^2+2c^2+a^2-b^2-c^2)+\tfrac{2}{\sqrt{3}}area(\Delta ABC) = \tfrac{1}{6}(a^2+b^2+c^2)+\tfrac{2}{\sqrt{3}}area(\Delta ABC).$$

However, $\overline{GI}$ is an arbitrary side of ΔGHI and it is symmetrical in a, b, and c, so $HI^2 = \tfrac{1}{6}(a^2+b^2+c^2)+\tfrac{2}{\sqrt{3}}area(\Delta ABC)$ and $GH^2 = \tfrac{1}{6}(a^2+b^2+c^2)+\tfrac{2}{\sqrt{3}}area(\Delta ABC)$. Thus $GI = IH = GH$. QED

Remark

Why is the previous result called Napoleon's Theorem? It appeared in print, for the first time, in *The Ladies' Diary* (1826) wherein two different proofs were provided; the proofs were answers to a question posed in the 1825 edition of the magazine. Some decades later the name 'Napoleon' was added to the theorem, implying that the French Emperor had some connection to the theorem. It is not outlandish to think that Napoleon Bonaparte (1769-1824) might have stated the theorem, or even proved it, since he had mathematical ability and befriended notable mathematicians like Gaspard Monge, Joseph Fourier, and Joseph Louis Lagrange. But it is highly probable that Napoleon had nothing to do with the theorem that bears his name (Grunbaum, 2011)

Chapter 7

The Geometry of the Circle

We have been using circles in previous chapters, especially when discussing some constructions with a straightedge and a compass. It is now time to analyze the geometry of the circle in greater depth.

7.1 Some preliminary ideas

Given $r > 0$ and a point O on the plane we define a circle Ω, with center O and radius r, as the set of all points P such that $OP = r$, that is, $\Omega = \{P : OP = r\}$. The word **radius** is also used to denote any segment $\overline{OQ}$ with $OQ = r$. A **chord** is any segment whose endpoints lie on a circle, while a **secant** is any line that contains a chord, that is, any line that intersects the circle at two points. A **diameter** is any chord that passes through the center of a circle.

Proposition 7.1.1

Let $\overline{AB}$ be a chord of a circle with center O such that a line L, which passes through O, bisects $\overline{AB}$. Then L is perpendicular to $\overline{AB}$.

Proof

Build $\overline{OB}$ and $\overline{OA}$ and let C be the midpoint of $\overline{AB}$ (figure 7.1). Then $\Delta ACO \cong \Delta BCO$ thanks to the SSS congruence property; in particular $\angle ACO \cong \angle BCO$. Let α be their measure. Since both angles form a linear pair we can conclude that $\alpha + \alpha = 180$, so $\alpha = 90$. QED

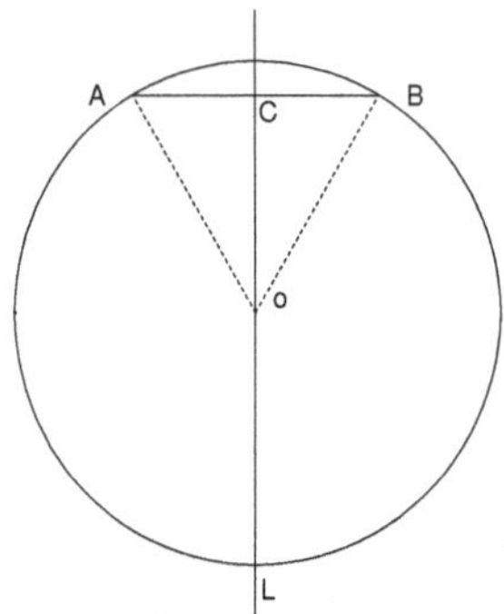

Figure 7.1: Bisecting a chord

Proposition 7.1.2

Let L be a line that passes through the center of a circle and is perpendicular to a chord $\overline{AB}$. Then L bisects $\overline{AB}$.

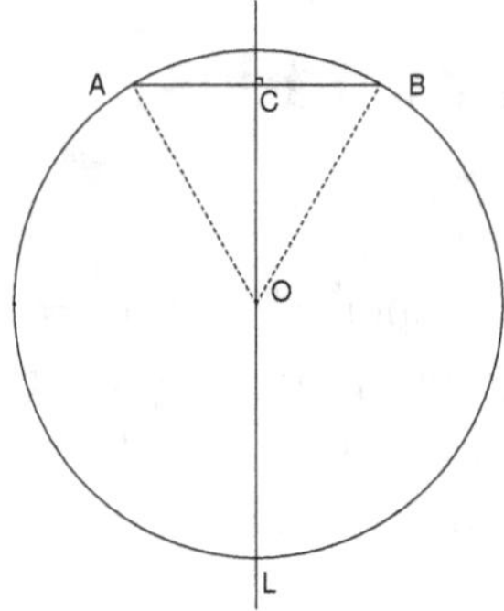

Figure 7.2: Line perpendicular to a chord

Proof

Let C be the point where L intersects $\overline{AB}$ (Figure 7.2). Join O with A and with B. Using HL congruence we can assert that $\Delta ACO \cong \Delta BCO$. In particular, $AC = BC$. QED

Proposition 7.1.3

Let L be a line that bisects a chord $\overline{AB}$ and makes a right angle with it (Figure 7.3). Then L must pass through the point O, the center of the circle.

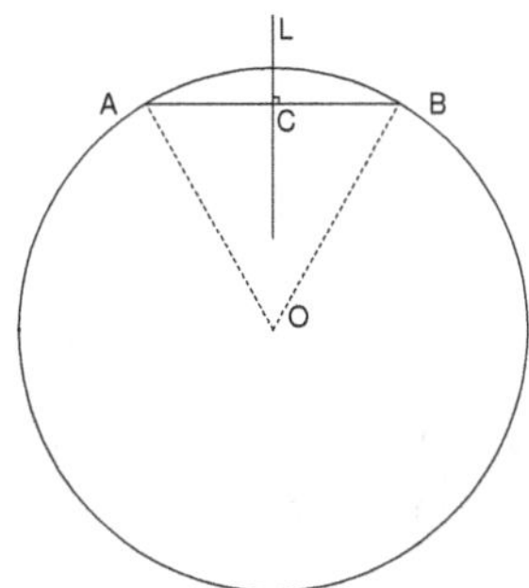

Figure 7.3: The perpendicular bisector of a chord

Proof

Build $\overline{OA}$ and $\overline{OB}$. The line L happens to be the perpendicular bisector of $\overline{AB}$. Since $OA = OB$, the characterization theorem of perpendicular bisectors implies that O belongs to L. QED

Proposition 7.1.4

Suppose that $\overline{AB}$ is a diameter of a circle with center O. Let C be any point on the circle. Then $\angle ACB$ is a right angle.

Proof

Just draw the radius $\overline{OC}$ and observe that both ΔAOC and ΔBOC are isosceles, hence their respective base angles (call them α and β) are congruent. That is, $m\angle OAC = m\angle OCA = \alpha$ while $m\angle OCB = m\angle OBC = \beta$. On the other hand, the sum of the measures of the internal angles of ΔACB is 180, consequently $\alpha + (\alpha + \beta) + \beta = 180$. Then $m\angle ACB = \alpha + \beta = 90$. QED

Two corollaries

1. Suppose that A, B, C lie on a circle with center at O and $m\angle ABC = 90$. Then $\overline{AC}$ is a diameter. Why is this so? Proceeding by contradiction assume that $\overline{AC}$ does not pass through O. Let us build the segment $\overline{CD}$ that passes through O and intercepts the circle at D (Figure 7.4). If D lies in the interior of $\angle ABC$ we can conclude that $m\angle DBC < m\angle ABC = 90$. But proposition 7.1.4 implies that $m\angle DBC = 90$; thus, $90 < 90$, a clear impossibility.

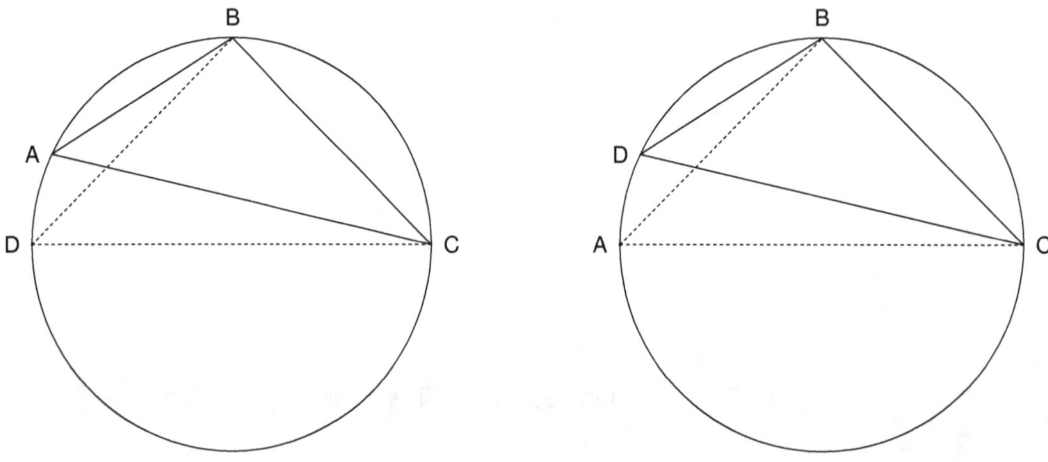

Figure 7.4: Proving the first corollary

Next assume that D is not in the interior of $\angle ABC$, so A is in the interior of $\angle DBC$ (Figure 7.4). Since A is in the interior of $\angle DBC$ it follows that $90 = m\angle ABC < m\angle DBC$. However, proposition 7.1.4 implies that $m\angle DBC = 90$. Therefore $90 < 90$, a flagrant contradiction again.

2. Assume that ΔABC has a right angle at C and, using $\overline{AB}$ as a diameter, we build a circle. Does this circle pass through C? Suppose that C lies in the interior of the circle (Figure 7.5). Then we can prolong $\overline{BC}$ until it meets the circle at D. According to proposition 7.1.4, $m\angle BDA = 90$. But the Exterior Angle Theorem, applied to ΔACD, implies that $m\angle BCA > m\angle BDA$, so $90 > 90$. We have reached an impossibility. If C were to lie on the exterior of the circle, a similar argument leads to a contradiction.

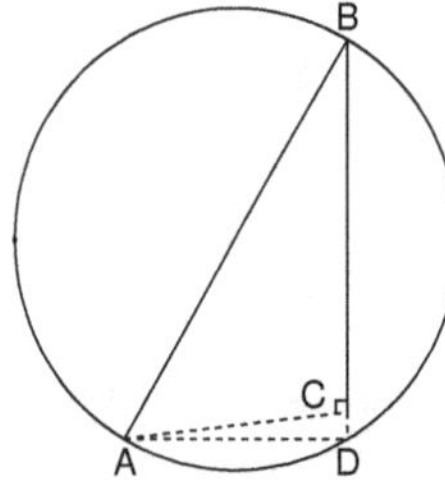

Figure 7.5: The second corollary

Remarks

1. Suppose we lose the center of a circle. Is it possible to find it? Start by building two chords, say $\overline{AB}$ and $\overline{CD}$ and let L_1 and L_2 be their corresponding perpendicular bisectors. According to proposition 7.1.3, both L_1 and L_2 must pass through the center of the circle. Thus the intersection of L_1 and L_2 is precisely the lost center.

2. Can we always build a circle that passes through three non-collinear points? Let A, B, C be the three points that do not lie on a line. Build $\overline{AB}$ and $\overline{BC}$. Thereafter build the perpendicular bisector L_1 of $\overline{AB}$ and the perpendicular bisector L_2 of $\overline{BC}$. They must intersect at a point because if they were to be parallel then the line that contains $\overline{AB}$ and the line that contains $\overline{BC}$ would be parallel, which is an impossibility since they meet at point B. Let O be the point of intersection of L_1 and L_2. By the characterization theorem of perpendicular bisectors we have that $OB = OC$ and $OB = OA$. Let $r = OC = OB = OA$. We can see[1] that the circle with center O and radius r passes through A, B, C. This circle is called the **circumcircle** of ΔABC.

3. We just learned that given three non-collinear points, P_1, P_2, P_3, it is possible to build a circle, with center O and radius r, that contains the three of them. Is it the only circle that passes through the three given points? Suppose that there exists a circle with center O' and radius r' that passes through P_1, P_2, P_3. Thereafter construct the chords $\overline{P_1P_2}$ and $\overline{P_2P_3}$, and the corresponding perpendicular bisectors L_1 and L_2. Due to proposition 7.1.3 we know that L_1 and L_2 pass through O'. Since the point of intersection of two lines is unique we can assert that $O = O'$ and since $OP_1 = r$ it follows that $r = r'$. But a circle is completely determined by its center and radius, consequently uniqueness has been proven.

7.2 Tangent line to a circle

A line L is **tangent** to a circle if it intersects the circle at just one point, which is called the point of tangency. The next proposition provides a path to build a tangent at any point on the circle: just draw a segment from the center to the point and then draw a line perpendicular to the segment at the above-mentioned point. This line is precisely the tangent to the circle at the given point! Unfortunately, this simple way of building tangent lines does not work for other conics.

[1]Compare this remark with the last proposition of section 2.6.

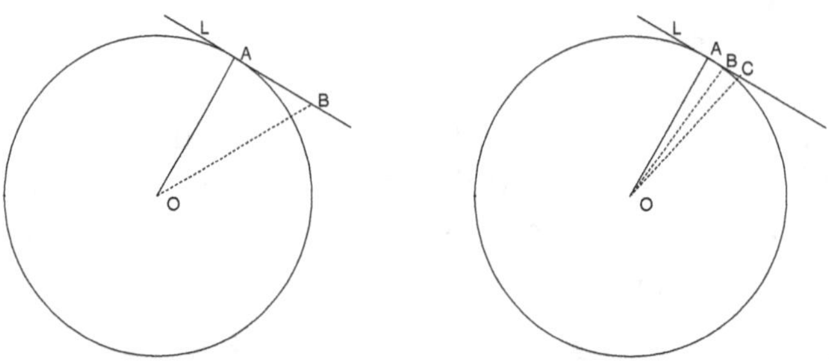

Figure 7.6: Perpendicularity and tangency

Proposition 7.2.1

Let A be an arbitrary point on a circle with center O and radius r. Then the line L is tangent to the circle at A if and only if L is perpendicular to $\overline{OA}$.

Proof

Assume that $\overline{OA}$ is perpendicular to L at A (Figure 7.6). Let B be any point on L, $B \neq A$. Since ΔOAB is a right triangle we can conclude that $OB > OA = r$. Hence B does not lie on the circle. That is to say, L and the circle only share the point A. Consequently, L is the required tangent.

Next we have to prove that if L is the tangent to the circle at A then $\overline{OA}$ is perpendicular to L. Actually, we will prove the logical contrapositive of it; namely, if $\overline{OA}$ is not perpendicular to L then L is not the tangent to the circle at A. With this purpose in mind let us assume that $\overline{AB}$ is not perpendicular to L at A. Draw the perpendicular from O to L and call it $\overline{OB}$ (Figure 7.6); without loss of generality let us suppose that B lies to the right of A. Then build $\overline{BC}$, with C on L, so that $BC = AB$. The SAS property of congruency implies that $\Delta OAB \cong \Delta OCB$, hence $OC = OA = r$. Thus C, which is different from A, lies on L and the circle. Therefore L is not the tangent to the circle at A. QED

Proposition 7.2.2

Let P be a point on the outside of a circle and let the lines that contain $\overline{PA}$ and $\overline{PB}$ be tangent to the circle at A and B respectively (Figure 7.7). Then $PA = PB$.

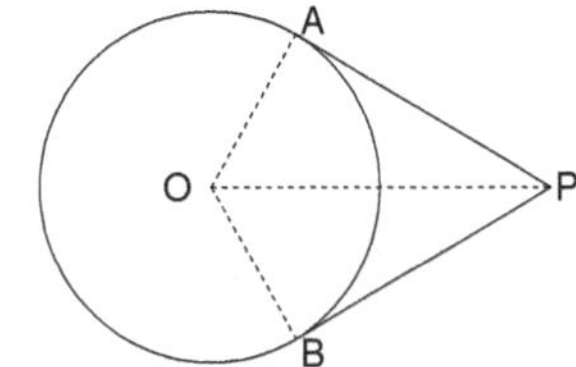

Figure 7.7: Two tangents to a circle from a point on the outside

Proof

Build $\overline{OA}$ and $\overline{OB}$. From the previous proposition we can assert that $\overline{PA}$ is perpendicular to $\overline{OA}$ and $\overline{PB}$ is perpendicular to $\overline{OB}$. After building $\overline{OP}$ we notice that $\Delta AOP \cong \Delta BOP$ thanks to HL congruency. In particular, $AP = BP$. QED

A construction

Let O be the center of a circle Ω_1 and P a point on its exterior. Let us discuss how to construct two lines that pass through P and are tangent to the given circle: Start by building a circle Ω_2 with diameter $\overline{OP}$ and let Q and R be the points of intersection between both circles. We claim that the lines $\overleftrightarrow{PQ}$ and $\overleftrightarrow{PR}$ are tangent to Ω_1. Indeed, proposition 7.1.4, applied to Ω_2, implies that $\overline{OQ} \perp \overline{QP}$. Then $\overleftrightarrow{PQ}$ is tangent to Ω_1 at Q. In a similar fashion, we can assert that $\overline{OR} \perp \overline{RP}$, thus $\overleftrightarrow{PR}$ is tangent to Ω_1 at R.

Inscribing a circle inside a triangle

We will say that a circle is inscribed in ΔABC if it lies inside the triangle and the sides are tangent to the circle. Is it always possible to build an inscribed circle? We already know that we can always circumscribe a circle around a triangle in the sense that the vertices of the latter lie on the former: build the perpendicular bisectors of each side, which meet at a point that happens to be the center of the circle that passes through the three vertices. This center, quite appropriately, is called the **circumcenter**[2]. Next we will see that it is always possible to build an inscribed circle in any triangle.

Let us recall that in section 2.9 we proved that the three angle bisectors meet at a point I (the **incenter**) in the interior of ΔABC. Moreover, the three perpendicular

[2]The corresponding radius is called the **circumradius.**

segments $\overline{IA}$, $\overline{IB}$, and $\overline{IC}$, from I to the three sides of the triangle, are congruent. Next we build the circle with center I and radius r, where $r = IA = IB = IC$. This will be the inscribed circle that we were looking for, which is called the **incircle** . It should be noted that the radius r is called the **inradius**.

We should emphasize that if a circle of radius r is inscribed in a triangle with semiperimeter s then the area of the triangle, usually denoted K, is given by

$$\boxed{K = s \times r}.$$

The proof of this result is left as an exercise.

Can we calculate the radius of the Earth?

Assuming that the Earth is a perfect sphere, which it is not, we climb a mountain that is m miles high. From point A, at the top of the mountain, using adequate instruments we find the horizon line (Figure 7.8) and measure $\angle OAB$. Then $\overline{OB}$ is perpendicular to $\overline{AB}$, where O is the center of the Earth. We observe that

$$\sin\alpha = \tfrac{r}{r+m},$$

where r is the unknown and α =m$\angle OAB$. Therefore $r\sin\alpha + m\sin\alpha = r$, so

$$r = \tfrac{m\sin\alpha}{1-\sin\alpha}.$$

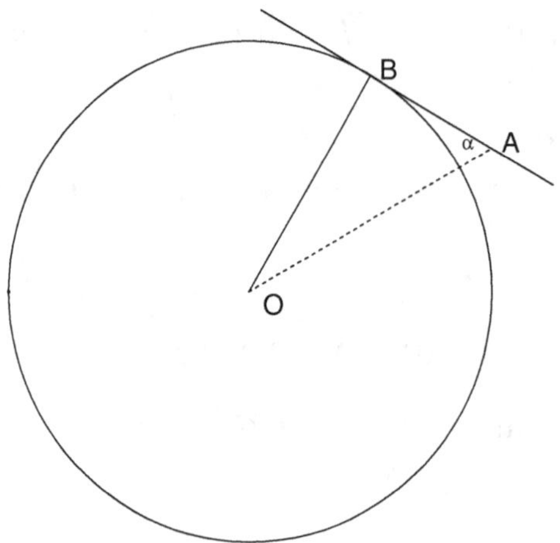

Figure 7.8: Calculating the radius of the Earth

We should point out that this ingenious method comes from ancient Greece, some 2,100 years ago. Modern measurements provide the following data: when $m = 3$ miles, $\alpha = 87.7666^o$. Then $r = 3,944.368421$ miles.

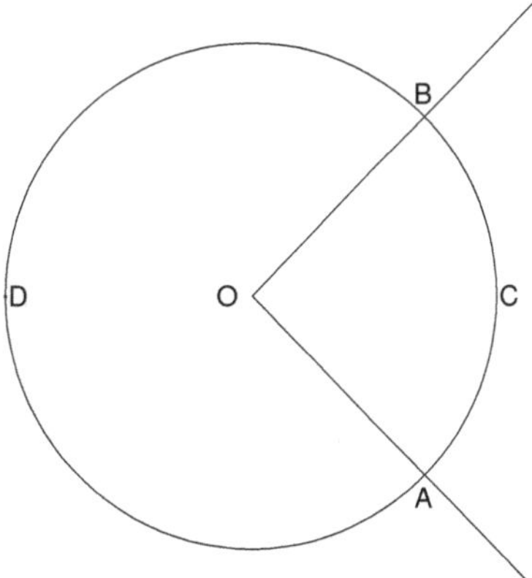

Figure 7.9: Central angle, major and minor arcs

7.3 Inscribed angles

Before we state the theorem on inscribed angles we need some definitions. A **central angle** of a circle is any angle whose vertex is at the center of the given circle (Figure 7.9). On the other hand, a **minor arc** $\overset{\frown}{AB}$ (that is, $\overset{\frown}{ACB}$) is the subset of the circle that lies on the interior of a central angle, as well as the points A and B. A **major arc** $\overset{\frown}{ADB}$ is the subset of the circle that lies on the exterior of a central angle, including the points A and B. An **inscribed angle**, $\angle ABC$, is any angle such that its vertex B lies on the circle and its sides 'cut' the circle at A and C (Figure 7.10). Often we will say that the inscribed angle 'subtends' the arc $\overset{\frown}{AC}$.

Measure of an arc

Given any minor arc $\overset{\frown}{AB}$, we define $m\ \overset{\frown}{AB}= \alpha$ where α is the measure of the corresponding central angle. Often we will omit the letter m and write simply $\overset{\frown}{AB}$ to denote the measure of the arc under consideration. From the context it should be clear whether we mean the arc or the measure of it. On the other hand, we should agree that if $\overset{\frown}{AB}$ happens to be a semi-circle then its measure is 180.

If $\overset{\frown}{AB}$ happens to be a major arc (to be more precise we should write $\overset{\frown}{ADB}$) then we define $m\ \overset{\frown}{AB}= 360 - \alpha$, where α is the measure of the corresponding minor arc.

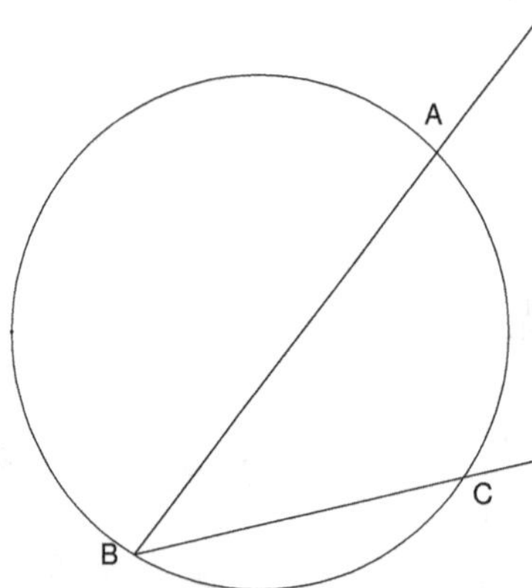

Figure 7.10: Inscribed angle

Lemma

For any three points A, B, C on a circle with center O,

$$\overset{\frown}{ABC} = \overset{\frown}{AB} + \overset{\frown}{BC}.$$

Proof

Assume that $\overset{\frown}{ABC}$ is a minor arc. Then

$$\overset{\frown}{ABC} = m\angle AOC = m\angle AOB + m\angle BOC = \overset{\frown}{AB} + \overset{\frown}{BC}.$$

If $\overset{\frown}{ABC}$ happens to be a semi-circle we proceed as before:

$$\overset{\frown}{ABC} = 180 = m\angle AOB + m\angle BOC = \overset{\frown}{AB} + \overset{\frown}{BC}.$$

Suppose that $\overset{\frown}{ABC}$ is a major arc such that A and C lie on the same side of the diameter that passes through B. Then $\overset{\frown}{ABC} = 360 - \alpha$, where $\alpha = m\angle AOC$, while $\overset{\frown}{AB} = 360 - (\alpha + \beta)$ and $\overset{\frown}{BC} = \beta$. Then $\overset{\frown}{ABC} = \overset{\frown}{AB} + \overset{\frown}{BC}$. There are other cases to consider (see list of problems at the end of the chapter.) QED

Theorem

Let AOB be an inscribed angle that subtends an arc $\overset{\frown}{AB}$. Then

$$\beta = \tfrac{1}{2}\,\overset{\frown}{AB}\,,$$

where $\beta = m\angle AOB$. That is to say, the measure of an inscribed angle is exactly half the measure of the corresponding central angle.

Proof

We have to consider three cases separately.

1. Assume that one of the sides of the inscribed angle is a diameter[3] (Figure 7.11). We start by joining the center P of the circle with the point A and let $\alpha = \overset{\frown}{AB}$. Since ΔOPA is isosceles it follows that $m\angle PAO = \beta$, hence $\alpha = \beta + \beta$ by the SEAT property of triangles. Therefore $\beta = \frac{1}{2}\alpha$ and we are done.

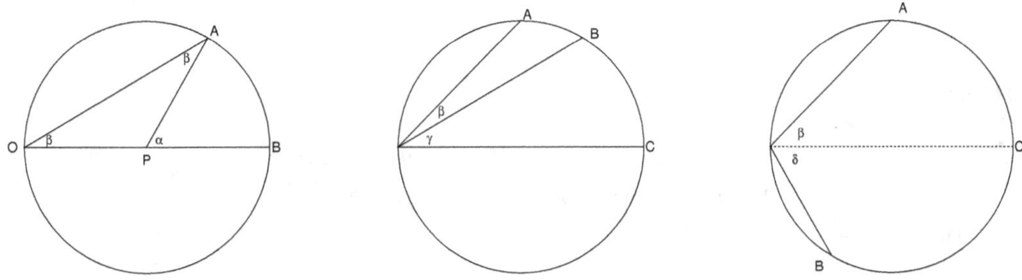

Figure 7.11: Inscribed Angle Theorem (three cases)

2. Assume that both sides of the inscribed angle lie on one side of a diameter $\overline{OC}$ (Figure 7.11). Letting $\gamma = m\angle BOC$ and applying case 1, as well as the lemma, we get $\overset{\frown}{BC} = 2\gamma$ and $\overset{\frown}{AC} = 2(\beta + \gamma)$. Therefore

$$\overset{\frown}{AB} + 2\gamma = \overset{\frown}{AB} + \overset{\frown}{BC} = \overset{\frown}{AC} = 2(\beta + \gamma) = 2\beta + 2\gamma,$$

   thus $\overset{\frown}{AB} = 2\beta$, which in turn leads to $\beta = \frac{1}{2}\overset{\frown}{AB}$.

3. Assume that the sides of the inscribed angle lie on opposite sides of a diameter $\overline{OC}$ (Figure 7.11). Letting $\delta = m\angle BOC$, $\gamma = m\angle AOC$, and applying case 1, as well as the lemma, we get

$$\overset{\frown}{AB} = \overset{\frown}{AC} + \overset{\frown}{CB} = 2\gamma + 2\delta = 2(\gamma + \delta) = 2\beta,$$

   thus $\beta = \frac{1}{2}\overset{\frown}{AB}$. QED

[3]Strictly speaking we would have to talk about 'rays' of an angle instead of sides.

Corollary

Let $\angle ABC$ and $\angle ADC$ be inscribed angles that subtend the same arc $\overset{\frown}{AC}$ of a circle. Then $\angle ABC \cong \angle ADC$.

The proof takes only two lines: Thanks to the theorem on inscribed angles, $m\angle ABC = \frac{1}{2}\overset{\frown}{AC}$ and $m\angle ADC = \frac{1}{2}\overset{\frown}{AC}$; consequently $\angle ABC \cong \angle ADC$. QED

We should point out that the corollary of the theorem on inscribed angles will play a crucial role in chapter 8 when discussing a proof of Ptolemy's Theorem.

Remarks

1. The theorem on inscribed angles, and its corollary, are very important tools in geometry. To illustrate the usefulness of the theorem let us provide an alternative proof of the half-angle formula for the cosine, which was proven in chapter 6 using the addition formula for cosines. Let us build a unit circle with center O and let us construct the central angle EOB so that $m\angle EOB = \alpha$ (we are only assuming that $0 < \alpha < 90$). Then extend the segment $\overline{OE}$ to the left so that it becomes the diameter $\overline{AE}$ and build $\overline{AB}$ (Figure 7.12). According to the above-mentioned theorem, $m\angle OAB = \alpha/2$.

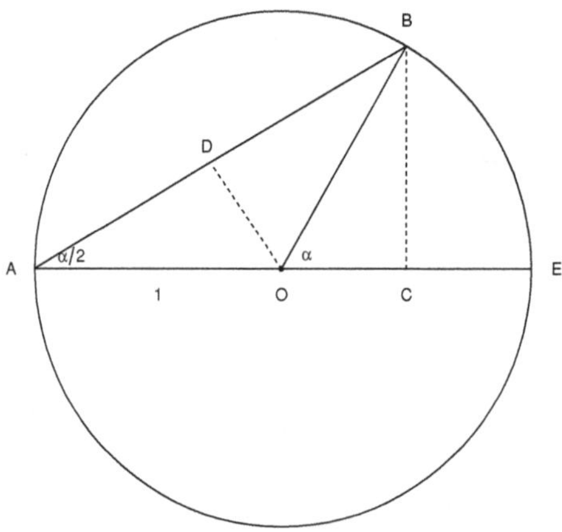

Figure 7.12: The cosine of a half-angle formula

Thereafter drop the heights $\overline{OD}$ and $\overline{BC}$. Using ΔADO we get $\cos\frac{\alpha}{2} = AD$

and using ΔACB it follows that

$$\cos\tfrac{\alpha}{2} = \tfrac{1+OC}{AB}.$$

But $AB = 2AD$ since ΔAOB is isosceles[4]. Therefore

$$\cos\tfrac{\alpha}{2} = \tfrac{1+OC}{2AD} = \tfrac{1+\cos\alpha}{2\cos\frac{\alpha}{2}},$$

and consequently $\cos^2\frac{\alpha}{2} = \frac{1+\cos\alpha}{2}$. Thus, $\cos\frac{\alpha}{2} = \sqrt{\frac{1+\cos\alpha}{2}}$. QED

2. Whenever an inscribed angle subtends a semi-circle, its measure is exactly 90. Moreover, according to the theorem on inscribed angles, if $\overline{AB}$ is a diameter then $\overset{\frown}{AB}= 90 \times 2 = 180$. That is to say, the measure of any semi-circle is precisely 180. This result is in agreement with proposition 7.1.4 and the convention made in the subsection on the measure of an arc.

3. Suppose that $\angle ABC$ and $\angle DEF$ are inscribed angles in a circle. We assert that if $m\angle ABC < m\angle DEF$ then $AC < DF$. Indeed, the measure of the corresponding central angles are $2 \times m\angle ABC$ and $2 \times m\angle DEF$, respectively; thus, $AC < DF$ thanks to the Hinge Theorem.

4. The theorem on inscribed angles and its corollary are also very useful when solving concrete problems in geometry. For instance, suppose that a rectangle $ACDE$ is inscribed in a circle (Figure 7.13), with $AE = AF = 5$ and $FC = 3$, and we wish to calculate AB. Right away we note that $EC = \sqrt{8^2 + 5^2} = \sqrt{89}$. Since $\angle ABE$ and $\angle ACE$, both inscribed angles, subtend the same arc, it follows that $\angle ABE \cong \angle ACE$. Then $\Delta ABF \sim \Delta ECF$, thus

$$\tfrac{AB}{\sqrt{89}} = \tfrac{5}{\sqrt{50}}.$$

Consequently $AB = \frac{5\sqrt{89}}{\sqrt{50}} = \sqrt{\frac{89}{2}}$.

A basic property of chords

Given two chords, $\overline{AB}$ and $\overline{CD}$, which intersect at a point E (Figure 7.14), the following equality is true:

$$AE \times EB = CE \times ED.$$

[4] In any isosceles triangle a height is also a median.

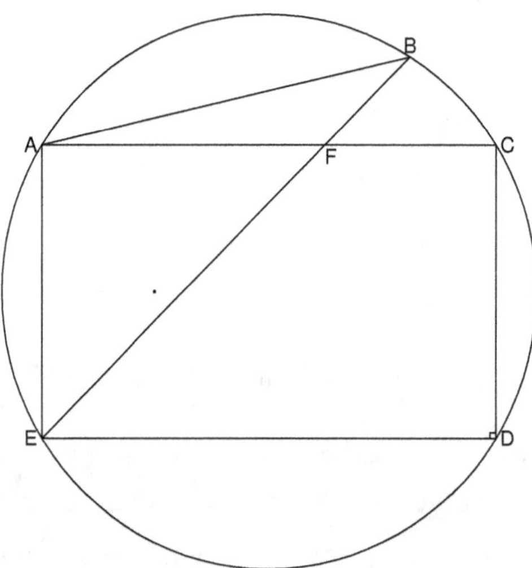

Figure 7.13: A concrete problem

Indeed, build segments $\overline{AC}$ and $\overline{BD}$. Since the inscribed angles CAB and CDB subtend the same arc, it follows that $\angle CAB \cong \angle CDB$. Using the fact that $\angle AEC \cong \angle DEB$ we can conclude that

$$\Delta AEC \sim \Delta DEB.$$

Therefore

$$\frac{AE}{DE} = \frac{CE}{EB},$$

which in turn leads to $AE \times EB = DE \times CE$. QED

A geometrical construction

Suppose that the semiperimeter s and the area a of a rectangle are given and we wish to find its length and width through a geometrical construction. Denoting the unknowns as x and y, we have to solve the system

$$x + y = s, \quad xy = a.$$

Letting $b = \sqrt{a}$, our system of equations becomes

$$x + y = s, \quad xy = b^2. \tag{7.1}$$

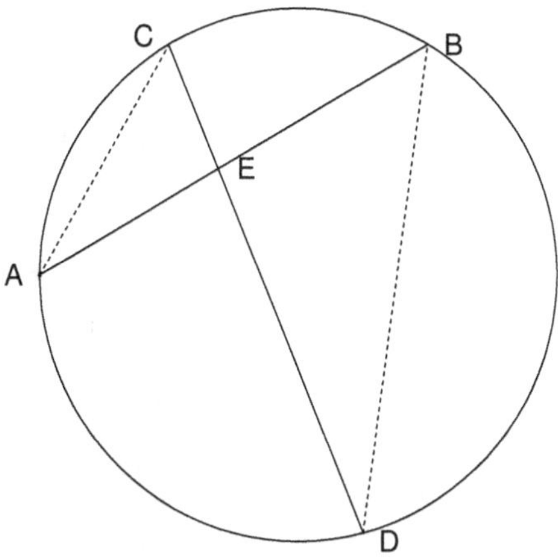

Figure 7.14: A property of chords

Next build a circle with diameter $\overline{AB}$ such that $AB = s$. Assuming that $b < \frac{s}{2}$, draw a segment $\overline{AC}$, perpendicular to $\overline{AB}$, so that $AC = b$ (Figure 7.15). Then build $\overline{CD}$ in such a way that $\overline{CD} \perp \overline{AC}$ and drop the perpendicular $\overline{DE}$, which we prolong until it meets the circle at F. Using HL congruence it follows that $\Delta ODE \cong \Delta OFE$, hence $DE = EF$ and therefore $EF = ED = b$.

Then we define $x = AE$ and $y = EB$. Using the basic property of chords we can assert that $xy = b \times b = b^2 = a$, while obviously $x + y = s$. The construction under consideration does not work if $b > \frac{s}{2}$, but we might analyze the case when $b = \frac{s}{2}$. Under these circumstances, $x = y = \frac{s}{2}$ satisfy the system of equations (7.1) since $xy = \frac{s^2}{4} = b^2$.

Extending the basic formula for inscribed angles

Suppose that a line L is tangent to a circle at a point A, let B be any other point on the circle ($B \neq A$), build the chord $\overline{AB}$ and let α be the measure of the angle formed by $\overline{AB}$ and the line L (Figure 7.16). Then

$$\alpha = \frac{1}{2}\overset{\frown}{AB}.$$

The justification can be provided readily. Indeed, start by building the diameter $\overline{AC}$. If $C = B$ then $\alpha = 90$ and $\overset{\frown}{AB} = 180$, thus the proof comes to an end. Assuming

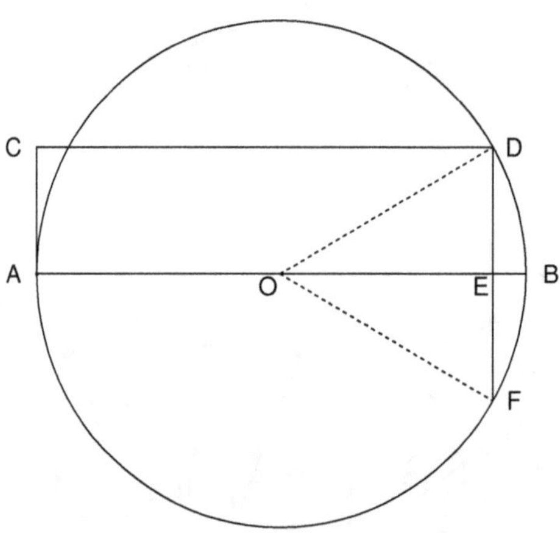

Figure 7.15: A geometrical construction

that $\alpha < 90$ let $\beta = m\angle BAC$ (Figure 7.16). Hence $\alpha + \beta = 90$, which in turn leads to $2\alpha = 180 - 2\beta$. But $\overset{\frown}{AB} + \overset{\frown}{BC} = 180$ and $\beta = \frac{1}{2}\overset{\frown}{BC}$, therefore

$$\overset{\frown}{AB} = 180 - 2\beta = 2\alpha,$$

consequently $\alpha = \frac{1}{2}\overset{\frown}{AB}$.

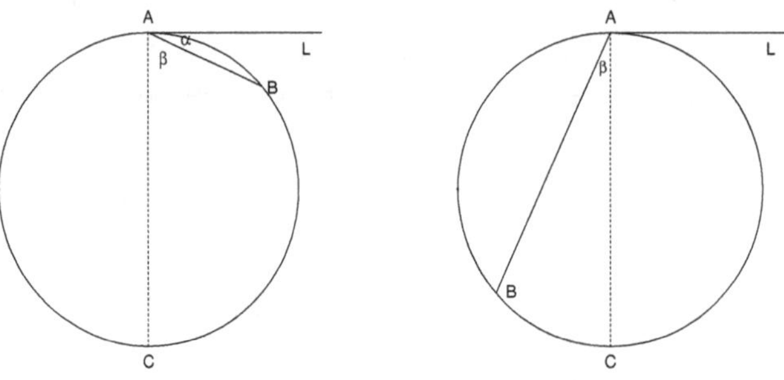

Figure 7.16: Extended formula for inscribed angles

Next assume that $\alpha > 90$ (Figure 7.16). We will show that $\alpha = \frac{1}{2}\overset{\frown}{ACB}$, where $\overline{AC}$ is a diameter. Letting $\beta = m\angle BAC$ we will have that $\beta = \frac{1}{2}\overset{\frown}{BC}$ and

$$\overset{\frown}{ACB} = 180 + \overset{\frown}{BC} = 180 + 2\beta.$$

Hence $\alpha = 90 + \beta = \frac{1}{2}\overset{\frown}{ACB}$. QED

The extended version of the basic formula for inscribed angles has many applications. For instance, assume that A, B, C belong to a circle and that the secant $\overleftrightarrow{BC}$ and the tangent at A intersect at a point D (Figure 7.17). Then $AD^2 = DC \times DB$. Indeed, build $\overline{AB}$ and $\overline{AC}$. Since $m\angle DAB = \frac{1}{2}\overset{\frown}{AB} = m\angle ACB$ we can assert that $\Delta ADB \sim \Delta CDA$, hence $\frac{BD}{AD} = \frac{AD}{DC}$, which in turn leads to $AD^2 = DC \times DB$.

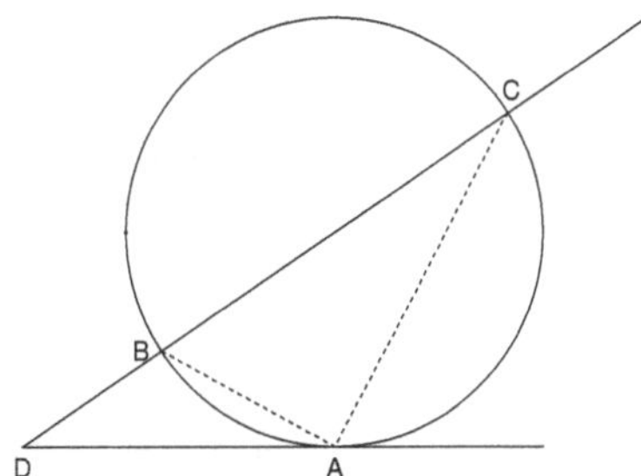

Figure 7.17: A secant and a tangent problem

7.4 Two visions

As we found before, sometimes there is not just one way to solve a problem. When dealing with circles this dichotomy appears more often than not, especially when the challenge is not straightforward. Let us discuss in detail the following problem: Suppose a point A on a circle with center O is given, a circle with diameter OA is built, and a line through A meets the smaller circle at B and the bigger circle at C. Our task is to show that $AB = BC$ (Figure 7.18). We can follow two radically distinct approaches.

1. Build $\overline{OC}$ and $\overline{OB}$. Since $\overline{OA}$ is a diameter we can conclude that $m\angle OBA = 90$, hence $m\angle OBC = 90$. But $OC = OA$, so HL congruence implies that $\Delta OBC \cong \Delta OBA$. In particular $AB = BC$.

2. Right away we note that ΔAPB is isosceles since $PA = PB$. Similarly, ΔAOC is isosceles since $OC = OA$. Therefore, by ITT, $m\angle OCA = m\angle BPA = m\angle OAC$. Then $\overline{OC} \parallel \overline{PB}$ because corresponding angles, with respect to the secant $\overleftrightarrow{AC}$, are congruent. But $PO = PA$, so TT1 implies that $AB = BC$.

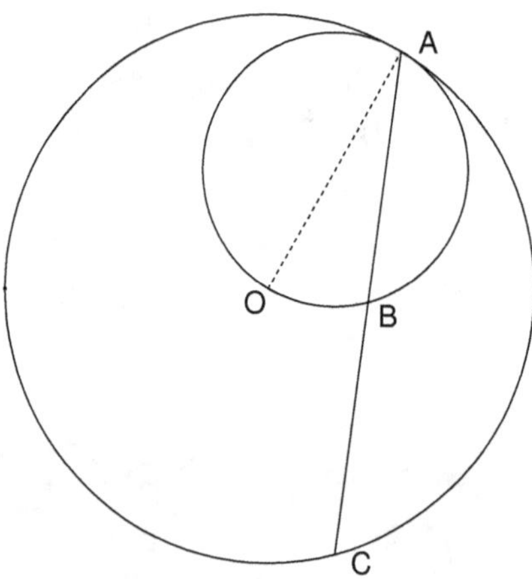

Figure 7.18: A problem about circles

7.5 Length of an angle bisector

At the end of section 6.8 we were able to find the length of the median of any triangle in terms of its sides. A natural question to ask is whether there is a similar formula for the length of any angle bisector of a triangle. Well, we mean the length of the cevian that lies on an angle bisector. Let ABC be an arbitrary triangle. As we know by

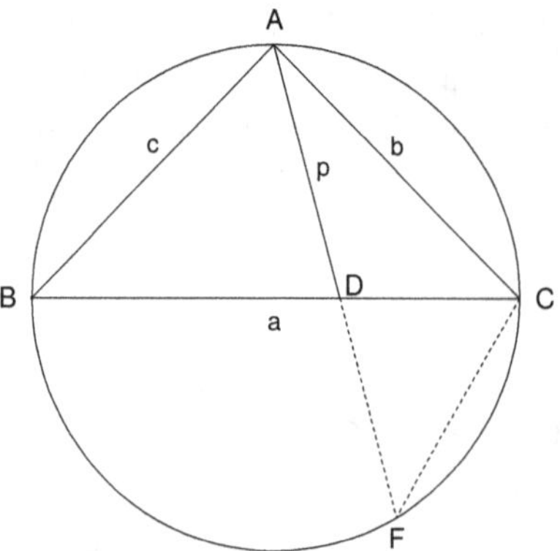

Figure 7.19: Finding the length of an angle bisector

now, there is a circle that passes through its three vertices: the circumcircle (Figure

7.19). Assuming that p denotes the length of an arbitrary angle bisector ($p = AD$) we will show that

$$p^2 = \frac{bc}{(b+c)^2}((b+c)^2 - a^2).$$

Indeed, the Angle Bisector Theorem (section 4.9) implies that $\frac{BD}{DC} = \frac{c}{b}$. Then $\frac{BD}{DC} + \frac{DC}{DC} = \frac{c}{b} + \frac{b}{b}$, thus $\frac{a}{DC} = \frac{b+c}{b}$. Therefore

$$DC = \frac{ab}{b+c}.$$

In identical fashion we reach the equality

$$BD = \frac{ac}{b+c}.$$

We will use these two equalities at the end of the proof.

Next we extend $\overline{AD}$ until it reaches the circumcircle at F, and then build $\overline{FC}$. Since $\angle ABD \cong \angle AFC$ (both subtend the same arc) and $\angle BAD \cong \angle FAC$ we can conclude that $\Delta ABD \sim \Delta AFC$. Hence $\frac{AF}{AB} = \frac{AC}{AD}$, that is,

$$\frac{AF}{c} = \frac{b}{p}.$$

Thus

$$\frac{c}{d} = \frac{AF}{b} = \frac{p+DF}{b}.$$

A basic property of chords (section 7.3) implies that

$$BD \times DC = p \times DF.$$

Consequently

$$\frac{c}{p} = \frac{p+\frac{BD\times DC}{p}}{b},$$

so $bc = p^2 + BD \times DC$. Therefore

$$p^2 = bc - \frac{ac}{b+c} \times \frac{ab}{b+c} = \frac{bc(b+c)^2 - a^2bc}{(b+c)^2} = \frac{bc((b+c)^2 - a^2}{(b+c)^2}. \text{ QED}$$

Another proof of Steiner-Lehmus Theorem

Let $\overline{AD}$ and $\overline{BE}$ be two angle bisectors of ΔABC, with $p = AD$ and $q = BE$. Steiner-Lehmus Theorem asserts that if $p = q$ then $a = b$ (Figure 7.20).

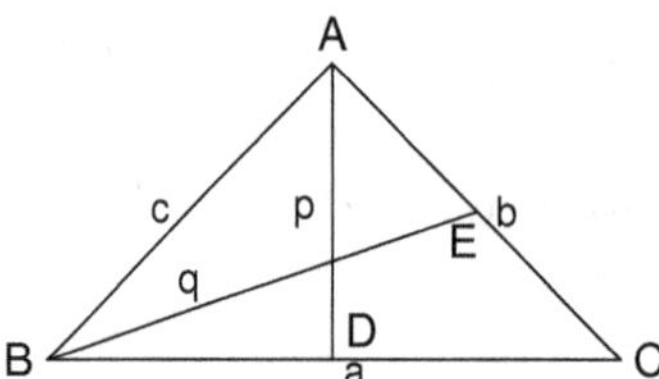

Figure 7.20: An algebraic proof of Steiner-Lehmus Theorem

Proof

The proof that we are about to present is algebraic in nature and appeared in print in (Isaacs, 2001; Xenos, 2006). Let us assume that $p = q$. Since $p^2 = \frac{bc}{(b+c)^2}((b+c)^2 - a^2)$ and $q^2 = \frac{ac}{(a+c)^2}((a+c)^2 - b^2)$, we will have

$$\frac{bc((b+c)^2-a^2)}{(b+c)^2} = \frac{ac((a+c)^2-b^2)}{(a+c)^2}.$$

Hence

$$bc(a+c)^2((b+c)^2 - a^2) = ac(b+c)^2((a+c)^2 - b^2),$$

so

$$(a+c)^2(b+c)^2bc - a^2bc(a+c)^2 = ac(b+c)^2(a+c)^2 - ab^2c(b+c)^2.$$

Therefore

$$(a+c)^2(b+c)^2(bc - ac) = abc(a(a+c)^2 - b(b+c)^2),$$

which in turn leads to

$$(a+c)^2(b+c)^2(b-a) = ab(a(a^2+2ac+c^2) - b(b^2+2bc+c^2)) =$$
$$ab(ac^2 - bc^2 + 2ca^2 - 2b^2c + a^3 - b^3) =$$

$$= ab(c^2(a-b) + 2c(a+b)(a-b) + (a-b)(a^2+ab+b^2)) =$$
$$ab(c^2 + 2c(a+b) + a^2 + ab + b^2)(a-b).$$

Let $\Phi = (a+c)^2(b+c)^2$ and $\Psi = ab(c^2 + 2c(a+b) + a^2 + ab + b^2)$, both of which are evidently positive numbers. We have just shown that

$$\Phi \times (b-a) = \Psi \times (a-b),$$

consequently

$$(\Phi + \Psi) \times (a-b) = 0.$$

Since $\Phi + \Psi \neq 0$ we can conclude that $a - b = 0$, that is, $a = b$. QED

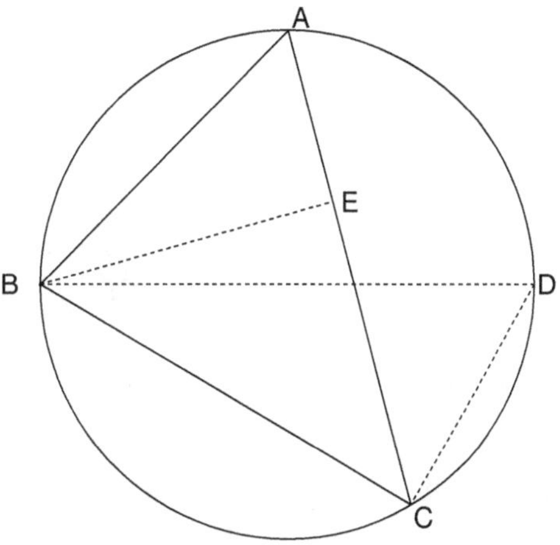

Figure 7.21: Proving Brahmagupta's Theorem

7.6 Brahmagupta's Theorem

Brahmagupta (598-668), a famous Indian mathematician, is credited with a remarkable theorem that provides the area of an inscribed triangle in terms of its sides and the circumradius[5]: Given ΔABC

$$\text{area}(\Delta ABC) = \tfrac{1}{4R}abc,$$

where a, b, c are the lengths of the three sides and R is the circumradius. A proof goes as follows. Letting O be the circumcenter, build the diameter $\overline{BD}$ and drop the perpendicular $\overline{BE}$ from the vertex B to the side $\overline{AC}$ (Figure 7.21). Since $\overline{BD}$ is a diameter we can assert that $\angle BCD$ is a right angle. Moreover, $\angle BAC \cong \angle BDC$ since both angles subtend the same arc. Then AA similarity implies that $\Delta AEB \cong \Delta DCB$. Consequently

$$\tfrac{BD}{AB} = \tfrac{BC}{BE}.$$

Thus, $\frac{2R}{c} = \frac{a}{BE}$. Therefore

$$\text{area}(\Delta ABC) = \tfrac{1}{2}AC \times BE = \tfrac{1}{2}b \times \tfrac{ac}{2R} = \tfrac{1}{4R}abc. \text{ QED}$$

---

[5]Recall that the circumradius of a triangle is the radius of the circumscribed circle. The center of the latter is the circumcenter, which happens to be the point of concurrence of the corresponding perpendicular bisectors.

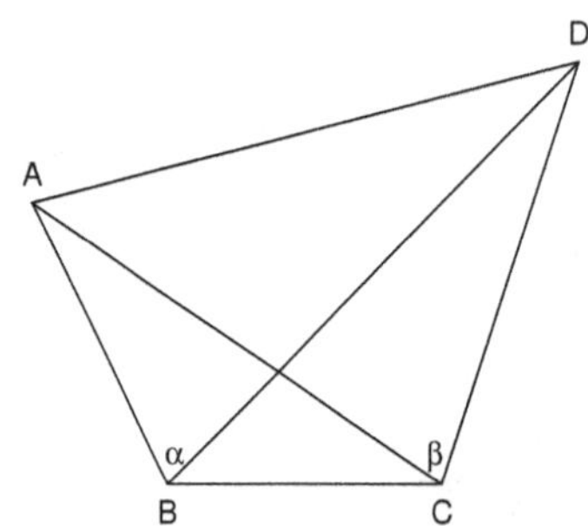

Figure 7.22: A condition under which the four vertices of a quadrilateral are concyclic

7.7 Concyclic points

Points $P_1, P_2, \ldots, P_n$ are said to be concyclic if they lie on a circle. It is easy to show that three non-collinear points are concyclic since we can draw the corresponding segments, build a triangle whose sides are the segments, and then the perpendicular bisectors of each segment. As we know by now, the latter meet at a point that happens to be the center of the circle that contains the three given points. Is there a proposition that guarantees when any four non-collinear points are concyclic? Next we will state and prove a proposition that answers the question.

Proposition

Let A, B, C, and D be the vertices of a quadrilateral (Figure 7.22). Assume that $\alpha = \beta$, where $\alpha = m\angle ABD$ and $\beta = m\angle ACD$. Then A, B, C, and D are concyclic.

Proof

Points A, B, and D lie on a circle. We claim that C lies on the circle too. Proceeding by contradiction suppose that C does not lie on the circle. There are two possibilities:

1. *C lies inside the circle.* Prolong $\overrightarrow{AC}$ and let E be its intersection with the circle (Figure 7.23). Next build $\overline{ED}$. We note that $\angle AED \cong \angle ABD$ because both are inscribed angles that subtend the same arc, namely $\overset{\frown}{AD}$. Thus $\gamma = \beta$, where $\gamma = m\angle AED$. But the Exterior Angle Theorem, applied to ΔECD, implies that $\beta > \gamma$. We have reached a contradiction.

2. *C lies outside the circle.* The ray $\overrightarrow{AC}$ (Figure 7.23) will meet the circle at a certain point E. Next we construct $\overline{ED}$ and let $\gamma = m\angle AED$. Since $\angle ABD \cong \angle AED$ (both subtend $\overset{\frown}{AD}$) we can conclude that $\gamma = \beta$. But the Exterior Angle

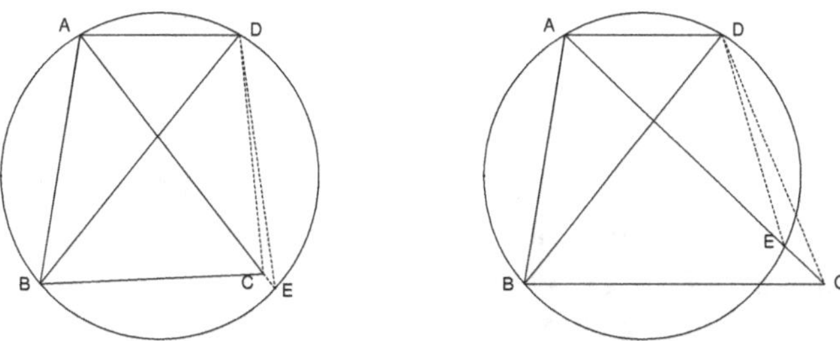

Figure 7.23: Proposition about concyclic points (two cases)

Theorem, applied to ΔCED, implies that $\gamma > \beta$. Again, we have reached a contradiction. QED

Yet another proof of Steiner-Lehmus Theorem

The reader might recall that in section 3.4 we provided a proof, based in part on the basic theory of parallelograms, of Steiner-Lehmus Theorem; that is, if the angle bisectors are congruent then the triangle has to be isosceles. Moreover, in section 7.5 we provided a second proof of this theorem. Using the preceding proposition, a necessary condition under which four points are concyclic, we will be able to provide a rather short proof of the above-mentioned theorem (Gilbert and MacDonnell, 1963).

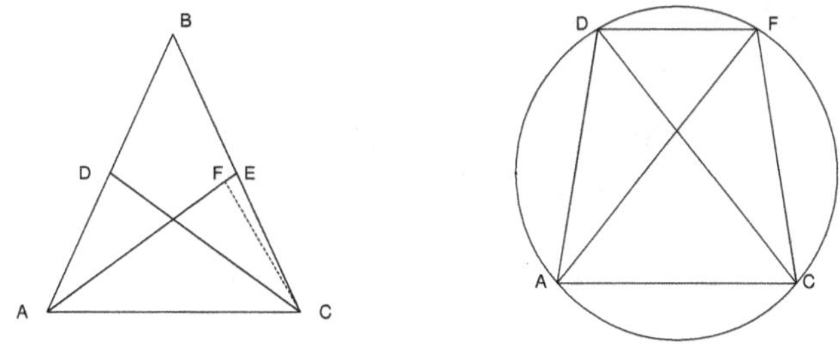

Figure 7.24: A third proof of Steiner-Lehmus Theorem

Proof

Let ΔABC have two congruent angle bisectors, namely $\overline{AE}$ and $\overline{CD}$ (Figure 7.24). Proceeding by contradiction assume that ΔABC is not isosceles. Letting $2\alpha =$

$m\angle BAC$ and $2\beta = m\angle BCA$ we can then assume, without loss of generality, that $\alpha < \beta$. Next build $\angle DCF$ so that $m\angle DCF = \alpha$. It should be noted that, from the construction, it is clear that $AF < AE$.

Since $m\angle DAE = m\angle DCF$ (their common measure is α), the preceding proposition implies that there is a circle that contains A, C, F and D. Since $\alpha + \alpha < \alpha + \beta$, and we are dealing with measures of inscribed angles, it follows that $DC < AF$. But $AF < AE$, consequently $DC < AE$. We have arrived at a contradiction since, by hypothesis, $AE = CD$. QED

7.8 Inscribed quadrilaterals

In the previous section we discussed a necessary condition under which a quadrilateral can be inscribed in a circle. Now we will analyze a necessary and sufficient condition under which a quadrilateral can be inscribed in a circle.

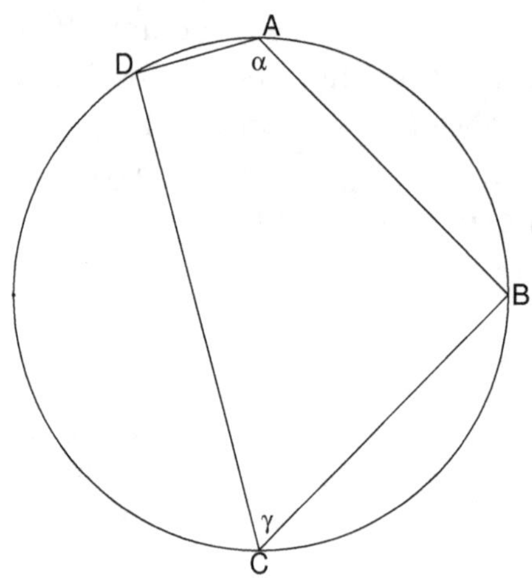

Figure 7.25: Inscribed quadrilateral

Proposition

Suppose that a quadrilateral is inscribed in a circle. Then both pairs of opposite angles are supplementary, that is, the sum of their measures is 180.

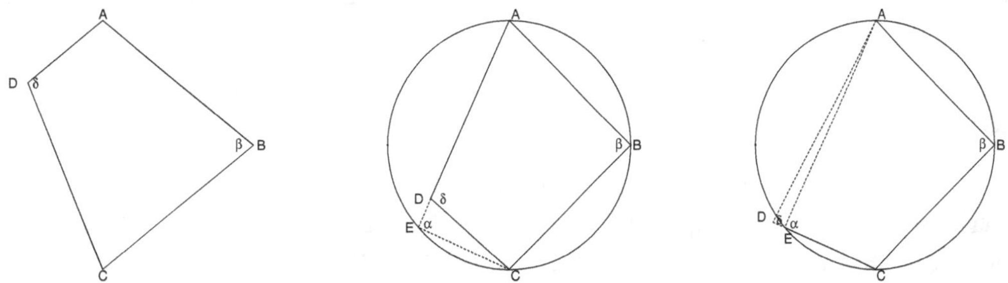

Figure 7.26: Proving a proposition about inscribed quadrilaterals

Proof

Let the quadrilateral $ABCD$ be inscribed in a circle (Figure 7.25) and define $\alpha = \frac{1}{2}\, \overset{\frown}{BCD}$, $\gamma = \frac{1}{2}\, \overset{\frown}{DAB}$. Hence

$$\alpha + \gamma = \frac{1}{2}\, \overset{\frown}{BCD} + \frac{1}{2}\, \overset{\frown}{DAB} = \frac{1}{2}(\overset{\frown}{BCD} + \overset{\frown}{DAB}) = \frac{1}{2}(360) = 180.$$

QED

Proposition

If a pair of opposite angles of a quadrilateral are supplementary then the quadrilateral can be inscribed in a circle.

Proof

Let $ABCD$ be a quadrilateral and assume that $\beta + \delta = 180$, where $\beta = m\angle ABC$ and $\delta = m\angle ADC$ (Figure 7.26). Let us draw a circle through the points A, B, C . We claim that D has to lie on this circle too. If it were not to lie on it, there are two possibilities:

1. Suppose that D is in the interior of the circle (Figure 7.26) and extend $\overline{AD}$ until it touches the circle at E. Letting $\alpha = m\angle AEC$, thanks to the previous proposition it follows that $\alpha + \beta = 180$. Hence $\alpha + \beta = \delta + \beta$, and consequently $\alpha = \delta$. This is an impossibility since $\delta > \alpha$ due to the Exterior Angle Theorem.

2. Another possibility is that D lies on the exterior of the circle (Figure 7.26). After building $\overline{AE}$ and defining $\alpha = m\angle AEC$, the preceding proposition implies that $\alpha + \beta = 180$. Since $\beta + \delta = 180$ it follows that $\alpha + \beta = \delta + \beta$, so $\alpha = \delta$. Again, this is an impossibility since the Exterior Angle Theorem implies that $\alpha > \delta$.

In summary, the circle that passes through A, B, C also has to pass through D because otherwise we reach a contradiction! QED

Corollary

Let ΔABC be an *equilateral* triangle and suppose that P is a point on the exterior of the triangle (say, on the opposite side of $\overline{BC}$) (Figure 7.27). Then P belongs to the circumcircle of ΔABC if and only if $m\angle BPC = 120$.

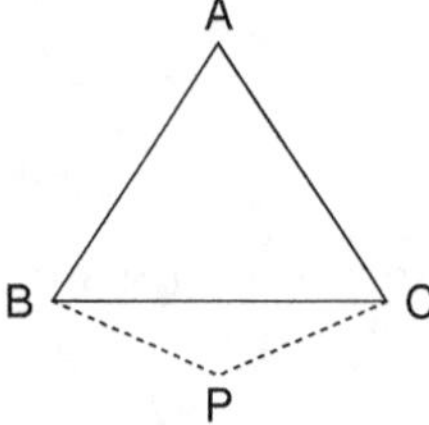

Figure 7.27: A property of equilateral triangles

The Fermat Point

Let us consider an arbitrary acute triangle ABC. A point P, in the interior of ΔABC, is called a Fermat point if

$$m\angle BPA = m\angle CPA = m\angle CPB = 120^o$$

Two questions come immediately to our mind: Does every acute triangle have a Fermat point? Is a Fermat point unique? Let us answer these questions separately[6].

Existence

Given ΔABC, an acute triangle, build an equilateral triangle on $\overline{AB}$ and an equilateral triangle on $\overline{AC}$. Thereafter construct the circumcircles of both equilateral triangles; they meet at A and at a point P inside[7] ΔABC. By the preceding corollary we have that $m\angle APB = 120^o$ and $m\angle APC = 120^o$. Therefore $m\angle BPC = 120^o$ too.

[6]The concept of a Fermat point could be defined for every triangle provided that none of the internal angles is bigger than or equal to 120°. But we have chosen to work only with acute triangles in order to simplify the proof of existence and uniqueness.

[7]We use the hypothesis that the triangle is acute in order to make sure that P lies inside the triangle.

Uniqueness

Let Q be any point in the interior of ΔABC, an acute triangle, such that $m\angle AQB = m\angle AQC = m\angle BQC = 120^o$. Then build equilateral triangles on $\overline{AB}$ and $\overline{AC}$, and construct the circumcircles Ω_1 and Ω_2 of both equilateral triangles. Applying the preceding corollary it follows that Q belongs to Ω_1 and Q belongs to Ω_2. Then Q belongs to the intersection of Ω_1 and Ω_2. Thus, Q is precisely the point found when discussing the proof of existence of a Fermat point. Having proven existence and uniqueness we can speak about *the* Fermat point, which from now on will be denoted with the letter F.

Fermat-Torricelli Theorem

Pierre de Fermat (1601-1665), an outstanding mathematician, posed the following problem: Among all interior points P of an acute triangle ABC, which one minimizes the sum $PA+PB+PC$? The answer was given by Evangelista Torricelli (1608-1647), who found that the solution was to choose what we nowadays call the Fermat point!

Theorem

Let ΔABC be an arbitrary acute triangle. Among all interior points P, if P is the Fermat point then the sum $PA + PB + PC$ is a minimum[8].

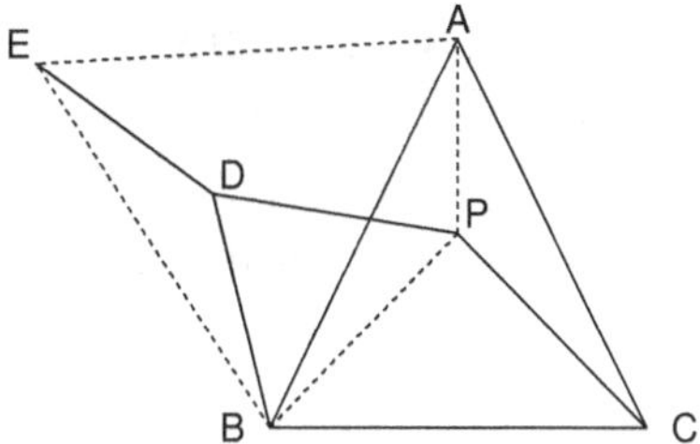

Figure 7.28: Proving Fermat-Torricelli Theorem

Proof

Let P be any interior point of ΔABC. Since $m\angle APB + m\angle APC + m\angle BPC = 360^o$ it must be true that at least one among $\angle APB$, $\angle APC$, and $\angle BPC$ has to be bigger

[8]The proof that we will discuss follows Berele and Goldman (2001).

than or equal 120° because otherwise $m\angle APB + m\angle APC + m\angle BPC < 360^o$. Without loss of generality assume that $m\angle APB \geq 120^o$ (Figure 7.28). Then $m\angle ABP < 60^o$. Next build an equilateral triangle on $\overline{AB}$, namely ΔABE, and construct ΔBDE in such a way that $m\angle EBD = m\angle ABP$ and $BD = BP$. Then SAS congruence implies that $\Delta EBD \cong \Delta ABP$, hence $ED = AP$.

After building $\overline{DP}$ we notice that $m\angle DBP = 60^o$ and since ΔDBP is isosceles we can conclude that the base angles have a measure of 60° too. Therefore ΔDBP is equilateral, consequently $DP = BD = BP$. We need to minimize $ED + DP + PC$. Evidently, this sum will attain a minimum if $\overline{ED}$, $\overline{DP}$ and $\overline{PC}$ lie on a line.

We will bring the proof to an end by showing that the choice $P = F$, where F is the Fermat point, is the best option, in the sense that it minimizes the sum of the distances to the three vertices. Indeed, the Fermat point has the property that $m\angle BFA = m\angle BFC = 120^o$ and since $\Delta EBD \cong \Delta ABF$ it follows that $m\angle EDB = m\angle AFB = 120^o$, hence $m\angle EDB + m\angle BDF = 180^o$. That is to say, the segments $\overline{ED}$ and $\overline{DF}$ lie on the same line.

In a similar fashion, we can observe that $m\angle BFC + m\angle BFD = 120^o + 60^o = 180^o$. So, $\overline{DF}$ and $\overline{FC}$ lie on the same line. In summary, we have shown that the points E, D, F, and C lie on the same line. QED

Remark

The notion of Fermat point is related to triangles, not circles. We discuss it in chapter 7 because the chosen approach to existence and uniqueness of the Fermat point uses circles. On the other hand, the reader must have noticed the practical utility of Fermat-Torricelli Theorem: it provides an answer to the question about where a power plant should be built if we wish to provide electricity to three cities, minimizing the length of the transmission lines. The theorem works well if the triangle formed by the three cities happens to be acute, or more generally, if none of the interior angles is bigger than or equal to 120°.

7.9 Inscribing regular polygons in a circle

The idea of inscribing regular polygons inside a circle will play an important role when defining and approximating the number π in the next section. So, let us spend some effort studying the relationship between the length of the side of an inscribed regular polygon and the radius of the circle. Let us work with a circle with center O and radius r. How do we build a regular triangle inside the circle? Quite simple, just notice that $360/3 = 120$.

Then choose points A, B, D on the circle in such a way that $m\angle AOB = m\angle BOD = m\angle AOD = 120^o$ (Figure 7.29). Obviously $\Delta AOB \cong \Delta DOB \cong \Delta AOD$ (SAS congru-

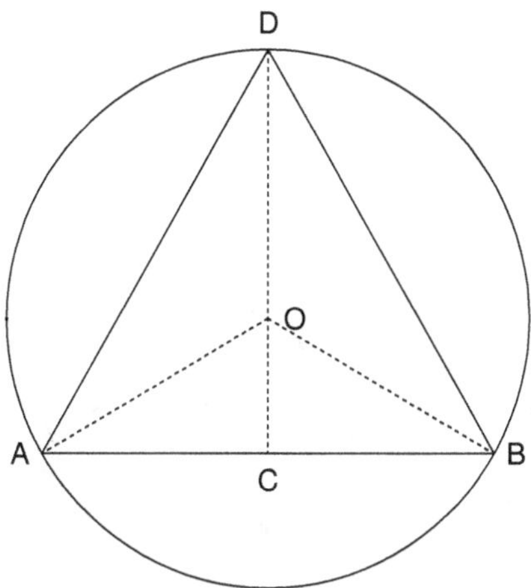

Figure 7.29: Inscribed equilateral triangle

ency). Let $x = AB = BD = AD$ and draw the height OC. Since ΔAOB is isosceles we will have that $m\angle BOC = 60^o$, hence

$$\frac{\sqrt{3}}{2} = \sin 60 = \frac{\frac{x}{2}}{r}.$$

Therefore $x = r\sqrt{3}$.

Our next task is to work with the regular hexagon. This time $\frac{360}{6} = 60$, so we choose A and B on the circle in such a way that $m\angle AOB = 60^o$. Since the triangle is isosceles we must have that $m\angle OAB = m\angle OBA = 60^o$ (recall that, in Euclidean geometry, the sum of the measures of the internal angles of any triangle is 180°). Therefore ΔAOB is an equilateral triangle, consequently $x = r$ where x is the length of the side of the regular hexagon inscribed in the circle with radius r and center O.

How about the regular 12-gon? This time $\frac{360}{12} = 30$. Let ΔAOB be the triangle corresponding to the hexagon, thus $AB = r$ (Figure 7.30). Then drop the height OC and extend it to the point D on the circle. Since ΔAOB is isosceles we can conclude that $m\angle AOD = 30^o$; moreover, $AC = \frac{r}{2}$ and $m\angle CAD = 15^o$. We notice that

$$\cos 15^o = \sqrt{\frac{1+\cos 30^o}{2}} = \sqrt{\frac{1+\frac{\sqrt{3}}{2}}{2}} = \frac{1}{2}\sqrt{2+\sqrt{3}}.$$

But, working with ΔACD, we get $\cos 15^o = \frac{\frac{r}{2}}{x}$ where x is the length of the side of the regular 12-gon (also known as a 'dodecagon'). Therefore

$$\frac{1}{2}\sqrt{2+\sqrt{3}} = \frac{\frac{r}{2}}{x},$$

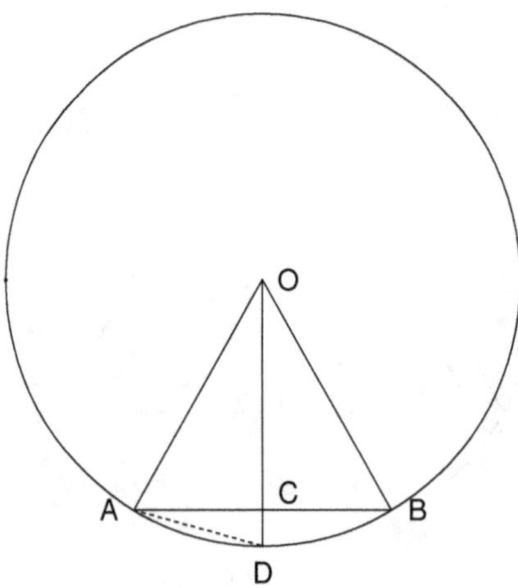

Figure 7.30: Length of side of a regular inscribed hexagon and a 12-gon

hence

$$x = \frac{1}{\sqrt{2+\sqrt{3}}}r = \sqrt{2-\sqrt{3}}r.$$

What could be done if the half-angle formula is not available? There is an alternative procedure; let $s = OC$ and apply Pythagoras' Theorem to ΔOCA. Then $s^2 = r^2 - \frac{r^2}{4}$, so $s = \frac{\sqrt{3}}{2}r$. Next we apply Pythagoras' Theorem to ΔACD and obtain

$$x^2 = \frac{r^2}{4} + (r-s)^2 = \frac{r^2}{4} + (r - \frac{\sqrt{3}}{2}r)^2 = \frac{r^2}{4} + r^2\frac{(2-\sqrt{3})^2}{4} = r^2(2-\sqrt{3}),$$

therefore $x = \sqrt{2-\sqrt{3}}r$.

The preceding approach can be used to construct the 24-gon, 48-gon, etc. Our next objective is to work with the pentagon and the decagon. The latter will be simpler than the former, so let us start with the decagon. As before, let us denote the length of its side by x. Since $\frac{360}{10} = 36$ we have to deal with ΔOAB (Figure 7.31), with $OA = OB = r$ and $AB = x$. Then bisect $\angle OAB$ and let D be the point where the angle bisector intersects $\overline{OB}$. Since $m\angle OAB = m\angle OBA = 72^o$ we can assert that $m\angle OAD = m\angle BAD = 36^o$. On the other hand, obviously $m\angle BDA = 72^o$; thus, ΔBAD is isosceles with $AD = AB = x$. But ΔODA is isosceles too, hence $OD = x$ and consequently $BD = r - x$.

AA similarity implies that $\Delta OBA \sim \Delta ABD$, hence

$$\frac{x}{r-x} = \frac{r}{x},$$

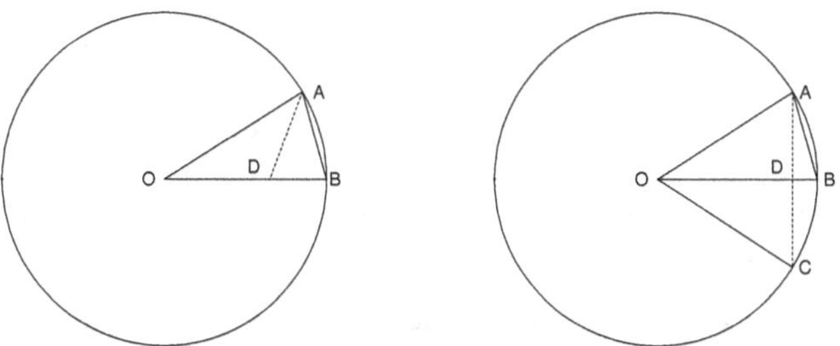

Figure 7.31: Length of side of a regular decagon and a regular pentagon

which in turn leads to $x^2 + rx - r^2 = 0$. The only positive root of this equation is $x = (\frac{\sqrt{5}-1}{2})r$. We have succeeded in expressing x in terms of the radius of the circle.

How could we use the preceding formula to solve the equivalent problem for the pentagon? Since $\frac{360}{5} = 72$ we build ΔAOC with $x = AC$ the length of the side of the regular pentagon (Figure 7.31), and bisect $\angle AOC$. The angle bisector intersects $\overline{AC}$ at D and intersects the circle at B. Letting $s = OD$ we will have that $DB = r - s$. Moreover, we observe that $AB = (\frac{\sqrt{5}-1}{2})r$ since it is the length of the side of the regular decagon. Next we apply Pythagoras' Theorem to ΔODA and ΔADB, so

$$r^2 - s^2 = AD^2 = \tfrac{r^2}{4}(\sqrt{5}-1)^2 - (r-s)^2.$$

Simplifying this expression we get $s = (\frac{\sqrt{5}+1}{4})r$. But $\frac{x}{2} = AD$; thus, applying Pythagoras' Theorem to ΔODA we arrive at

$$\tfrac{x}{2} = \sqrt{r^2 - s^2} = \sqrt{r^2 - \tfrac{(\sqrt{5}+1)^2}{4^2}r^2}.$$

After a little bit of arithmetic we reach the answer, namely $x = \frac{r}{2}\sqrt{10 - 2\sqrt{5}}$. We trust that by now the reader has understood some of the techniques that need to be followed in order to deal with any regular inscribed polygon.

As a side result we can see that

$$\cos 36^o = \tfrac{OD}{OA} = \tfrac{\sqrt{5}+1}{4},$$

$$\sin 36^o = \tfrac{AD}{OA} = \sqrt{1 - (\tfrac{\sqrt{5}+1}{4})^2} = \tfrac{1}{4}\sqrt{16 - (\sqrt{5}+1)^2}.$$

Then

$$\cos 72^o = \cos^2 36^o - \sin^2 36^o = (\tfrac{\sqrt{5}+1}{4})^2 - \tfrac{1}{4^2}(16 - (\sqrt{5}+1)^2) = \tfrac{1}{4}(\sqrt{5}-1).$$

7.10 The number π

Perimeter of a circle

Given a circle, let p_n be the perimeter of the inscribed regular polygon of 3×2^n sides ($n = 0, 1, 2, ...$). Using the triangle inequality we can see that $p_1 < p_2 < p_3 < ....$ The sequence (p_n) is bounded above by the perimeter of the circumscribed equilateral triangle. Since every increasing and bounded above sequence is convergent we can assert that there exists a unique number p such that $\lim_{n\to\infty} p_n = p$. We adopt p as the definition of the perimeter of the given circle.

Proposition

Given any two circles with centers O, O^*, perimeters p, p^*, and radii r, r^* respectively, the following equality is true:

$$\frac{p}{r} = \frac{p^*}{r^*}.$$

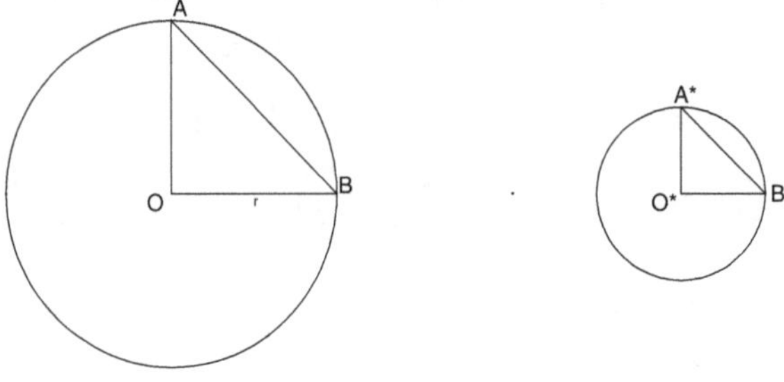

Figure 7.32: Preliminary steps in the quest for a proper definition

Proof

Let s_n be the length of the side of the regular polygon of 3×2^n sides, inscribed in the circle with center O and radius r. Similarly, let s_n^* be the length of the side of the inscribed regular polygon of 3×2^n sides, inscribed in the circle with center O^* and radius r^* (Figure 7.32). Then we build ΔAOB with $AO = BO = r$ and $AB = s_n$, and we build $\Delta A^*O^*B^*$ with $A^*O^* = B^*O^* = r^*$ and $A^*B^* = s_n^*$. Evidently

$$m\angle AOB = m\angle A^*O^*B^* = \frac{360}{3\times 2^n}.$$

Since $\frac{r}{r^*} = \frac{r}{r^*}$ and $\angle AOB \cong \angle A^*O^*B^*$, the SAS property of similarity (section 5.2) implies that $\Delta AOB \sim \Delta A^*O^*B^*$. Then

$$\frac{s_n}{r} = \frac{s_n^*}{r^*},$$

consequently

$$\frac{3\times 2^n\times s_n}{r} = \frac{3\times 2^n\times s_n^*}{r^*}.$$

That is,

$$\frac{p_n}{r} = \frac{p_n^*}{r^*}. \tag{7.2}$$

Since $\lim_{n\to\infty}\frac{p_n}{r} = \frac{p}{r}$ and $\lim_{n\to\infty}\frac{p_n^*}{r^*} = \frac{p^*}{r^*}$, from (7.2) it follows that[9]

$$\frac{p}{r} = \frac{p^*}{r^*}.$$

QED

Definition

Given any circle with perimeter p and radius r we define $\pi = \frac{p}{2r}$. The preceding proposition makes this a well-defined number. From the definition it follows that given the radius r of a circle, its perimeter follows from the formula

$$p = 2\pi r.$$

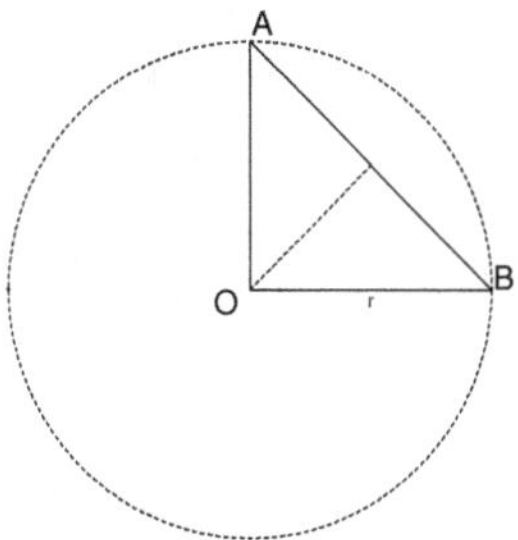

Figure 7.33: Looking for a formula for the area of a circle

[9]Keep in mind that the limit of a sequence is unique.

Area of a circle

Let us consider a circle with center O and radius r. We start by inscribing a regular polygon of $3{\times}2^n$ sides (Figure 7.33). In the usual way, the inscribed polygon is divided into 3×2^n congruent triangles like ΔAOB, where $AO = BO = r$ and $AB = s_n$. Let q_n be the height that starts at the vertex O. We note, right away, that $\frac{1}{2}s_n q_n$ is the area of ΔAOB. Thus

$$\tfrac{1}{2}(3 \times 2^n s_n q_n),$$

that is $\frac{1}{2}p_n q_n$, is the area of the inscribed polygon of 3×2^n sides. Our intuition (see remark below) tells us that $\lim_{n\to\infty} q_n = r$. Since $\lim_{n\to\infty} p_n = p$ it follows that

$$\lim_{n\to\infty} \tfrac{1}{2}p_n q_n = \tfrac{1}{2}pr.$$

We can then adopt the number $\frac{1}{2}pr$ as the area of the circle with radius r. Note that

$$\text{area of circle}{=}\tfrac{1}{2}pr = \tfrac{1}{2}(2\pi r)r = \pi r^2.$$

Remark

Can we prove that $\lim_{n\to\infty} q_n = r$? Figure 7.34 , where $OB = r$, $BC = \frac{s_n}{2}$, and $OC = q_n$, will serve us a guide to the proof. First of all we will prove that $\lim_{n\to\infty} s_n = 0$. Indeed, $0 < 3 \times 2^n \times s_n < p$, so

$$0 < s_n < \tfrac{p}{3\times 2^n}.$$

Since $\lim_{n\to\infty} \frac{p}{3\times 2^n} = 0$ we can conclude that $\lim_{n\to\infty} s_n = 0$. Next note that $q_n < r$ (a leg is always smaller than the hypotenuse) while the triangle inequality implies that $r < q_n + \frac{s_n}{2}$. Hence

$$r - \tfrac{s_n}{2} < q_n < r.$$

But $\lim_{n\to\infty} r - \frac{s_n}{2} = r$, therefore $\lim q_n = r$.

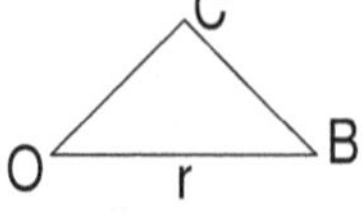

Figure 7.34: Justifying a limit

Approximations of π

Let us consider a circle with unit radius (Figure 7.35). As usual, let s_n be the length of the side of the regular polygon with 3×2^n sides inscribed in the circle. Observe that s_{n+1} corresponds to the polygon of $3 \times 2^{n+1}$ sides, twice as many sides as the previous polygon. Inside the unit circle build ΔDOC with $DC = s_n$ and prolong the height OA until it meets the circle at B. Then $AC = \frac{s_n}{2}$ and $BC = s_{n+1}$. Applying Pythagoras Theorem to ΔBAC we get

$$s_{n+1}^2 = \tfrac{s_n^2}{4} + (1 - OA)^2.$$

A second application of Pythagoras' Theorem, this time to ΔOAC, leads to $OA = \sqrt{1 - \frac{s_n^2}{4}}$. Then

$$s_{n+1}^2 = \tfrac{s_n^2}{4} + (1 - \sqrt{1 - \tfrac{s_n^2}{4}})^2,$$

consequently

$$s_{n+1}^2 = \tfrac{s_n^2}{4} + 1 - 2\sqrt{1 - \tfrac{s_n^2}{4}} + 1 - \tfrac{s_n^2}{4}.$$

Therefore

$$s_{n+1} = \sqrt{2 - \sqrt{4 - s_n^2}}, \quad n = 0, 1, 2, ... \tag{7.3}$$

Since s_0 was calculated at the beginning of section 7.9 ($s_0 = \sqrt{3}$), the recursive process defined by (7.3) can get started. For instance,

$$s_1 = \sqrt{2 - \sqrt{4 - 3}} = \sqrt{2 - 1} = 1,$$

$$s_2 = \sqrt{2 - \sqrt{4 - 1}} = \sqrt{2 - \sqrt{3}},$$

$$s_3 = \sqrt{2 - \sqrt{4 - (2 - \sqrt{3})}} = \sqrt{2 - \sqrt{2 + \sqrt{3}}},$$

$$s_4 = \sqrt{2 - \sqrt{4 - (2 - \sqrt{2 + \sqrt{3}})}} = \sqrt{2 - \sqrt{2 + \sqrt{2 + \sqrt{3}}}},$$

$$s_5 = \sqrt{2 - \sqrt{4 - (2 - \sqrt{2 + \sqrt{2 + \sqrt{3}}})}} = \sqrt{2 - \sqrt{2 + \sqrt{2 + \sqrt{2 + \sqrt{3}}}}}.$$

Furthermore,

$$\lim_{n \to \infty} p_n = p = 2\pi$$

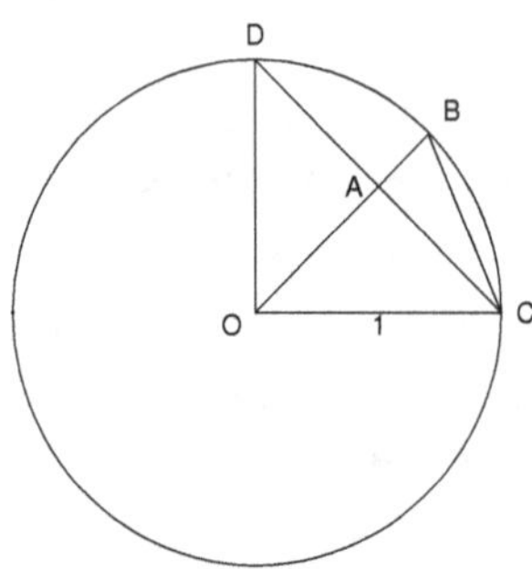

Figure 7.35: Approximation of π

where, let us recall, $p_n = 3 \times 2^n \times s_n$. Hence

$$\pi \approx \tfrac{p_n}{2} = 3 \times 2^{n-1} \times s_n$$

when n is 'large enough'. Using a calculator we get

$$\tfrac{p_3}{2} = 3 \times 2^2 \times s_3 = 3.132628613,$$

$$\tfrac{p_4}{2} = 3 \times 2^3 \times s_4 = 3.139350203,$$

$$\tfrac{p_5}{2} = 3 \times 2^4 \times s_5 = 3.141031951.$$

The last approximation, which involves a regular polygon of $3 \times 2^5 = 96$ sides, is already pretty good because it has allowed us to find the first three 'true' decimals of π. We could find other decimals of π if we were to choose larger values of n, but handling s_n becomes burdensome, so it is advisable to build a program using the capabilities of a graphing calculator (TI-83 or similar machines). This task is accomplished in chapter 15, where we are able to find the first eight 'true' decimals of π.

7.11 Measure in radians

The main goal in this section is to lay the foundations behind the concept of radian. In the process we will define, without ambiguity, the notions of length of an arc and area of a sector, as well as related mathematical ideas. By the way, a **sector** of a circle is the region bounded by an arc and two radii.

Length of an arc

Let us consider an arc $\overset{\frown}{AB}$, which stems from a circle centered at O and radius $r = OA = OB$ (Figure 7.36). Then drop a height from O to $\overline{AB}$. Since ΔOAB is isosceles we can conclude that $AP_1 = BP_1$, where P_1 is the point where the height meets the segment $\overline{AB}$. Next prolong $\overline{AP_1}$ until it touches the arc at C. Thanks to the SAS congruence property we can assert that $\Delta AP_1O \cong \Delta BP_1O$; thus, $\angle AOP_1 \cong \angle BOP_1$. Hence, again by SAS congruence, $\Delta AOC \cong \Delta BOC$, so $AC = BC$.

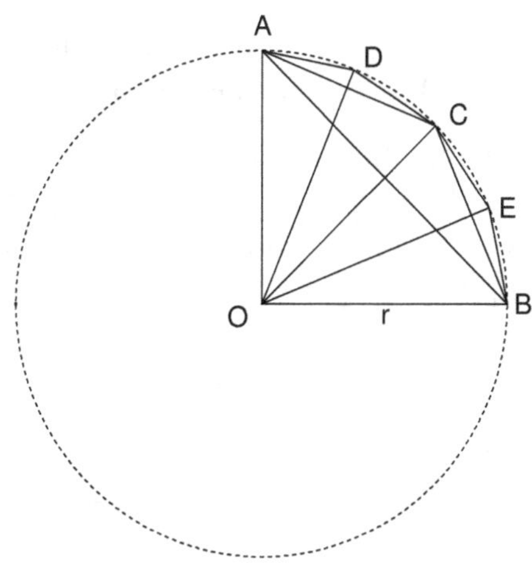

Figure 7.36: Length of an arc

Thereafter drop the height OP_2 from O to $\overline{AC}$ and the height OP_3 from O to $\overline{BC}$ and prolong them until they touch the arc at D and E respectively. The same reasoning, which allowed us to conclude that $AC = BC$, leads to $\Delta AP_2O \cong \Delta CP_2O$ and $\Delta CP_3O \cong \Delta BP_3O$. Therefore $\angle AOP_2 \cong \angle COP_2$ and $\angle COP_3 \cong \angle COP_3$. But, let us recall, $\angle AOC \cong \angle BOC$, consequently $\angle AOP_2 \cong \angle COP_2 \cong \angle COP_3 \cong \angle COP_3$. Then SAS congruence implies that $\Delta AOD \cong \Delta DOC \cong \Delta COE \cong \Delta EOB$, therefore $AD = DC = CE = EB$. Evidently, this process can be continued indefinitely.

Let $b_0 = AB$, $b_1 = AC = BC$, $b_2 = AD = DC = CE = EB$, and so on. Define $p_0 = b_0$, $p_1 = 2b_1$, $p_2 = 4b_2$,... Thus

$$p_n = 2^n b_n \quad n = 0, 1, 2, 3, ...$$

Using the triangle inequality we can readily conclude that $p_0 < p_1 < p_2 < ...$ Besides, the sequence (p_n) is bounded above (inscribe the circle, which contains the arc, inside a square; then the perimeter of the square is an upper bound of (p_n)). Since the sequence is increasing, and bounded above, it must converge to a unique number p. That is,

$$\lim_{n\to\infty} p_n = p.$$

We adopt p as the length of the arc $\overset{\frown}{AB}$.

Area of a sector

Let us recall that a subset of a circle, limited by $\overset{\frown}{AB}$ and the segments $\overline{OA}$ and $\overline{OB}$, is called a sector of the circle with center O and radius $r = OA = OB$ (Figure 7.36). We subdivide the sector following the procedure used to define the length of $\overset{\frown}{AB}$. At the nth step, a typical triangle has base b_n and height a_n; thus, its area is $\frac{1}{2}a_n b_n$. At this step there are 2^n congruent triangles, consequently an approximation to the area of the sector at the nth step is

$$\tfrac{1}{2}a_n b_n \times 2^n.$$

But[10] $\lim_{n\to\infty} a_n = r$ and $\lim_{n\to\infty} 2^n b_n = p$, where p is the length of $\overset{\frown}{AB}$. Then

$$\lim_{n\to\infty} \tfrac{1}{2}a_n b_n \times 2^n = \tfrac{1}{2}pr.$$

We can then adopt $\frac{1}{2}pr$ as the area of the sector with arc length p and radius r.

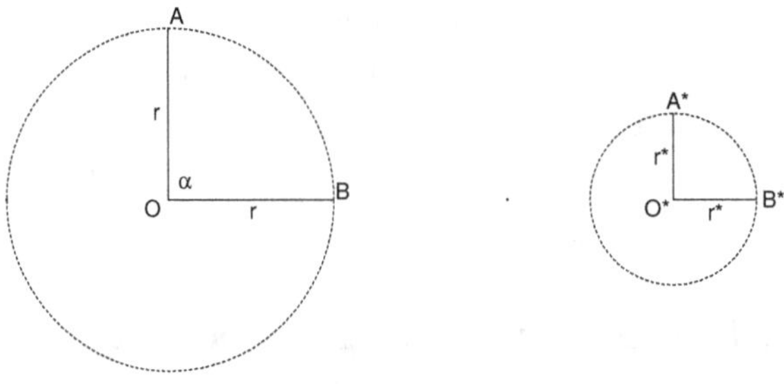

Figure 7.37: Towards the definition of radian

[10]We could provide a formal proof of the fact that the sequence (a_n) does indeed converge to r, following an argument similar to the one we employed in the last remark (section 7.10.)

Proposition

Let $\overset{\frown}{AB}$ and $\overset{\frown}{A^*B^*}$ be arcs corresponding to circles with centers O and O^*, and radii r and r^* respectively (Figure 7.37). Moreover $m\angle AOB = m\angle A^*O^*B^* = \alpha$. Then

$$\frac{p}{r} = \frac{p^*}{r^*},$$

where p =length $\overset{\frown}{AB}$, p^* =length $\overset{\frown}{A^*B^*}$.

Proof

For each sector we follow the procedure that led to the definition of length of an arc. At the nth step, a typical triangle for one sector will have base b_n, the other two sides r and angle in between with measure $\frac{\alpha}{2^n}$, while the typical triangle for the other sector will have base b_n^*, the other two sides r^* and angle in between with measure $\frac{\alpha}{2^n}$. SAS similarity allows us to conclude that both triangles are similar. Thus

$$\frac{b_n}{b_n^*} = \frac{r}{r^*},$$

which in turn implies that

$$\frac{2^n b_n}{2^n b_n^*} = \frac{r}{r^*}.$$

Hence

$$\frac{p_n}{p_n^*} = \frac{r}{r^*}.$$

But $\lim_{n\to\infty} p_n = p$ and $\lim_{n\to\infty} p_n^* = p^*$. Therefore $\frac{p}{p^*} = \frac{r}{r^*}$, that is, $\frac{p}{r} = \frac{p^*}{r^*}$. QED

Definition (measure of an angle in radians)

Given any angle, with vertex at O, choose A on one side and B on the other side so that $OA = OB = r$. Then build $\overset{\frown}{AB}$ and let p be its length. The quotient $\theta = \frac{p}{r}$ is known as the **measure in radians** of the given angle. This is a well-defined mathematical concept thanks to the preceding proposition.

A basic formula

Suppose that we have a circle with center O and radius r, and a sector AOB such that $m\angle AOB = \theta$ radians. Then, from the very definition of radian we get

$$p = r\theta,$$

where p is the length of $\overset{\frown}{AB}$. This formula is remarkable for its simplicity and plays an important role in the development of calculus. No wonder that in calculus and real analysis we prefer to work with radians.

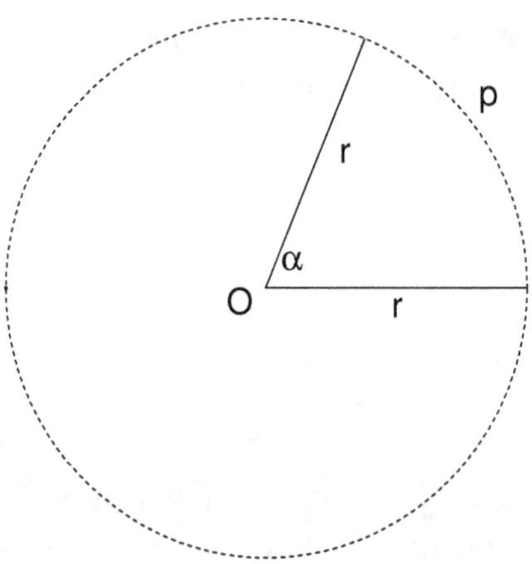

Figure 7.38: Conversion to and from radians

Using radians

Consider a right angle and choose A and B so that $OA = OB = 1$. Then the measure in radians of the given angle is $\frac{1}{4}(2\pi)$. Thus, a right angle has a measure of $\frac{\pi}{2}$ radians. We might then be tempted to set up the proportion $\frac{\theta}{\pi/2} = \frac{\alpha^o}{90^o}$, that is,

$$\frac{\theta}{\pi} = \frac{\alpha^o}{180^o}, \tag{7.4}$$

where θ is the measure in radians of an angle whose measure in degrees is α. This proportion sounds reasonable and it is usually accepted without any qualms. However, a justification is warranted. It follows from

$$\frac{p}{2\pi r} = \frac{\alpha^o}{360^o} \tag{7.5}$$

where p is the length of an arc, which comes from a circle with radius r and corresponding central angle with measure α^o (Figure 7.38). It is evident that (7.5) implies (7.4) since $\frac{p}{r} = \theta$ radians. A proof of (7.5) can be found, as an enrichment note, at the end of the chapter.

To convert the measure in degrees of an angle into radians, or vice-versa, we just need to use (7.4). For instance, an angle with measure 45° corresponds to $\pi/4$ radians, while $\pi/7$ radians correspond to an angle measure $180^o/7$ and 1 radian corresponds to an angle with measure $180^o/\pi$. Thus, we would say that 30 degrees (30°) correspond to $\frac{\pi}{6}$ radians.

Remark

Suppose we are given an arc $\overset{\frown}{AB}$ that comes from a circle with center O and radius r, and such that $m\angle AOB = \alpha^o$. From (7.5) we get:

$$p = \frac{\pi\alpha^o}{180^o}r.$$

This way of computing p is more complicated than the simple formula $p = r\theta$ that we would use if we were to work with radians.

Measure of the circumference of the Earth

Erathostenes (276 BCE-194 BCE) was an outstanding Alexandrian sage who excelled in many fields, especially mathematics, astronomy, and geography. Besides measuring the distance Earth-Sun, as well as the distance to the moon, he found an approximation to the circumference C of the Earth. We do not know the details on how he did it since his work *On the measurement of the Earth* has been lost. However, other contemporary authors credited him with a clever procedure that we will discuss schematically in the next paragraph.

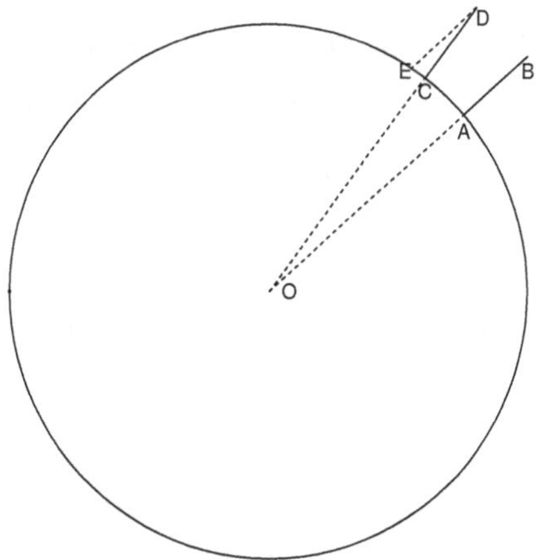

Figure 7.39: Measuring the circumference of the Earth

Assume that the Earth is a sphere and that the rays of light coming from the sun do so as parallel lines. This assumption seems reasonable since the Earth is far away from the sun and our planet is small compared to its star. According to tradition, Erathostenes observed that during the northern hemisphere summer solstice (June

21), at noon a stick cast no shadow in the city of Syene (modern day Aswan, Egypt) while, on the same day and time, a stick in Alexandria (north of Syene) cast a shadow. Since Alexandria and Syene are on the same meridian[11] and the distance between both cities is about 500 miles (see Figure 7.39, obviously not built on a proper scale, where $\overline{ED}$ is parallel to $\overline{OB}$ and $m\angle AOC = m\angle EDC = 7.5^o$; the rays of light are parallel to $\overline{OB}$ and the measure of $\angle EDC$ is experimentally determined) we will have

$$\frac{C}{500} = \frac{360^o}{7.5^o}.$$

Hence $C = 24,000$ miles. At the end of section 7.2 we mentioned that $r = 3,944.368421$ is an accepted modern value for the radius of the Earth, hence $C = 2\pi \times 3,944.368421 = 24,782.2$ miles is the length of the circumference, a value that is not far away from Erathostenes' calculations[12].

Exercises

1. Using propositions 7.1.4 and 7.1.2 provide a third proof of the problem described in section 7.4.

2. Complete the proof of the lemma that precedes the theorem on inscribed angles (Hint: Two further possibilities need to be considered, namely (i) when $\overset{\frown}{ABC}$ is a major arc and A and C lie on opposite sides of the diameter that passes through B; (ii) when $\overset{\frown}{ABC}$ is a major arc and $\overset{\frown}{AB}$ or $\overset{\frown}{BC}$ is a semi-circle).

3. Given a square with side x, inscribed in a circle with radius r, calculate x in terms of r.

4. Given a regular octagon with side x, inscribed in a circle with radius r, calculate x in terms of r.

5. Given a regular 16-gon with side x, inscribed in a circle with radius r, calculate x in terms of r (Hint: Use the half-angle formula to calculate $\cos \frac{45}{2}$).

6. A rectangle $ABCD$ is inscribed in a circle (Figure 7.40) and it is given that $DC = DE = 6$ and EA=2. Calculate DF.

7. Given a rectangle with sides p and q, make use of a geometrical construction to find the length of the side of a square with area $p \times q$. In other words, solve geometrically the equation $x^2 = pq$ (Hint: Build a segment $\overline{AB}$ of length p and

[11]This is not exactly true since Alexandria is a little west of Syene.

[12]Neither the radius nor the circumference of the Earth are exact values since our planet is not a perfect sphere and there are always measurement errors.

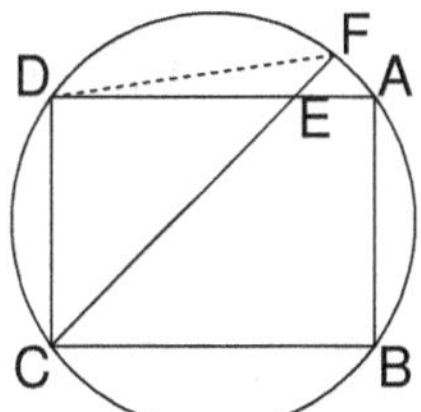

Figure 7.40: Exercise 6

prolong it in such a way that a new segment $\overline{BC}$, of length q, is built. Then construct a circle with diameter $p+q$ and draw a chord $\overline{DE}$ that passes through B and is perpendicular to $\overline{AC}$.)

8. Let two chords, $\overline{AB}$ and $\overline{CD}$, intersect at a point E inside the circle. Prove that $m\angle AEC = \frac{1}{2}(\overset{\frown}{AC} + \overset{\frown}{BD})$ (Hint: Build chord $\overline{AC}$ and observe that $m\angle AEC = 180 - (m\angle CAE + m\angle ACE)$.)

9. Suppose that A, B, C, D are points on a circle and that $\overleftrightarrow{AB}$ and $\overleftrightarrow{CD}$ intersect at a point E outside of the circle. Show that $AE \times BE = DE \times EC$ and $m\angle E = \frac{1}{2}(\overset{\frown}{AD} - \overset{\frown}{BC})$ (Hint: Construct chords $\overline{AC}$ and $\overline{BD}$ to prove the first equality. To prove the second equality just build $\overline{AD}$.)

10. Assume that A, B, C belong to a circle and that the secant $\overleftrightarrow{BC}$ and the tangent at A intersect at a point D. Show that $m\angle D = \frac{1}{2}(\overset{\frown}{AC} - \overset{\frown}{AB})$ (Hint: Build $\overline{AC}$ and keep in mind that the sum of the internal angles of ΔCAD is 180°.)

11. Assume that A and B belong to a circle and that the tangents at A and B meet at a point C and that $\overset{\frown}{ARB} > \overset{\frown}{ASB}$. Show that $m\angle C = \frac{1}{2}(\overset{\frown}{ARB} - \overset{\frown}{ASB})$.

12. Show that, for any equilateral triangle, the circumradius is exactly twice the length of the inradius.

13. Assume that $\Delta ABC \sim \Delta A'B'C'$ with ratio of similarity k, that is, $\frac{AB}{A'B'} = \frac{BC}{B'C'} = \frac{AC}{A'C'} = k$. Show that the circumradius and inradius of ΔABC are k times the circumradius and inradius of $\Delta A'B'C'$.

14. Referring to the proof of Brahmagupta's Theorem, find the measure of $\angle EBD$ in terms of the measures of $\angle BAC$ and $\angle BCA$.

15. A circle with radius 2 cm. is inscribed inside a right triangle with hypotenuse 10 cm. Calculate the area of the triangle. Solve the same problem when the radius is r and the hypotenuse is b. By the way, we are assuming that the corresponding right triangle exists (see section 11.4 and exercise 26, chapter 11).

16. Let us consider a semicircunference with radius r and the isosceles triangle ABC (Figure 7.41). Using $\overline{AB}$ as a diameter, we build the semicircle ADB. Calculate, in terms of the radius r, the area of the region limited by the arcs $\overset{\frown}{ADB}$ and $\overset{\frown}{AEB}$.

17. Let us consider a circle with radius r and draw two diameters, perpendicular to each other (Figure 7.42). With center at D and radius AD we build the arc $\overset{\frown}{AFC}$. Calculate, in terms of r, the area of the region between the arcs $\overset{\frown}{ABC}$ and $\overset{\frown}{AFC}$.

18. A circle, with center at O and radius r, is tangent to the base $\overline{BC}$ of the isosceles triangle ABC at D(Figure 7.43). The segments $\overline{AB}$ and $\overline{AC}$ are extended and are tangent to the circle at E and F, respectively. Prove that, if $\overline{OD}$ is extended, it will pass through A (Hint: Build $\overline{OB}$ and $\overline{OC}$.)

19. In the preceding problem, assuming that the isosceles triangle ABC has a right angle at A, calculate the area of ΔABC in terms of r (Hint: Notice that $\Delta ADC \sim \Delta AFO$.)

20. Let a circle be inscribed in a triangle and draw the three cevians to the respective points of tangency. Prove that the above-mentioned cevians are concurrent.

Enrichment Note: Proving a Basic Equality

We will show that given any circle with radius r, and any arc $\overset{\frown}{AB}$,

$$\frac{p}{2\pi r} = \frac{\alpha}{360},$$

where p is the length of $\overset{\frown}{AB}$ and α is the measure, in degrees, of the corresponding central angle.

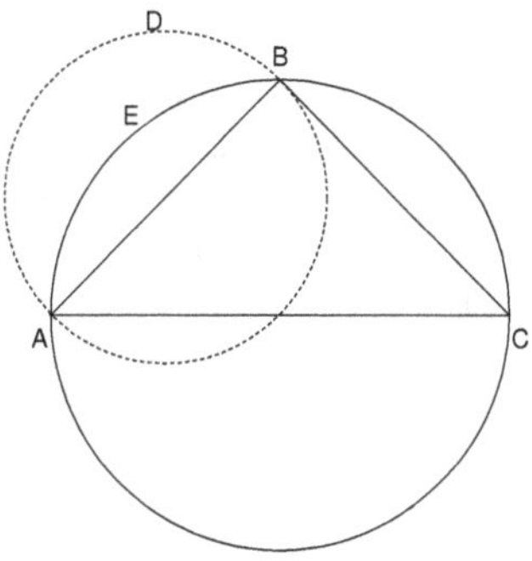

Figure 7.41: Exercise 16

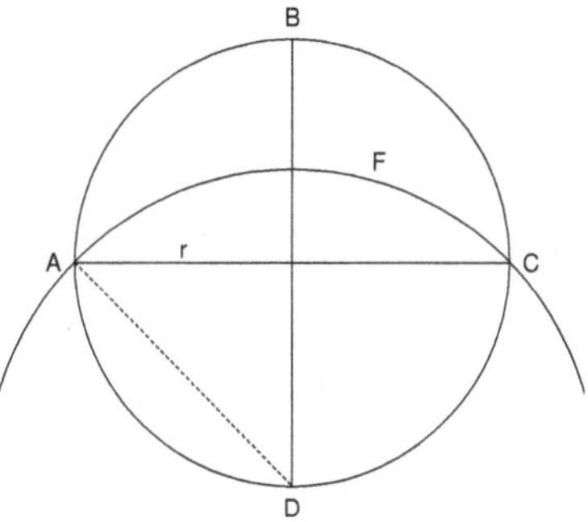

Figure 7.42: Exercise 17

Proof

Let n be an arbitrary positive integer. We fix it and then define $\beta = \frac{360}{n}$. A basic property of the real number system allows us to conclude that there exists a positive integer m such that

$$m \leq \frac{\alpha}{\beta} < m + 1.$$

We will consider two possibilities:

1. Let us assume that $m = \frac{\alpha}{\beta}$, that is, $\alpha = \beta m$. If we denote by t the length of the arc subtended by a central angle of measure β, then $p = mt$. Since $360 = \beta n$ we will have $2\pi r = nt$, consequently

$$\frac{p}{2\pi r} = \frac{m}{n} \text{ while } \frac{\alpha}{360} = \frac{m}{n}.$$

   Therefore $\frac{p}{2\pi r} = \frac{\alpha}{360}$ and we are done.

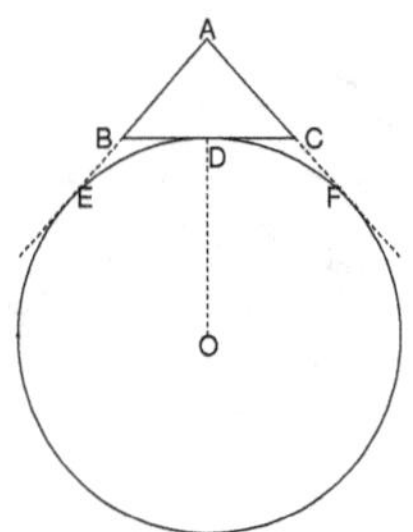

Figure 7.43: Exercise 18

2. Let us assume that $m < \frac{\alpha}{\beta} < m+1$. Thus

$$m\beta < \alpha < (m+1)\beta, \tag{7.6}$$

so $m\frac{360}{n} < \alpha < (m+1)\frac{360}{n}$. That is to say,

$$\frac{m}{n} < \frac{\alpha}{360} < \frac{m+1}{n}. \tag{7.7}$$

From (7.6) we get $mt < p < (m+1)t$, hence

$$\tfrac{m}{n} < \tfrac{p}{nt} < \tfrac{m+1}{n}.$$

But $360 = n\beta$, therefore $2\pi r = nt$. Consequently

$$\frac{m}{n} < \frac{p}{2\pi r} < \frac{m+1}{n}. \tag{7.8}$$

From (7.7) and (7.8) we get

$$-\tfrac{1}{n} < \tfrac{p}{2\pi r} - \tfrac{\alpha}{360} < \tfrac{1}{n},$$

which in turn leads to

$$|\tfrac{p}{2\pi r} - \tfrac{\alpha}{360}| < \tfrac{1}{n}$$

for any n. Since $\lim_{n\to\infty} \frac{1}{n} = 0$ we can finally conclude that

$$\tfrac{p}{2\pi r} - \tfrac{\alpha}{360} = 0,$$

that is, $\frac{p}{2\pi r} = \frac{\alpha}{360} = 0$. QED

Chapter 8

More on Trigonometry

8.1 Ptolemy's Theorem

Claudius Ptolemy (100-170 CE) was a noted Alexandrian Greek astronomer, whose great work *Syntaxis Mathematica* , also known as *Almagest*, exerted a deep influence up to the 16th century. It was in this book that a truly remarkable theorem, describing a property of a quadrilateral inscribed in a circle, appeared for the first time. We will use it to prove trigonometric formulas.

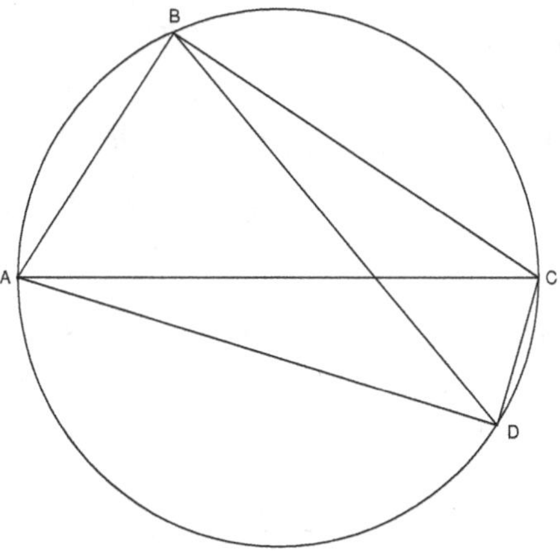

Figure 8.1: Ptolemy's Theorem

Theorem

Let $ABCD$ be a quadrilateral inscribed in a circle (Figure 8.1). Then the product of the diagonals of the quadrilateral equals the sum of the product of the opposite sides. That is,

$$BD \times AC = BC \times AD + AB \times DC.$$

Proof

Assume that $\angle ABD \cong \angle CBD$. Since $\angle ADB$ and $\angle ACB$ subtend the same arc AB we can conclude that both angles are congruent. Then AA similarity implies that $\Delta ABD \sim \Delta FBC$, where F is the point of intersection of both diagonals. Hence $\frac{BD}{BC} = \frac{AD}{FC}$, that is,

$$BD \times FC = BC \times AD. \tag{8.1}$$

In a similar fashion, since $\angle BAC$ and $\angle BDC$ subtend the same arc BC it follows that they must be congruent. Then AA similarity implies that $\Delta BAF \sim \Delta BDC$. Therefore $\frac{AF}{DC} = \frac{AB}{BD}$, thus

$$BD \times AF = AB \times DC. \tag{8.2}$$

Adding (8.1) and (8.2) we get the thesis of Ptolemy's Theorem, namely

$$BD \times AC = BC \times AD + AB \times DC.$$

What can be done if $\angle ABD$ is not congruent to $\angle CBD$? Without loss of generality let us assume that m$\angle CBD <$m$\angle ABD$. Draw $\overline{BE}$ so that $\angle ABE \cong \angle CBD$ (Figure 8.2). We note that $\angle BAE \cong \angle BDC$ (both subtend the same arc BC). Thanks to AA similarity we can conclude that $\Delta BAE \sim \Delta BDC$. Thus $\frac{AB}{BD} = \frac{AE}{DC}$, hence

$$AB \times DC = BD \times AE. \tag{8.3}$$

Next we realize that $\angle ADB \cong \angle BCA$ (both angles subtend the same arc AB). Moreover, it is evident that $\angle ABD \cong \angle CBE$. Then AA similarity implies that $\Delta DBA \sim \Delta CBE$, consequently $\frac{AD}{EC} = \frac{BD}{BC}$, thus

$$AD \times BC = BD \times EC. \tag{8.4}$$

Adding (8.3) and (8.4) we get

$$AB \times DC + AD \times BC = BD \times (AE + EC) = BD \times AC.$$

QED

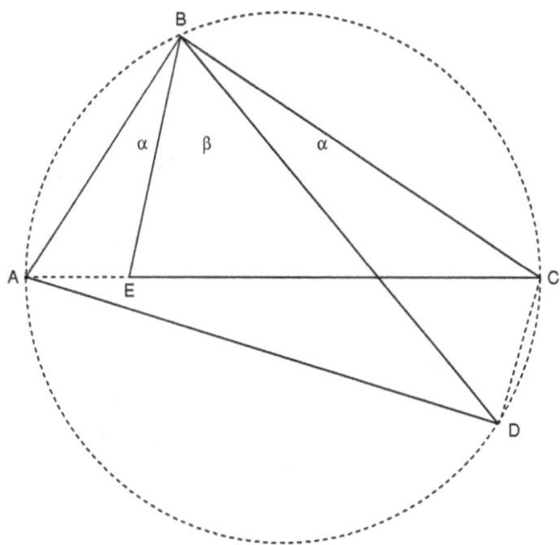

Figure 8.2: Proof of Ptolemy's Theorem

A second proof of Ptolemy's Theorem

Having used algebra to prove Heron's Theorem, we might wonder whether an algebraic approach could lead to a proof of Ptolemy's Theorem. The answer is affirmative provided that a little bit of trigonometry is employed. Referring to Figure 8.3 we note that $m\angle C + m\angle A = 180$, therefore $\cos C = \cos(180 - A) = -\cos A$. The law of cosines, applied to ΔABD, letting $x = BD$, implies that

$$x^2 = a^2 + d^2 - 2ad\cos A. \tag{8.5}$$

Similarly, the law of cosines, applied to ΔBDC, leads to $x^2 = b^2 + c^2 - 2bc\cos C$. Since $\cos C = -\cos A$ we will have

$$x^2 = b^2 + c^2 + 2bc\cos A. \tag{8.6}$$

From (8.5) and (8.6) it follows that $(a^2 - c^2) + (d^2 - b^2) = 2(ad + bc)\cos A$, thus

$$2\cos A = \frac{(a^2-c^2)+(d^2-b^2)}{bc+ad}.$$

Replacing this expression for $2\cos A$ in (8.5) we get

$$x^2 = a^2 + d^2 - ad[\frac{(a^2-c^2)+(d^2-b^2)}{bc+ad}] = \frac{ab(ac+db)+cd(bd+ac)}{bc+ad} = \frac{(ab+cd)(ac+db)}{bc+ad}.$$

Therefore

$$x^2 = \frac{(ab+cd)(ac+db)}{bc+ad}. \tag{8.7}$$

In a similar fashion, letting $y = AC$, it can be proven that

$$y^2 = \frac{(bc + ad)(bd + ac)}{cd + ab}. \tag{8.8}$$

Multiplying the expressions given in (8.7) and (8.8) we get $x^2y^2 = (ac + bd)^2$, consequently $xy = ac + bd$. QED

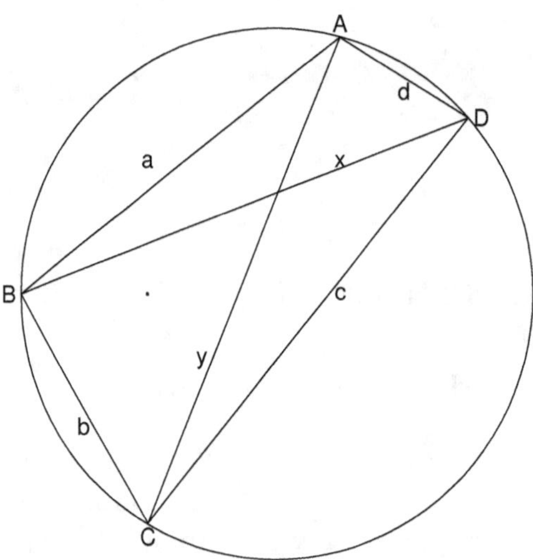

Figure 8.3: A second proof of Ptolemy's Theorem

8.2 Proving two trigonometric formulas

To show the power of Ptolemy's Theorem we will prove the formulas for the sine of the sum and the difference of two angles. The reader might remember that the former was proven in chapter 6, but no proof has so far been given for the latter.

1. Let us start by providing an alternative proof of the well-known formula

$$\sin(\alpha + \beta) = \sin\alpha\cos\beta + \cos\alpha\sin\beta,$$

assuming that α and β are the measures of two acute angles.

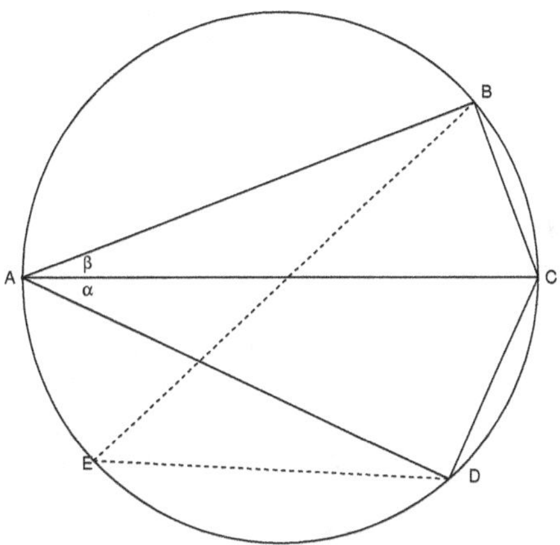

Figure 8.4: Sine of the sum of two angles

We build a circle of unit diameter (Figure 8.4). Let $\overline{AC}$ be a diameter and draw $\angle BAC$ and $\angle CAD$ with measure β and α respectively. Next construct the quadrilateral $ABCD$. We observe that m$\angle ABC$=m$\angle ADC$ since $\overline{AC}$ is a diameter, thus $\sin\alpha = CD$, $\cos\alpha = AD$, $\sin\beta = BC$, and $\cos\beta = AB$. Ptolemy's Theorem implies that $AB \times CD + BC \times AD = 1 \times BD$, that is

$$\cos\beta\sin\alpha + \sin\beta\cos\alpha = BD. \tag{8.9}$$

Now draw diameter $\overline{BE}$. We note that $\angle BAD \cong \angle BED$ (both inscribed angles subtend arc BCD), so m$\angle BED = \alpha + \beta$. But

$$\sin(\alpha + \beta) = BD. \tag{8.10}$$

Finally, combining (8.9) and (8.10) we get what we wanted to prove. QED

2. This time our purpose is to prove the formula

$$\sin(\alpha - \beta) = \sin\alpha\cos\beta - \cos\alpha\sin\beta,$$

assuming that α and β are acute angles ($\alpha > \beta$).

Let us start the proof by building a circle of unit diameter (Figure 8.5). Thereafter draw $\angle BAD$ with measure α and $\angle CAD$ with measure β, and build quadrilateral $ABCD$. Right away we notice that m$\angle ABD = 90$ and m$\angle ACD = 90$

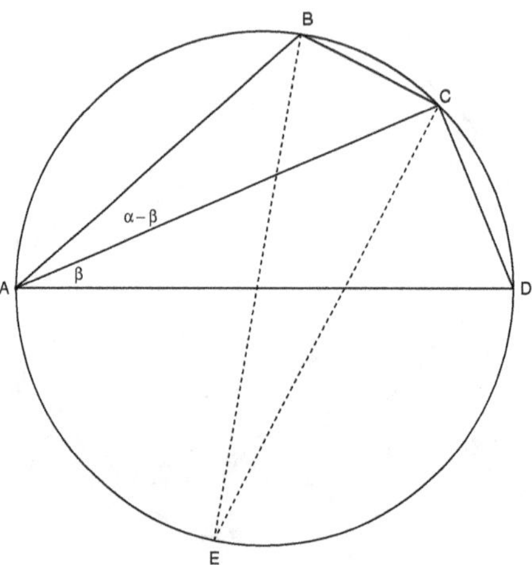

Figure 8.5: Sine of the difference of two angles

since $\overline{AD}$ is a diameter, hence $\sin\beta = CD$, $\cos\beta = AC$, $\sin\alpha = BD$, and $\cos\alpha = BC$. Ptolemy's Theorem implies that

$$AC \times BD = AB \times CD + BC \times 1. \tag{8.11}$$

Next draw diameter $\overline{BE}$. We note that $\angle BEC \cong \angle BAC$ since both inscribed angles subtend arc BC, therefore $\text{m}\angle BEC = \alpha - \beta$. But $\text{m}\angle BCE = 90$ since $\overline{BE}$ is a diameter, consequently

$$\sin(\alpha - \beta) = BC. \tag{8.12}$$

Replacing (8.12) in (8.11) we bring the proof to an end. QED

Another approach to the solution of a problem

In section 7.3 (4th remark) we considered a problem related to a rectangle of length 8 and width 5 inscribed in a circle, and a chord $\overline{EB}$ that intersects $\overline{AC}$ at F (Figure 8.6). Our task is to calculate AB. With this purpose in mind we note that $\overline{EC}$, whose length is $\sqrt{89}$, happens to be a diameter since the inscribed angle EDC is a right angle. Then $\angle EBC$ is also a right angle. But the right triangle EAF is isosceles, hence $45 = m\angle AFE = m\angle BFC = m\angle BCF$; thus, the right triangle FBC is isosceles too, a fact that leads to the equality $2BC^2 = 9$. That is, $BC = \frac{3}{\sqrt{2}}$. On the other hand, $EB = \sqrt{89 - \frac{9}{2}} = \frac{13}{\sqrt{2}}$. We are now ready to apply Ptolemy's Theorem to the inscribed quadrilateral $ABCE$:

$$AB\sqrt{89} + 5 \times \tfrac{3}{\sqrt{2}} = 8 \times \tfrac{13}{\sqrt{2}}.$$

Therefore $AB = \frac{89}{\sqrt{89}\sqrt{2}} = \sqrt{\frac{89}{2}}$.

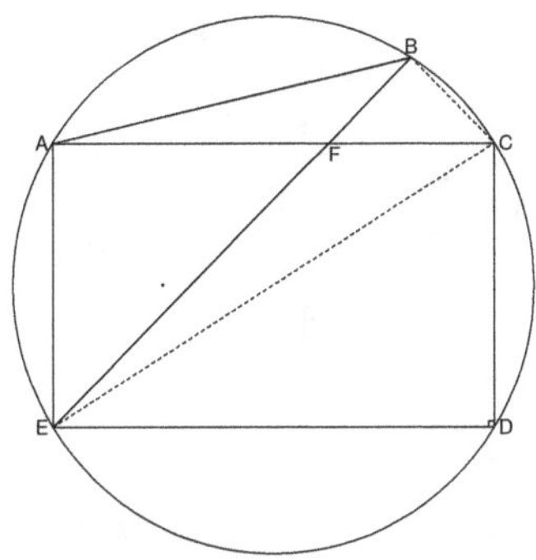

Figure 8.6: A second approach

8.3 Extending the trigonometry of acute angles

Let us work with radians. So far we have only studied the sine, cosine, and tangent for values between 0 and $\frac{\pi}{2}$. We can start by defining $\sin 0 = 0$, $\cos 0 = 1$. Let θ be any number between 0 and 2π, that is, $0 < \theta < 2\pi$, and consider a circle of *unit radius* (Figure 8.7). Then draw coordinate axes whose origin is at the center of the circle. Letting $\theta = m\angle AOB$ drop the perpendicular $\overline{BC}$.

Define

$$\sin\theta = \text{ordinate of } B, \quad \cos\theta = \text{abscissa } B.$$

We observe that $\sin\theta > 0$ if $0 < \theta < \pi$, while $\sin\theta < 0$ if $\pi < \theta < 2\pi$. Obviously, $\sin\pi = \sin 2\pi = 0$, $\sin\frac{\pi}{2} = 1$, and $\sin\frac{3\pi}{2} = -1$. In a similar fashion we observe that $\cos\theta > 0$ when $0 < \theta < \frac{\pi}{2}$ or $\frac{3\pi}{2} < \theta < 2\pi$, while $\cos\theta < 0$ when $\frac{\pi}{2} < \theta < \frac{3\pi}{2}$. From the very definition of cos we can see that $\cos\pi = -1$, $\cos\frac{\pi}{2} = \cos\frac{3\pi}{2} = 0$ and $\cos 2\pi = 1$.

As the point B does a complete run around the circle, counterclockwise, clearly

$$\sin(\theta + 2\pi) = \sin\theta, \quad \cos(\theta + 2\pi) = \cos\theta.$$

That is, both the sine and cosine are periodic functions with period 2π. The reader may observe that the given definitions for $\sin\theta$ and $\cos\theta$ (θ any real number) coincide with the definitions of $\sin\theta$ and $\cos\theta$, $0 < \theta < \frac{\pi}{2}$, presented in chapter 6. Moreover, for any θ we have

$$\cos^2\theta + \sin^2\theta = 1 \tag{8.13}$$

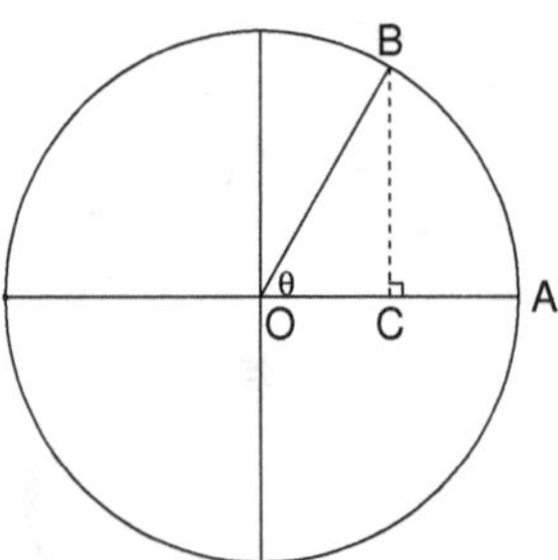

Figure 8.7: The sine and cosine of any angle

since we are working with the unit circle and the coordinates of any point on the unit circle are precisely $(\cos\theta, \sin\theta)$.

Can we extend the tangent function to the whole real line? Let B be any point on the unit circle, with coordinate axes whose origin is at the center O of the circle (Figure 8.8). Let $0 < \theta < 2\pi$ and draw a line L, perpendicular to the circle at A. We prolong $\overline{OB}$ in such a way that it intersects L at the point D. Then we define $\tan\theta$ as the ordinate of the point D. This definition coincides with the one given in chapter 6 when $0 < \theta < \frac{\pi}{2}$, and we observe that $\tan\theta$ is not defined when $\theta = \frac{\pi}{2}$ or $\theta = \frac{3\pi}{2}$. Moreover, $\tan 0 = 0$, $\tan 2\pi = 0$.

So far the point B moves counterclockwise on the unit circle. If B were to move clockwise we would be dealing with angles that have a negative measure. Keeping in mind that the ordinate of any point on the first and second quadrant is positive while the ordinate of any point on the third and fourth quadrant is negative, we can conclude that $\sin(-\alpha) = -\sin\alpha$. Similarly, since the abscissa of any point on the first and fourth quadrant is positive and the abscissa of any point on the second and third quadrant is negative, we can assert that $\cos(-\alpha) = \cos\alpha$. Then $\tan(-\alpha) = -\tan\alpha$.

It is to be noted that $\tan\theta$ is always increasing on any interval where it is defined. Furthermore, we can see that $\tan\theta$ is a periodic function with period π. If we draw a pair of coordinate axes and project on them the values of $\cos\theta$, $\sin\theta$, and $\tan\theta$, we obtain the three well-known graphs of the sine, cosine and tangent function.

In chapter 6 we *defined* $\sin(180-\alpha) = \sin\alpha$ and $\cos(180-\alpha) = -\cos\alpha$ whenever $\frac{\pi}{2} < \alpha < \pi$. Having developed the cosine and sine for any real number, the two

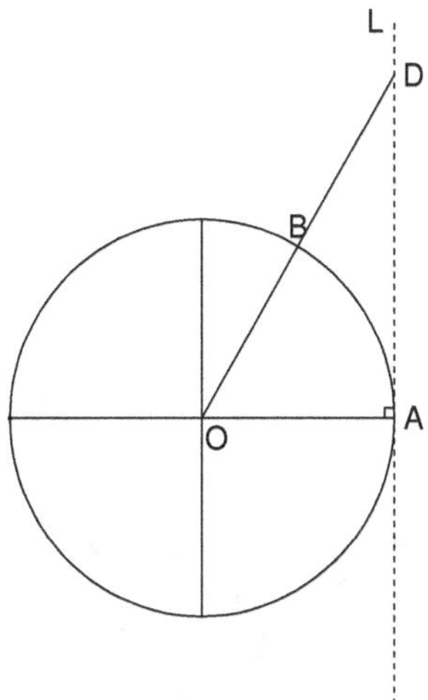

Figure 8.8: The tangent of any angle

preceding equalities follow from the definition. Indeed, let B be the point on the unit circle corresponding to α and draw a line that passes through B and is parallel to the x-axis, which meets the circle at D (Figure 8.9). Then build the segments $\overline{BC}$ and $\overline{DE}$, both perpendicular to the x-axis. We observe that $BC = DE$ since the distance between two parallel lines is a constant, consequently[1] $\Delta BOC \cong \Delta DOE$. In particular $\angle BOC \cong \angle DOE$, thus $m\angle COD = 180 - \alpha$. Therefore

$$\sin(180 - \alpha) = \sin\alpha,$$

$$\cos(180 - \alpha) = -\cos\alpha.$$

Extension of trigonometric identities

We have already proven that

$$\sin(\alpha \pm \beta) = \sin\alpha\cos\beta \pm \cos\alpha\sin\beta, \tag{8.14}$$

$$\cos(\alpha \pm \beta) = \cos\alpha\cos\beta \mp \sin\alpha\sin\beta \tag{8.15}$$

when $0 < \alpha < \frac{\pi}{2}$, $0 < \beta < \frac{\pi}{2}$, and $\alpha + \beta < \pi$.

A natural question to ask is whether these trigonometric identities are valid for any real numbers α, β. To answer this question we will follow an approach that appeared

[1]Recall the HL property of congruence among right triangles.

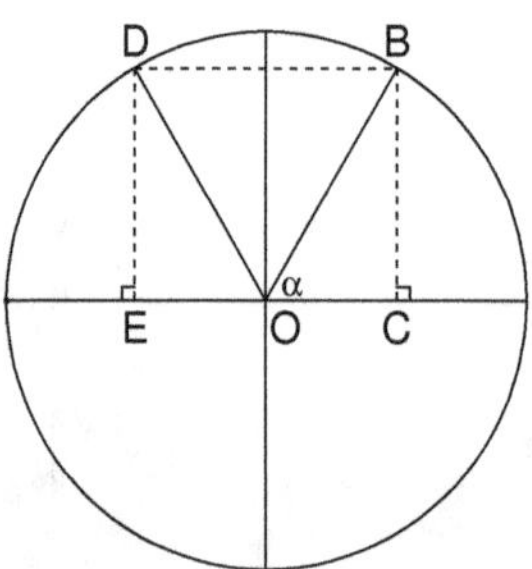

Figure 8.9: Justifying two identities

in a paper by Edward J. McShane (1904-1989) some 70 years ago (McShane, 1941). In the first place we will prove that, given any real numbers α, β with $\alpha > \beta$,

$$\cos(\alpha - \beta) = \cos\beta\cos\alpha + \sin\alpha\sin\beta. \tag{8.16}$$

Referring to the unit circle, with the origin of the Cartesian axes at the center, the coordinates of β are $B = (\cos\beta, \sin\beta)$ while the coordinates of α are $A = (\cos\alpha, \sin\alpha)$ (Figure 8.10). Letting d be the distance between both points, we will have[2]

$$d^2 = (\cos\alpha - \cos\beta)^2 + (\sin\alpha - \sin\beta)^2. \tag{8.17}$$

We have in front of us a triangle with vertices $(\cos\alpha, \sin\alpha)$, $(\cos\beta, \sin\beta)$, and $(0, 0)$. The angle opposite the side of length d has measure $\alpha - \beta$, which we denote with the letter γ. Next we rotate the above-mentioned triangle clockwise in such a way that the side next to the angle with measure β is made to coincide with the x-axis. The new coordinates of the triangle will be $(1, 0)$, $(\cos\gamma, \sin\gamma)$ and $(0, 0)$ (Figure 8.11). Since we have a congruent copy of the original triangle, it is evident that the length of the segment with endpoints $(\cos\gamma, \sin\gamma)$ and $(1, 0)$ is still d. Therefore

$$d^2 = (\cos\gamma - 1)^2 + \sin^2\gamma. \tag{8.18}$$

From (8.17) and (8.18) it follows that

$$(\cos\alpha - \cos\beta)^2 + (\sin\alpha - \sin\beta)^2 = (\cos\gamma - 1)^2 + \sin^2\gamma.$$

[2]We trust that the reader remembers, in the context of Cartesian geometry, the formula for the distance between two points.

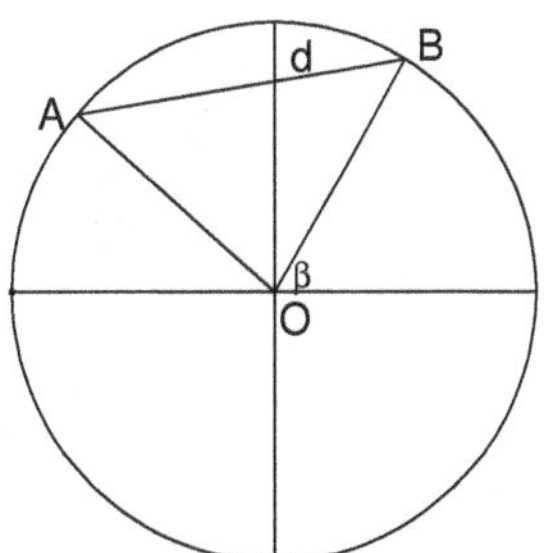

Figure 8.10: On the process of extending trigonometric identities

Simplifying both sides of the expression we get

$$2 - 2\cos\alpha\cos\beta - 2\sin\alpha\sin\beta = 2 - 2\cos\gamma,$$

therefore $\cos\alpha\cos\beta + \sin\alpha\sin\beta = \cos\gamma$, that is, $\cos(\alpha-\beta) = \cos\beta\cos\alpha + \sin\alpha\sin\beta$.

Having proven (8.16), right away we see that

$$\cos(-\beta) = \cos\beta, \tag{8.19}$$

$$\cos(\frac{\pi}{2} - \beta) = \sin\beta. \tag{8.20}$$

Then

$$\sin(\frac{\pi}{2} - \beta) = \cos(\frac{\pi}{2} - (\frac{\pi}{2} - \beta)) = \cos\beta, \tag{8.21}$$

and consequently

$$\sin(\alpha+\beta) = \cos(\tfrac{\pi}{2} - (\alpha+\beta)) = \cos((\tfrac{\pi}{2} - \alpha) - \beta) = \cos(\tfrac{\pi}{2} - \alpha)\cos(-\beta) + \sin(\tfrac{\pi}{2} - \alpha)\sin\beta,$$

so

$$\sin(\alpha + \beta) = \sin\alpha\cos\beta + \cos\alpha\sin\beta. \tag{8.22}$$

On the other hand, from (8.20) we get

$$\sin(-\alpha) = \cos(\tfrac{\pi}{2} - (-\alpha)) = \cos(\tfrac{\pi}{2} + \alpha) = \cos(\alpha - (-\tfrac{\pi}{2})) =$$
$$\cos\alpha\cos(-\tfrac{\pi}{2}) + \sin\alpha\sin(-\tfrac{\pi}{2}),$$

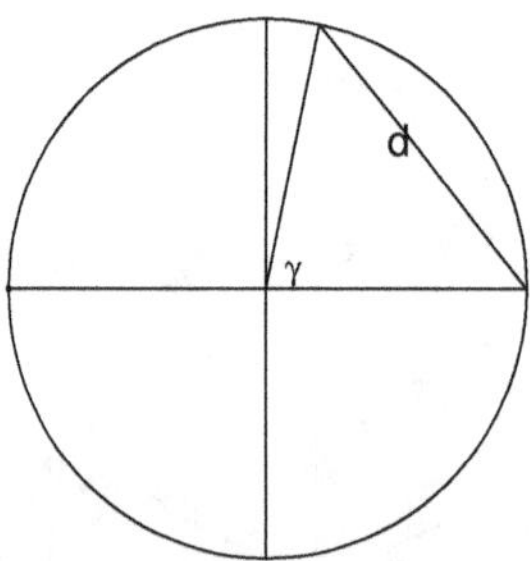

Figure 8.11: Rotating a triangle for a definite purpose

hence

$$\sin(-\alpha) = -\sin\alpha. \tag{8.23}$$

Then

$$\cos(\alpha+\beta) = \cos(\alpha-(-\beta)) = \cos\alpha\cos(-\beta)+\sin\alpha\sin(-\beta) = \cos\alpha\cos\beta-\sin\alpha\sin\beta. \tag{8.24}$$

Similarly

$$\sin(\alpha-\beta) = \sin(\alpha+(-\beta)) = \sin\alpha\cos(-\beta)+\cos\alpha\sin(-\beta) = sin\alpha\cos\beta-\cos\alpha\sin\beta. \tag{8.25}$$

Finally, for any integer n,

$$\sin(2n\pi+\alpha) = \sin(2n\pi)\cos\alpha+\cos(2n\pi)\sin\alpha = sin\alpha,$$

$$\cos(2n\pi+\alpha) = \cos(2n\pi)\cos\alpha-\sin(2n\pi)\cos\alpha = \cos\alpha.$$

8.4 Inverse trigonometric functions

The reader must have seen the main inverse trigonometric functions in pre-calculus and calculus; thus, this section should be merely a review. Let us restrict the sine function to the interval $[-\frac{\pi}{2}, \frac{\pi}{2}]$. On this interval the sine function is increasing, thus it is one-to-one. Hence, its inverse $\sin^{-1} : [-1,1] \to [-\frac{\pi}{2}, \frac{\pi}{2}]$ exists. For instance, $\sin^{-1}(\frac{1}{2}) = \frac{\pi}{6}$ since $\sin\frac{\pi}{6} = \frac{1}{2}$; $\sin^{-1}(\frac{\sqrt{2}}{2}) = \frac{\pi}{4}$ since $\sin(\frac{\pi}{4}) = \frac{\sqrt{2}}{2}$.

If we restrict the cosine function to $[0, \pi]$ we get a decreasing function; thus, it is one-to-one. Then we can define its inverse function $\cos^{-1} : [-1, 1] \to [0, \pi]$. For instance, $\cos^{-1}(\frac{1}{2}) = \frac{\pi}{3}$ since $\cos(\frac{\pi}{3}) = \frac{1}{2}$.

Next let us restrict the tangent function to the interval $(-\frac{\pi}{2}, \frac{\pi}{2})$. It is increasing everywhere, so it has to be one-to-one. Its inverse $\tan^{-1} \to (-\frac{\pi}{2}, \frac{\pi}{2})$ is well-defined. For example, $\tan^{-1}(1) = \frac{\pi}{4}$ because $\tan(\frac{\pi}{4}) = 1$. Luckily, all graphing calculators have the functions $\sin^{-1}$, $\cos^{-1}$, and $\tan^{-1}$ readily available.

Exercises

1. Given that $\angle A$ is in the third quadrant with $\sin A = -\frac{1}{4}$, and $\angle B$ is in the fourth quadrant with $\cos B = \frac{2}{3}$, calculate $\cos(A + B)$.
2. Find the solution set of $\cos^2\theta - 3\sin\theta + 2 = 0$, $0 \le \theta < 2\pi$.
3. Solve the equation $\tan x \sin^2 x = \tan x$, $0 \le x < 2\pi$.
4. Given that $x + y = \frac{\pi}{4}$ find the value of $(1 + \tan x)(1 + \tan y)$ (Hint: Show that $\tan(x + y) = \frac{\tan x + \tan y}{1 - \tan x \tan y}$.)
5. Show that $\cos(x + y)\cos(x - y) = \cos^2 x + \cos^2 y - 1$.
6. Show that $\cos(x + y)\cos(x - y) - \sin(x + y)\sin(x - y) = \cos^2 x - \sin^2 x$.
7. Show that $\cos x \cos y = \frac{1}{2}\cos(x + y) + \frac{1}{2}\cos(x - y)$.
8. Show that $\sin x \sin y = \frac{1}{2}\cos(x - y) - \frac{1}{2}\cos(x + y)$.
9. Show that $\sin x \cos y = \frac{1}{2}\sin(x + y) + \frac{1}{2}\sin(x - y)$.
10. Use the preceding exercise to show that $\sin x + \sin y = 2\sin(\frac{x+y}{2})\cos(\frac{x-y}{2})$.

Enrichment Note: Yet another proof of the addition formula

We will follow an approach credited to Jesse Douglas (1897-1965) (Schaumberger, 1962). Let us prove the trigonometric identity

$$\sin(\alpha + \beta) = \sin\alpha\cos\beta + \cos\alpha\sin\beta \tag{8.26}$$

for any real numbers α, β. With this purpose in mind we will divide the proof in six cases.

1. Assume that $\alpha+\beta<\pi$, $\alpha>0$, $\beta>0$. Under these circumstances we can think about a triangle with sides a, b, c and corresponding angles α, β, γ. Using the projection property of triangles (chapter 6) we get the equality $c = a\cos\beta + b\cos\alpha$, while the law of sines leads to $a = \frac{\sin\alpha}{\sin\gamma}c$, $b = \frac{\sin\beta}{\sin\gamma}c$. Then

$$c = \left(\tfrac{\sin\alpha}{\sin\gamma}c\right)\cos\beta + \left(\tfrac{\sin\beta}{\sin\gamma}c\right)\cos\alpha.$$

Hence $\sin\gamma = \sin\alpha\cos\beta + \sin\beta\cos\alpha$. But $\gamma = \pi - (\alpha+\beta)$, consequently

$$\sin\gamma = \sin(\pi - (\alpha+\beta)) = \sin(\alpha+\beta).$$

Therefore $\sin(\alpha+\beta) = \sin\alpha\cos\beta + \sin\beta\cos\alpha$.

2. Assume that $\alpha = 0$ or $\beta = 0$. If $\alpha = 0$ then $\sin(\alpha+\beta) = \sin\beta$ while

$$\sin\alpha\cos\beta + \sin\beta\cos\alpha = 0\times\cos\beta + \sin\beta\cos 0 = \sin\beta.$$

In a similar fashion the identity (8.26) is true when $\beta = 0$.

3. Assume that $\alpha+\beta = \pi$. Then $\sin(\alpha+\beta) = \sin\pi = 0$ while

$$\begin{aligned}\sin\alpha\cos\beta + \sin\beta\cos\alpha &= \sin\alpha\cos(\pi-\alpha) + \sin(\pi-\alpha)\cos\alpha = \\ &-\sin\alpha\cos\alpha + \sin\alpha\cos\alpha = 0.\end{aligned}$$

4. Assume that $0<\alpha<\pi$, $0<\beta<\pi$, and $\alpha+\beta>\pi$. We observe that $\pi-\alpha>0$ and $\pi-\beta>0$, and $(\pi-\alpha)+(\pi-\beta) = 2\pi-(\alpha+\beta)$. But $-(\alpha+\beta)<-\pi$, hence $(\pi-\alpha)+(\pi-\beta)<\pi$. Applying the first case we get

$$\sin(\pi-\alpha+\pi-\beta) = \sin(\pi-\alpha)\cos(\pi-\beta) + \sin(\pi-\beta)\cos(\pi-\alpha).$$

However, $\sin(2\pi-(\alpha+\beta)) = \sin(-(\alpha+\beta)) = -\sin(\alpha+\beta)$ while

$$\sin(\pi-\alpha)\cos(\pi-\beta) + \sin(\pi-\beta)\cos(\pi-\alpha) = -\sin\alpha\cos\beta - \sin\beta\cos\alpha.$$

Thus $-\sin(\alpha+\beta) = -\sin\alpha\cos\beta - \sin\beta\cos\alpha$, which in turn leads to

$$\sin(\alpha+\beta) = \sin\alpha\cos\beta + \cos\alpha\sin\beta.$$

Note that so far we have proven the preceding formula whenever $0\le\alpha<\pi$ and $0\le\beta<\pi$.

5. Assume that $\pi \leq \alpha < 2\pi$ and $0 \leq \beta < 2\pi$. We would have to consider three possibilities: either $\pi \leq \alpha < 2\pi$ and $0 \leq \beta < \pi$, or $0 \leq \alpha < \pi$ and $\pi \leq \beta < 2\pi$, or $\pi \leq \alpha < 2\pi$ and $\pi \leq \beta < 2\pi$.

   Let us suppose that $\pi \leq \alpha < 2\pi$ and $0 \leq \beta < \pi$. Then $0 \leq \alpha - \pi < \pi$, so

$$\sin(\alpha - \pi + \beta) = \sin(\alpha - \pi)\cos\beta + \cos(\alpha - \pi)\sin\beta.$$

   But

$$\sin(\alpha + \beta - \pi) = \sin(-(\pi - (\alpha + \beta))) = -\sin(\pi - (\alpha + \beta)) = -\sin(\alpha + \beta),$$

$$\sin(\alpha - \pi) = \sin(-(\pi - \alpha)) = -\sin(\pi - \alpha) = -\sin\alpha,$$

$$\cos(\alpha - \pi) = \cos(-(\pi - \alpha)) = \cos(\pi - \alpha) = -\cos\alpha.$$

   Thus $-\sin(\alpha + \beta) = -\sin\alpha\cos\beta - \cos\alpha\sin\beta$, consequently $\sin(\alpha + \beta) = \sin\alpha\cos\beta\cos\alpha\sin\beta$. The other two sub-cases can be proven in exactly the same way.

6. Finally, we claim that the formula for the sine of the sum of two angles holds true for any real numbers α and β. Indeed, choose a natural number p such that $0 \leq \frac{\alpha}{2\pi} - 1 < p$. Defining $m = -p$ we get $0 \leq \frac{\alpha}{2\pi} - 1 < -m$, consequently $0 \leq \alpha < 2\pi(1 - m)$, that is, $0 \leq \alpha + 2m\pi < 2\pi$. In a similar fashion it can be shown that there exists an integer n such that $0 \leq \beta + 2n\pi < 2\pi$. Applying the previous case we obtain

$$\sin(\alpha+2m\pi+\beta+2n\pi) = \sin(\alpha+2m\pi)\cos(\beta+2n\pi)+\cos(\alpha+2m\pi)\sin(\beta+2n\pi).$$

   Since both sin and cos have period 2π we can conclude that $\sin(\alpha + \beta) = \sin\alpha\cos\beta + \cos\alpha\sin\beta$. QED

Chapter 9

The Cartesian Alternative

9.1 A brief review of the main formulas of Cartesian geometry

The reader is surely acquainted, from pre-calculus mathematics, with the basic aspects of Cartesian geometry[1]. Let us recall some of the main results of Cartesian geometry, referred to lines and circles, which are valid once a pair of coordinate axes is brought into existence and the correspondence between points on the plane and ordered pairs is established.

1. The distance between two distinct points (x_1, y_1) and (x_2, y_2) is given by

$$\sqrt{(x_2 - x_1)^2 + (y_2 - y_1)^2}.$$

2. The slope m of the segment with endpoints (x_1, y_1) and (x_2, y_2) is defined by $\frac{y_2 - y_1}{x_2 - x_1}$, provided that $x_1 \neq x_2$. When $x_2 > x_1$ and $y_2 > y_1$ the slope is positive and we say that the segment has a positive inclination. If b denotes the distance between (x_2, y_2) and (x_2, y_1), and a denotes the distance between (x_1, y_1) and (x_2, y_1), we will have $m = \frac{b}{a}$. Similarly, when $x_2 > x_1$ and $y_2 < y_1$ the slope is negative and we say that the segment has a negative inclination. If b denotes the distance between (x_1, y_1) and (x_1, y_2), and a denotes the distance between (x_1, y_2) and (x_2, y_2), we will have

$$m = \frac{y_2 - y_1}{x_2 - x_1} = -\frac{y_1 - y_2}{x_2 - x_1} = -\frac{b}{a}.$$

[1]The word 'Cartesian' comes from 'Cartesius', the Latin version of René Descartes' last name. In 1637 Descartes published, as an appendix to his great work *The Method*, the first systematic application of algebra to the quest for solutions to geometric problems.

3. The slope of a non-vertical line L is the slope of any segment contained in L. This fact characterizes a line with slope m, which passes through a given point (x_1, y_1), and leads to the equation

$$y - y_1 = m(x - x_1)$$

where (x, y) is any point on L. Thus, the equation of a line that passes through two points (x_1, y_1) and (x_2, y_2), $x_1 \neq x_2$, is

$$y - y_1 = \frac{y_2 - y_1}{x_2 - x_1}(x - x_1).$$

A line has a positive or negative inclination depending on whether any segment contained in it has a positive or negative inclination.

4. The equation of a vertical line that passes through $(0, k)$ is simply $x = k$. The concept of slope is not defined for vertical lines.

5. The equation of a line with slope m and y-intercept $(0, b)$ is $y = mx + b$.

6. The equation of a horizontal line that passes through $(0, b)$ is $y = b$.

7. Two non-vertical lines L_1 and L_2, with slope m_1 and m_2 respectively, are parallel if and only if $m_1 = m_2$. Under these circumstances we write $L_1 \parallel L_2$.

8. Two non-vertical lines L_1 and L_2, with slope m_1 and m_2 respectively, are perpendicular ($L_1 \perp L_2$) if and only if $m_1 m_2 = -1$.

9. Let $\overline{PQ}$ be a segment with endpoints (x_1, y_1), (x_2, y_2). The coordinates of the midpoint of the segment are

$$\left(\frac{x_1 + x_2}{2}, \frac{y_1 + y_2}{2}\right).$$

10. The equation of a circle centered at the origin, and with radius r, is given by $x^2 + y^2 = r^2$. If the circle is centered at any point (a, b), its equation becomes

$$(x - a)^2 + (y - b)^2 = r^2.$$

As an illustration of the interplay between Euclidean and Cartesian geometry that is needed to justify certain propositions, let us provide a proof of item 8 above assuming, without loss of generality, that L_1 has a positive inclination and L_2 has a negative inclination. Suppose that $L_1 \perp L_2$, intercepting each other at a point O. Then build a line parallel to the x-axis, which crosses the lines L_1 and L_2 at C and D, respectively (Figure 9.1), and drop a perpendicular $\overline{OE}$ from O to $\overline{CD}$. We have built a right triangle OCD, thus $c^2 = ab$ where $c = OE$, $a = CE$, and $b = ED$ (recall exercise 6, chapter 4). Since $m_1 = \frac{c}{a}$ and $m_2 = -\frac{c}{b}$ it follows that

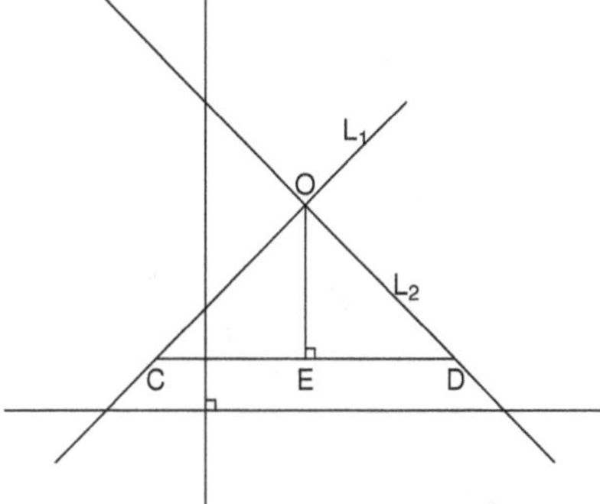

Figure 9.1: An important relationship linked to perpendicular lines

$$m_1 m_2 = -\frac{c^2}{ab} = -\frac{ab}{ab} = -1.$$

Next let us assume that $m_1 m_2 = -1$. Evidently, L_1 and L_2 cannot be parallel because parallelism leads to $m_1 = m_2$ and therefore $m_1 m_2 = m_1^2 \geq 0$. Consequently L_1 and L_2 meet at a point O. Referring to Figure 9.1 we will have $\frac{c}{a}(-\frac{c}{b}) = -1$, consequently $c^2 = ab$. Hence

$$OC^2 + OD^2 = a^2 + c^2 + c^2 + b^2 = a^2 + ab + ab + b^2 = (a + b)^2.$$

Then the converse of the Pythagorean theorem implies that $m\angle COD = 90$. QED

9.2 Several problems of increasing level of difficulty

To illustrate the power and versatility of Cartesian methods we will discuss five particularly interesting problems. The first three are well-known propositions from elementary geometry.

1. The diagonals of a square make a right angle at their point of intersection.

Proof

Let us consider an arbitrary square $ABCD$ of side a. Choose the coordinate axes as shown in Figure 9.2. Then the coordinates of A, B, C, D are $(0,0)$, $(0,a)$, (a,a) and $(a,0)$ respectively. Let L_1 and L_2 be the two diagonals The slope of L_1 is $m_1 = \frac{a-0}{a-0} = 1$ while the slope of L_2 is $m_2 = \frac{a-0}{0-a} = -1$. Therefore $m_1 m_2 = -1$ and we can conclude that $L_1 \perp L_2$.

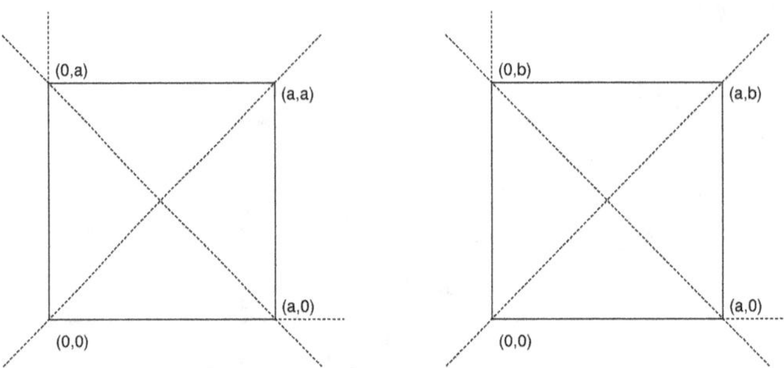

Figure 9.2: Two problems about diagonals and perpendicularity

2. Suppose that the diagonals of a rectangle make a right angle when they intersect. Then the rectangle has to be a square.

Proof

Let us consider a rectangle $ABCD$, with width b and length a, such that its perpendicular diagonals are labeled L_1 and L_2, with slopes m_1 and m_2. Choosing the coordinate axes as shown in Figure 9.2 we can see that the coordinates of A, B, C, D are $(0,0)$, $(0,b)$, (a,b) and $(a,0)$ respectively. Then $m_1 = \frac{b-0}{a-0} = \frac{b}{a}$ while $m_2 = \frac{b-0}{0-a} = -\frac{b}{a}$. Since $L_1 \perp L_2$ we must have $m_1 m_2 = -1$. Thus, $\frac{b}{a}(-\frac{b}{a}) = -1$. Hence $b^2 = a^2$, which in turn leads to $b = a$. That is to say, the rectangle has to be a square!

3. Let ABC be a triangle inscribed in a circle of radius r so that $\overline{AC}$ is a diameter. Then $m\angle ABC = 90$.

Proof

Choose the coordinate axes in such a way that the origin is located at the center of the circle (Figure 9.3). Then the coordinates of A and C are $(-r, 0)$ and $(r, 0)$ respectively. We do not know what the coordinates of B are so we denote them (x, y). Since (x, y) lies on the circle we will have $x^2 + y^2 = r^2$. Let L_1 be the line that contains $\overline{AB}$ and L_2 the line that contains $\overline{BC}$, with slopes m_1 and m_2 respectively. We then have $m_1 = \frac{y-0}{x+r} = \frac{y}{x+r}$ while $m_2 = \frac{y-0}{x-r} = \frac{y}{x-r}$. Hence

$$m_1 m_2 = \frac{y}{x+r}\frac{y}{x-r} = \frac{y^2}{x^2-r^2} = \frac{y^2}{-y^2} = -1.$$

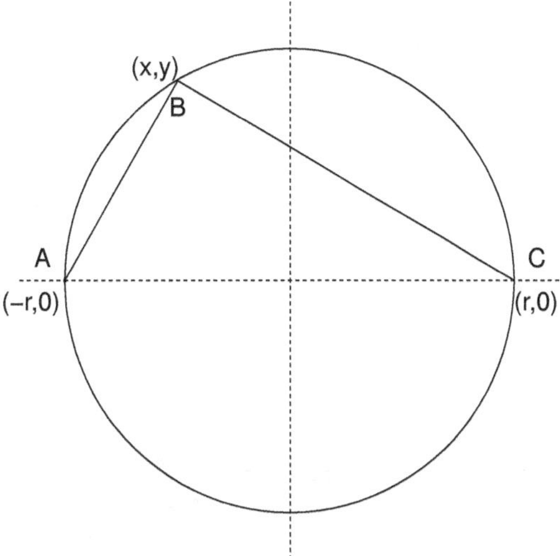

Figure 9.3: An inscribed angle that subtends a diameter

We have been able to show that $L_1 \perp L_2$, hence $m\angle ABC = 90$. The reader may compare this approach with proposition 7.1.4 (chapter 7).

4. Let ΔABC be a right triangle with $m\angle BCA = 90^o$ (Figure 9.4). The median $\overline{CM}$ is perpendicular to median $\overline{BN}$ and $BC = 5$. Calculate[2] BN.

A Euclidean approach

When two medians intersect, the respective segments are in a relation of 2 to 1 (recall the first remark after the proof of the concurrency of the medians in section 3.3). Thus, if $DN = x$ then $BD = 2x$. We observe that, thanks to AA similarity, $\Delta BCN \sim \Delta BDC$. Therefore $\frac{3x}{5} = \frac{5}{2x}$, then $x = \frac{5}{\sqrt{6}}$; thus,

$$BN = 3x = \frac{15}{\sqrt{6}} = \frac{5\sqrt{6}}{2}.$$

A Cartesian approach

Draw the coordinate axes as shown in Figure 9.5. Letting $a = CA$ we can assert that the coordinates of A are $(a, 0)$ while the coordinates of N are $(\frac{a}{2}, 0)$. On

[2]This problem and the next are small variations of problems developed in Taback (1990). The original problems were proposed in mathematical contests for High School students.

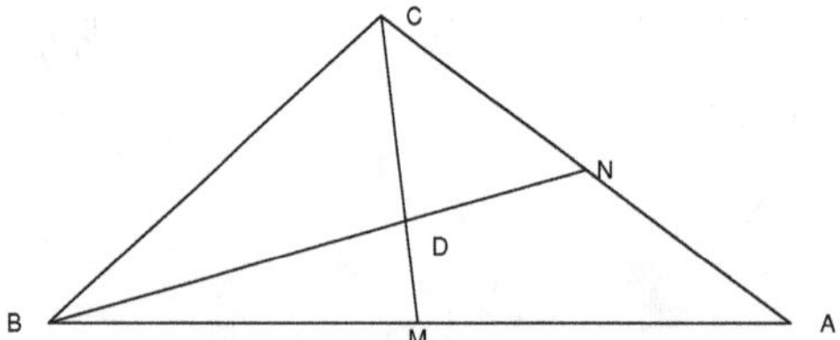

Figure 9.4: A Euclidean approach

the other hand, since M is the midpoint of $\overline{AB}$, its coordinates are $(\frac{0+a}{2}, \frac{5+0}{2})$, that is, $(\frac{a}{2}, \frac{5}{2})$. The slope of the median $\overline{CM}$ is

$$\frac{\frac{5}{2}-0}{\frac{a}{2}-0} = \frac{5}{a},$$

while the slope of the median $\overline{BN}$ is

$$\frac{5-0}{0-\frac{a}{2}} = -\frac{10}{a}.$$

But both medians are perpendicular, so $\frac{5}{a}(-\frac{10}{a}) = -1$; hence $a = \sqrt{50}$. Finally, we apply the formula for the distance between two points and obtain the answer:

$$BN = \sqrt{(\frac{a}{2})^2 + 5^2} = \sqrt{\frac{50}{4} + 25} = \frac{5\sqrt{6}}{2}.$$

Remark

The reader may have noticed that both approaches are radically different. The Euclidean approach seems simpler because we are using a powerful result, namely the 2 to 1 relationship when two medians intersect each other. In the preceding problem the Cartesian approach seems longer, but it is quite systematic.

5. A trapezoid has perpendicular diagonals, one of them of length 10. If the height of the trapezoid is 6, calculate its area

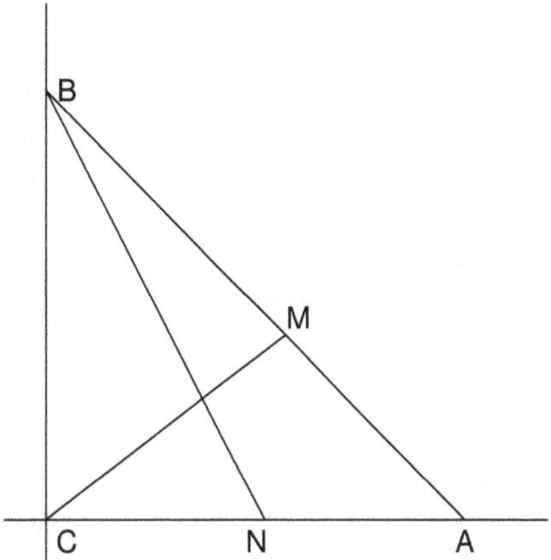

Figure 9.5: A Cartesian approach

Solution

A Cartesian approach seems more amenable in this problem. Let $ABCD$ be a trapezoid with $BD = 10$. Choose the coordinate axes in such a way that $\overline{AB}$ falls on the x-axis (Figure 9.6). Then the coordinates of A, B, C, D are $(0,0), (b,0), (c,6)$, and $(d,6)$ respectively (recall that the height of the trapezoid is 6). The slopes of the diagonals[3] are $\frac{6}{c}$ and $\frac{6}{d-b}$. Since the diagonals are perpendicular to each other we must have

$$\frac{6}{d-b}\frac{6}{c} = -1.$$

Hence

$$c(b-d) = 36. \tag{9.1}$$

On the other hand, applying the formula for the distance between two points we get $(d-b)^2 + 6^2 = 10^2$, so

$$b - d = 8. \tag{9.2}$$

From (9.1) and (9.2) it follows that $c = \frac{36}{8} = 4.5$. Finally,

$$\text{area}= \tfrac{6}{2}(c - d + b) = 3(b - d + c) = 3(8 + 4.5) = 37.5.$$

[3]Without loss of generality we are assuming that $b > d$.

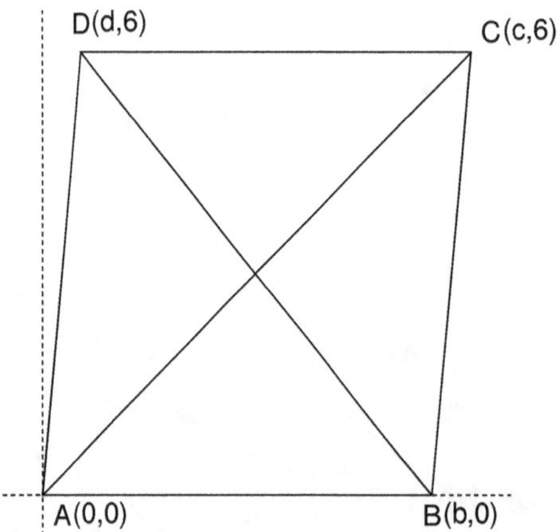

Figure 9.6: The problem of the trapezoid

Surprisingly, we have been able to solve the problem without having to find b and d. A Euclidean approach can be chosen (see exercise 7 at the end of the chapter), but it requires more work.

9.3 Algebra comes to the rescue

The five previous problems involved segments that were perpendicular to each other, a fact that seemed to suggest the use of a Cartesian approach. After all, there is a simple relationship involving slopes when perpendicularity takes place ($m_1m_2 = -1$). However, the range of applicability of Cartesian methods is vast as the following problem, which comes from the first Iberoamerican Mathematical Olympiad held in Colombia in 1985, will show. We will have to handle a considerable amount of algebraic work.

Let P be an interior point of an equilateral triangle ABC such that $PA = 5$, $PB = 7$ and $PC = 8$. Find the length of the side of ΔABC.

Solution

Let $a = AB = BC = CA$ and choose the coordinate axes as shown in Figure 9.7. Then the coordinates of A and B are $(0, 0)$ and $(a, 0)$ respectively. How

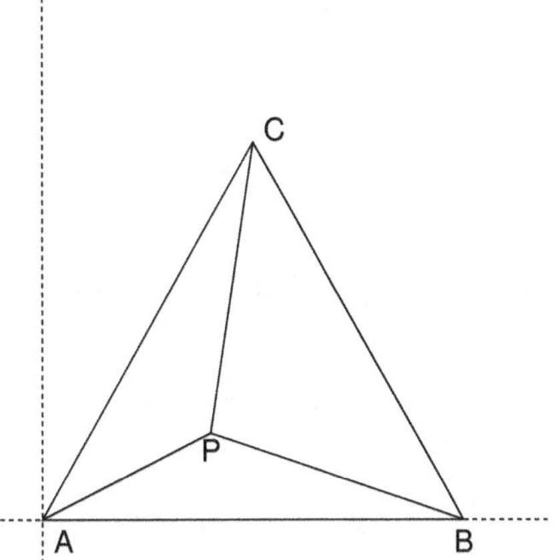

Figure 9.7: The equilateral triangle problem

about the coordinates of point C ? The height of the triangle, namely h, is

$$h = \sqrt{a^2 - \frac{a^2}{4}} = \frac{\sqrt{3}}{2}a.$$

Therefore, the coordinates of C are $(\frac{a}{2}, \frac{\sqrt{3}}{2}a)$. Let (x, y) be the coordinates of P. Using the formula for the distance between two points we get

$$\sqrt{x^2 + y^2} = 5, \sqrt{(x - \frac{a}{2})^2 + (y - \frac{\sqrt{3}}{2}a)^2} = 8, \sqrt{(x - a)^2 + y^2} = 7.$$

Thus,

$$x^2 + y^2 = 25, \tag{9.3}$$

$$x^2 - ax + \frac{a^2}{4} + y^2 - \sqrt{3}ay + \frac{3}{4}a^2 = 64, \tag{9.4}$$

$$x^2 - 2ax + a^2 + y^2 = 49. \tag{9.5}$$

This is a good start because we have obtained three equations and three unknowns (namely x, y, a). Introducing (9.3) into (9.4) and (9.5) we arrive at

$$-ax + a^2 - \sqrt{3}ay = 39, \tag{9.6}$$

$$-2ax + a^2 = 24. \tag{9.7}$$

From (9.7) we get

$$x = \frac{a^2 - 24}{2a}. \tag{9.8}$$

Replacing this value in (9.6) we obtain the equation $a^2 - 2\sqrt{3}ay = 54$. Hence

$$y = \frac{a^2 - 54}{2\sqrt{3}a}. \tag{9.9}$$

Substituting (9.8) and (9.9) into the equation $x^2+y^2 = 25$ we reach the equation

$$(\tfrac{a^2-24}{2a})^2 + (\tfrac{a^2-54}{2\sqrt{3}a})^2 = 25.$$

Simplifying, we get the biquadratic equation

$$a^4 - 138a^2 + 1161 = 0.$$

Letting $z = a^2$ leads to the quadratic equation $z^2 - 138z + 1161 = 0$. Solving this equation we get $z = 9$ or $z = 129$. Hence $a = 3$ or $a = \sqrt{129}$. The first possibility has to be discarded because the triangle inequality, applied to ΔAPC, does not hold true ($8 < 3 + 5$ happens to be false). Then $a = \sqrt{129} \approx 11.36$ is the solution to the problem. As expected, the Cartesian approach has replaced the need for a geometric argument with a mostly algebraic procedure.

Is a Euclidean approach feasible? It is, but it requires much more ingenuity and work. The interested reader can browse through the Bulletin of the Cuban Society of Mathematics (No. 6, 1986, 26-27) for such an approach.

Exercises

Use Cartesian geometry to prove statements 1 through 6 below.

1. The diagonals of a rectangle have the same length.
2. Given a right triangle with a right angle at A, let D be the midpoint of $\overline{BC}$. Then $AD = \frac{1}{2}BC$.
3. The median of any trapezoid is parallel to the bases.

4. Given triangle ABC (Figure 9.8), if $\overline{DE} \parallel \overline{AC}$ and $AD = DB$ then $BE = EC$.

5. Given triangle ABC (Figure 9.8), if $AD = DB$ and $BE = EC$ then $\overline{DE} \parallel \overline{AC}$.

6. We join the midpoints of a rhombus in order to build a quadrilateral. Then the quadrilateral is a rectangle.

7. Use only a Euclidean approach to solve the following problem: A trapezoid (Figure 9.8) has perpendicular diagonals, one of them of length 10. If the height of the trapezoid is 6, calculate its area (Hint: Draw the height $\overline{EF}$ that passes through G. Then notice that triangles AGF, FGB, DGE, and CGE are similar.)

8. Let ABC be a right triangle with $m\angle BCA = 90$. Let M be the midpoint of $\overline{AB}$ and let N be the midpoint of $\overline{AC}$. Assume that $\overline{CM}$ is perpendicular to $\overline{BN}$ and $BC = 10$. Calculate BN.

9. Let P be an interior point of an equilateral triangle ABC such that $PA = 6$, $PB = 8$, and $PC = 9$. Find the length of the side of ΔABC.

10. Consider a circle with radius 5 (Figure 9.8) centered at O. Given that $OC = 3$ and $AC = 4$, calculate BC using, separately, a Euclidean approach and a Cartesian approach. Compare both approaches.

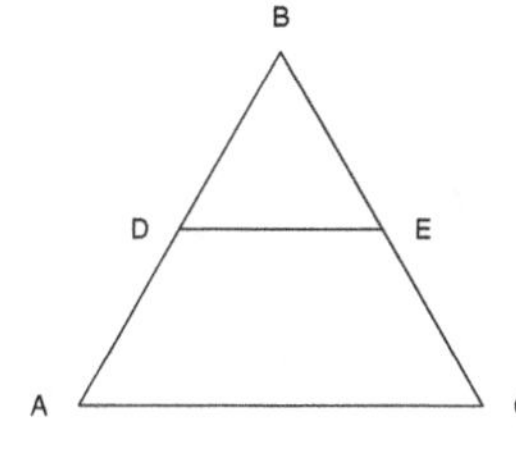

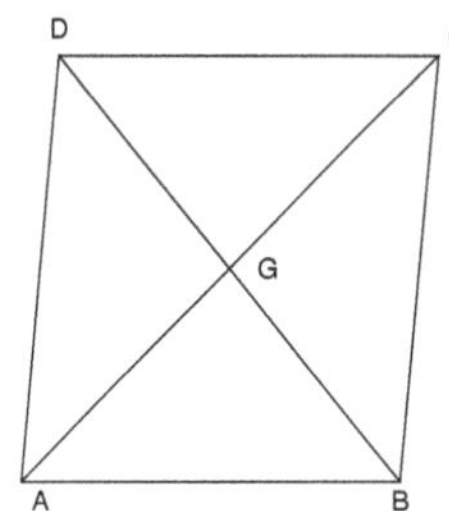

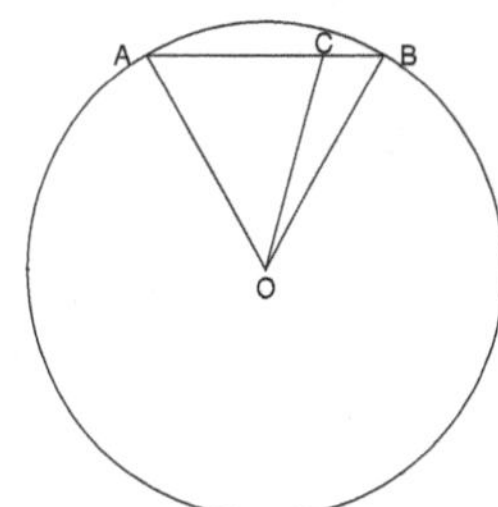

Figure 9.8: Diagrams related to exercises 4, 5, 7, 10

Chapter 10

Revisiting the Complex Numbers

The reader is surely acquainted with complex numbers from her/his previous mathematical experience. They are numbers like $3+5i$ or $7-2i$ and we can add them or multiply them keeping in mind that $i^2=-1$. For instance, $(3+5i)+(7-2i)=10+3i$ while $(3+5i)(7-2i)=21-6i+35i-10i^2=31+29i$. To be more precise, we should say that C, the set of complex numbers, that is, numbers of the form $a+bi$, where a,b are reals and i is a non-real number such that $i^2=-1$, has the reals as a subset and two operations (addition and multiplication) that are extensions of the corresponding operations among real numbers. Hopefully, no misunderstanding will arise from the fact that the same symbols are used whether we are adding or multiplying complex numbers or real numbers.

As we mentioned before, the number i has a particular property, namely $i^2=-1$. Thus, i is not a real number because the square of any real number is non-negative. Moreover, the operations of addition and multiplication are defined by

$$(a+bi)+(c+di)=(a+c)+(b+d)i,$$

$$(a+bi)(c+di)=(ac-bd)+(ad+bc)i.$$

With these two operations, the set C happens to be a field with the real numbers 0 and 1 as the additive and multiplicative identity, respectively. Indeed, it is not difficult to show that:

1. $(a+bi)+(c+di)=(c+di)+(a+bi)$.

2. $(a+bi)+[(c+di)+(u+vi)]=[(a+bi)+(c+di)]+(u+vi)$.

3. $(a+bi)(c+di)=(c+di)(a+bi)$.

4. $(a+bi)[(c+di)(u+vi)]=[(a+bi)(c+di)](u+vi)$.

5. $(a+bi)[(c+di)+(u+vi)]=(a+bi)(c+di)+(a+bi)(u+vi)$.

6. $(a+bi)+0=a+bi$, $(a+bi)1=a+bi$.

7. The equation $(a+bi)+z=0$ has a solution in C, namely $z=-a-bi$.

8. The equation $(a+bi)w=1$ ($a\neq 0$ or $b\neq 0$) has a solution in C, namely $w=\frac{a}{a^2+b^2}-\frac{b}{a^2+b^2}i$.

10.1 The real and imaginary parts of a complex number

If $r+si=0$ (r and s arbitrary real numbers) then $r=0$ and $s=0$. Why is this so? Since $r=-si$ we will have $r^2=-s^2$; thus $r^2+s^2=0$, which in turn leads to $r^2=0$ and $s^2=0$. Consequently $r=0$ and $s=0$. We can go a little farther and formulate the following question: Does $a+bi=c+di$ imply that $a=c$ and $b=d$? Indeed it does since $(a-c)+(b-d)i=0$. Then $a-c=0$ and $b-d=0$, so $a=c$ and $b=d$. We have just seen that given any complex number z there exist unique real numbers a,b such that $z=a+bi$. We call a the real part of z and b the imaginary part of z and write $a=Re(z)$, $b=Im(z)$. Thus $z=Re(z)+iIm(z)$.

Examples

1. $(7+i)(3-i)=22-4i$; thus, $Re(7+i)(3-i)=22$ and $Im(7+i)(3-i)=-4$.

2. $\frac{1}{3-2i}=\frac{3+2i}{(3-2i)(3+2i)}=\frac{3+2i}{9+4}=\frac{3}{13}+\frac{2}{13}i$. Hence $Re\frac{1}{3-2i}=\frac{3}{13}$, $Im\frac{1}{3-2i}=\frac{2}{13}$.

3. $\frac{5+4i}{1-i}=\frac{(5+4i)(1+i)}{(1-i)(1+i)}=\frac{1+9i}{1+1}=\frac{1}{2}+\frac{9}{2}i$. Hence $Re\frac{5+4i}{1-i}=\frac{1}{2}$, $Im\frac{5+4i}{1-i}=\frac{9}{2}$.

10.2 The isomorphism between C and $\Re\times\Re$

Let $\Re$ denote the set of real numbers. We will consider the Cartesian product $\Re\times\Re$ with two operations:

$$(a,b)\oplus(c,d)=(a+c,b+d),$$

$$(a,b)\otimes(c,d)=(ac-bd,ad+bc).$$

It can be shown that $\Re\times\Re$ becomes a field with $(0,0)$ as the identity with regard to $\oplus$ and $(1,0)$ as the identity with regard to $\otimes$. As an illustration, let us prove that, given (a,b) the equations $(a,b)\oplus(w,z)=(0,0)$ and $(a,b)\otimes(u,v)=(1,0)$ have a solution[1]. Indeed, we realize that $(a, b) \oplus (-a, -b) = (0, 0)$. This unique solution is denoted $-(a, b)$, that is, $-(a, b) = (-a, -b)$.

On the other hand, if we were to find real numbers u, v such that $(a, b) \otimes (u, v) = (1, 0)$ then $(au - bv, av + bu) = (1, 0)$, which in turn leads to

$$au - bv = 1, av + bu = 0$$

Since $(a, b) \neq (0, 0)$ we must have $a \neq 0$ or $b \neq 0$; assume, without loss of generality, that $b \neq 0$. Then $u = -\frac{a}{b}v$, thus

$$a(-\tfrac{a}{b}v) - bv = 1.$$

Therefore $v = -\frac{b}{a^2+b^2}$, and consequently

$$u = -\tfrac{a}{b}(\tfrac{-b}{a^2+b^2}) = \tfrac{a}{a^2+b^2}.$$

A simple calculation will allow us to assert that $(a, b) \otimes (\frac{a}{a^2+b^2}, -\frac{b}{a^2+b^2}) = (1, 0)$. We have succeeded in showing that the equation $(a, b) \otimes (u, v) = (1, 0)$ has a solution in $\Re \times \Re$. Since this solution is certainly unique we can denote it $(a, b)^{-1}$.

The ordered couple $(0, 1)$ has an important property, namely

$$(0, 1) \otimes (0, 1) = (-1, 0) = -(1, 0).$$

Moreover, since $(b, 0) \otimes (0, 1) = (0, b)$ we will have

$$(a, b) = (a, 0) \oplus (b, 0) \otimes (0, 1).$$

Next we define a function $h : \Re \times \Re \to C$, $h(a, b) = a + bi$. It is not hard to prove that h is one-to-one and onto. Furthermore

$$h[(a, b) \oplus (c, d)] = h(a, b) + h(c, d),$$

$$h[(a, b) \otimes (c, d)] = h(a, b)h(c, d).$$

The structures $(\Re \times \Re, \oplus, \otimes)$ and $(C, +, .)$ are thus isomorphic. Hence, for all practical purposes we may think of complex numbers as ordered pairs that obey certain properties when added or multiplied. In the next section we will make use of this representation. It should be noted that the above-mentioned isomorphism really shows that C is well-defined.

[1]For the second equation we are assuming that $(a, b) \neq (0, 0)$.

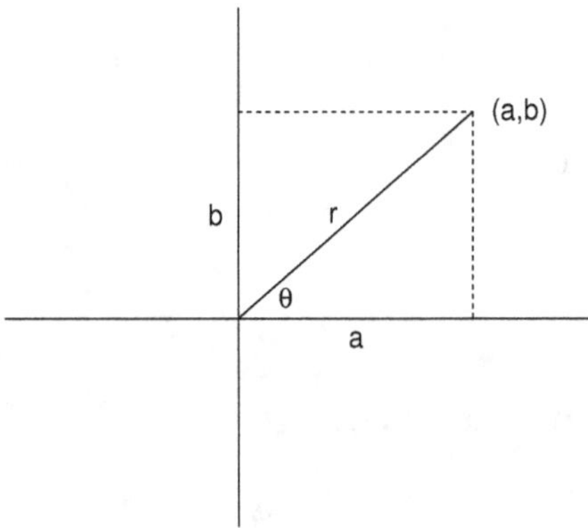

Figure 10.1: Graphic representation of $a + bi$

10.3 Graphic representation of complex numbers

Given any complex number $z = a + bi$ let us think of it as an ordered pair (a, b) on the Cartesian plane (Figure 10.1). Define

$$|a + bi| = \sqrt{a^2 + b^2},$$

which is called the 'modulus' or absolute value of z. In a similar fashion, we define θ (the 'argument' of z) as the unique number such that $0 \leq \theta < 2\pi$ and

$$\cos\theta = \tfrac{a}{r}, \qquad \sin\theta = \tfrac{b}{r},$$

where $r = |a + bi|$. Since $a = r\sin\theta$ and $b = r\sin\theta$ we will have

$$z = |z|(\cos\theta + i\sin\theta).$$

This expression is called the 'polar form of a complex number'.

Example

Suppose that we need to find the modulus and argument of $1 + \sqrt{3}i$. Right away we notice that $|1 + \sqrt{3}i| = \sqrt{1 + 3} = 2$. Next it is our task to determine the unique θ, $0 \leq \theta < 2\pi$, such that $\cos\theta = 1/2$ and $\sin\theta = \sqrt{3}/2$. Obviously $\theta = \pi/3$.

Often it is easier to look at $\tan\theta$. That is to say, we need to find θ in $[0, 2\pi]$ such that $\tan\theta = \sqrt{3}$, with θ in the first quadrant. Then $\theta = \pi/3$. Although $\tan(\frac{\pi}{3} + \pi) = \sqrt{3}$ too, but $\cos(\frac{\pi}{3} + \pi) = -\frac{1}{2}$ and $\sin(\frac{\pi}{3} + \pi) = -\frac{\sqrt{3}}{2}$. Thus, we have to rule out $\frac{\pi}{3} + \pi$ as the argument of $1 + \sqrt{3}i$.

10.4 The conjugate of a complex number

Given $z = a + bi$ we define $\overline{z} = a - bi$, and call $\overline{z}$ the conjugate of z. It can be easily verified that, for any complex numbers w, z, $\overline{w + z} = \overline{w} + \overline{z}$ and $\overline{wz} = \overline{wz}$. Moreover, using mathematical induction it can be proven that

$$\overline{w_1 + ... + w_n} = \overline{w_1} + ... + \overline{w_n},$$

$$\overline{w_1 ... w_n} = \overline{w_1} ... \overline{w_n}.$$

A particular case of the last property establishes that $(\overline{w})^n = \overline{w^n}$. We should also keep in mind that $z + \overline{z} = 2Re(z)$, $\overline{\overline{z}} = z$, and $|\overline{z}| = |z|$.

A property of the roots of a polynomial with real coefficients

Let $p(z) = a_n z^n + a_{n-1} z^{n-1} + ... + a_1 z + a_0$, where all the coefficients are real numbers. We will prove that if $p(w) = 0$ then $p(\overline{w}) = 0$, that is, if a complex number is a root of $p(z)$ then its conjugate is a root too[2]. Indeed, let us assume that $a_n w^n + ... + a_1 w + a_0 = 0$. Then $\overline{a_n w^n + ... + a_1 w + a_0} = \overline{0} = 0$. But

$$\overline{a_n w^n + ... + a_1 w + a_0} = \overline{a_n w^n} + ... + \overline{a_1 w} + \overline{a_0} = a_n \overline{w^n} + ... + a_1 \overline{w} + a_0 =$$

$$a_n (\overline{w})^n + ... + a_1 \overline{w} + a_0.$$

Therefore $p(\overline{w}) = 0$, as we wished to prove.

The conjugate comes to the help of the absolute value

Given any complex number $z = a + bi$ we note that $z\overline{z} = (a + bi)(a - bi) = a^2 + b^2$, so

$$|z|^2 = a^2 + b^2 = z\overline{z}.$$

This simple fact will allow us to justify, rather swiftly, two important properties of the absolute value.

1. $|zw| = |z||w|$.

   Indeed

   $$|zw|^2 = zw\overline{zw} = zw\overline{z}\overline{w} = z\overline{z}w\overline{w} = |z|^2|w|^2.$$

   Therefore $|zw| = |z||w|$. A particular case of this equality implies that

[2]This result will be very useful in later chapters when discussing the search of roots of polynomial equations.

$$|z^2| = |z|^2.$$

2. $|w + z| \leq |w| + |z|$.

   Indeed

$$|w + z|^2 = (w + z)(\overline{w} + \overline{z}) = w\overline{w} + w\overline{z} + z\overline{w} + z\overline{z}.$$

   But $w\overline{z} + z\overline{w} = w\overline{z} + \overline{w\overline{z}} = 2Re(w\overline{z}) \leq 2|w\overline{z}|$. Hence

$$|w + z|^2 \leq |w|^2 + |z|^2 + 2|w\overline{z}| = |w|^2 + |z|^2 + 2|w||z| = (|w| + |z|)^2.$$

   Consequently $|w + z| \leq |w| + |z|$.

10.5 The square root of a complex number

As we know, given a positive real number r, the symbol $\sqrt{r}$ denotes the unique positive number such that $(\sqrt{r})^2 = r$. We might ask whether given a complex number $a + bi$ there exists a complex number $x + iy$ such that $(x + iy)^2 = a + ib$. If such a number (or numbers) exists, we can assign a meaning to $\sqrt{a + ib}$. Actually, we will soon see that there are two distinct solutions to the equation

$$(x + iy)^2 = a + ib. \tag{10.1}$$

Suppose that $x + iy$ is a solution of (10.1). Then $x^2 - y^2 + 2xyi = a + ib$. Hence

$$x^2 - y^2 = a \tag{10.2}$$

and

$$xy = \frac{b}{2}. \tag{10.3}$$

But

$$x^2 + y^2 = |x + iy|^2 = |(x + iy)^2| = |a + ib| = \sqrt{a^2 + b^2}. \tag{10.4}$$

Adding (10.2) and (10.4) we get $2x^2 = a + \sqrt{a^2 + b^2}$. Therefore

$$x^2 = \tfrac{1}{2}(a + \sqrt{a^2 + b^2})$$

and consequently

$$x = \pm\sqrt{\frac{1}{2}(a + \sqrt{a^2 + b^2})}. \tag{10.5}$$

In a similar fashion, subtracting (10.2) from (10.4) we get $2y^2 = \sqrt{a^2 + b^2} - a$. Hence

$$y = \pm\sqrt{\frac{1}{2}(\sqrt{a^2 + b^2} - a)}. \tag{10.6}$$

From (10.5) and (10.6) we might think that there are four candidates for $x + iy$. However, since $xy = \frac{b}{2}$ (recall (10.3)) we will only have two candidates depending on whether $b > 0$ or $b < 0$.

1. Assume that $b > 0$. Then

$$x = \sqrt{\tfrac{1}{2}(a + \sqrt{a^2 + b^2})}, \quad y = \sqrt{\tfrac{1}{2}(\sqrt{a^2 + b^2} - a)}$$

   or

$$x = -\sqrt{\tfrac{1}{2}(a + \sqrt{a^2 + b^2})}, \quad y = -\sqrt{\tfrac{1}{2}(\sqrt{a^2 + b^2} - a)}.$$

   Indeed, in either case

$$xy = \sqrt{\tfrac{1}{4}(a^2 + b^2) - a^2} = \tfrac{1}{2}\sqrt{b^2} = \tfrac{1}{2}|b| = \tfrac{1}{2}b.$$

2. Assume that $b < 0$. Then

$$x = \sqrt{\tfrac{1}{2}(a + \sqrt{a^2 + b^2})}, \quad y = -\sqrt{\tfrac{1}{2}(\sqrt{a^2 + b^2} - a)}$$

   or

$$x = -\sqrt{\tfrac{1}{2}(a + \sqrt{a^2 + b^2})}, \quad y = \sqrt{\tfrac{1}{2}(\sqrt{a^2 + b^2} - a)}.$$

   In either case

$$xy = -\sqrt{\tfrac{1}{4}(a^2 + b^2 - a^2)} = -\tfrac{1}{2}\sqrt{b^2} = -\tfrac{1}{2}|b| = -\tfrac{1}{2}(-b) = \tfrac{1}{2}b.$$

The reader can easily verify that the two candidates are solutions of the equation (10.1).

Example

Suppose we need to find the solutions of $z^2 = 1 + i\sqrt{3}$, that is, we are interested in calculating $\sqrt{1 + i\sqrt{3}}$. Applying the formulas that we have just developed, we can assert that the two solutions are:

$$\sqrt{\tfrac{1}{2}(\sqrt{1+3}+1)} + i\sqrt{\tfrac{1}{2}(\sqrt{1+3}-1)}$$

and

$$-\sqrt{\tfrac{1}{2}(\sqrt{1+3}+1)} - i\sqrt{\tfrac{1}{2}(\sqrt{1+3}-1)}.$$

That is to say, $\sqrt{\frac{3}{2}} + i\frac{1}{\sqrt{2}}$ and $-\sqrt{\frac{3}{2}} - i\frac{1}{\sqrt{2}}$.

Remark

Finding the square root of a complex number required some work. It is to be expected that, if we were to follow the same approach, finding the nth root of a complex number might require even more work. In section 10.7 we will show an alternative approach: using the polar form of a complex number we will be able to implement a systematic procedure with the goal of solving the equation $z^n = c$, c any complex number. To nobody's surprise, the equation $z^n = c$ has n distinct solutions!

10.6 De Moivre's formula

Given any real number θ, and any natural number n,

$$(\cos\theta + i\sin\theta)^n = \cos(n\theta) + i\sin(n\theta).$$

The proof is by induction. The formula is obviously true for $n = 1$. Assume that it is true for n, and let us try to show that it has to be true for $n+1$. Indeed

$$\begin{gathered}(\cos\theta + i\sin\theta)^{n+1} = (\cos\theta + i\sin\theta)(\cos\theta + i\sin\theta)^n = (\cos\theta + i\sin\theta)(\cos(n\theta) + \\ i\sin(n\theta)) = [\cos\theta\cos(n\theta) - \sin\theta\sin(n\theta)] + i[(\sin\theta\cos(n\theta) + \cos\theta\sin(n\theta)] = \\ \cos(\theta + n\theta) + i\sin(\theta + n\theta) = \cos((n+1)\theta) + i\sin((n+1)\theta).\end{gathered}$$

We have succeeded in providing a proof. The formula, named after Abraham de Moivre (1667-1754), will play a vital role in the next section.

10.7 The solutions of $z^n = c$

Let $c = |c|(\cos\theta + i\sin\theta)$, $c \neq 0$, and let n be any natural number. We will show that the formula

$$z_k = |c|^{1/n}(\cos(\frac{\theta + 2k\pi}{n}) + i\sin(\frac{\theta + 2k\pi}{n})), \ k = 0, 1, 2, ..., n-1 \tag{10.7}$$

provides n distinct solutions of $z^n = c$.

The proof is as follows. First of all, thanks to de Moivre's formula we have

$$z_k^n = |c|(\cos(\theta + 2k\pi) + i\sin(\theta + 2k\pi)) = |c|(\cos\theta + i\sin\theta) = c.$$

Thus, we have shown that the complex numbers z_k, $k = 0, ..., n-1$, are solutions of $z^n = c$. Are these solutions distinct? We would have to prove that if $k \neq j$ then $z_k \neq z_j$, or the logical contrapositive, namely, if $z_k = z_j$ then $k = j$.

Assume that $z_k = z_j$. Then

$$|c|^{1/n}(\cos(\tfrac{\theta+2k\pi}{n}) + i\sin(\tfrac{\theta+2k\pi}{n})) = |c|^{1/n}(\cos(\tfrac{\theta+2j\pi}{n}) + i\sin(\tfrac{\theta+2j\pi}{n})).$$

So

$$\cos(\tfrac{\theta+2k\pi}{n}) + i\sin(\tfrac{\theta+2k\pi}{n}) = \cos(\tfrac{\theta+2j\pi}{n}) + i\sin(\tfrac{\theta+2j\pi}{n}).$$

But $0 \le \theta < 2\pi$, thus $0 \le \theta + 2k\pi < 2\pi + 2k\pi \le 2\pi + 2\pi(n-1) = 2\pi n$. Therefore

$$0 \le \tfrac{\theta+2k\pi}{n} < 2\pi.$$

In a similar fashion we can reach the inequalities

$$0 \le \tfrac{\theta+2j\pi}{n} < 2\pi.$$

Let $\alpha = \frac{\theta+2k\pi}{n}$ and $\beta = \frac{\theta+2j\pi}{n}$. Since $\cos\alpha + i\sin\alpha = \cos\beta + i\sin\beta$, where $0 \le \alpha < 2\pi$ and $0 \le \beta < 2\pi$, we can conclude[3] that $\alpha = \beta$. Hence

$$\frac{\theta+2k\pi}{n} = \frac{\theta+2j\pi}{n}.$$

Consequently $k = j$. QED

In the next section we will state a proposition, which will allow us to assert that the n different complex numbers z_k, $k = 0, 1, \ldots n-1$ are all the solutions of the equation $z^n = c$. None others can be found.

Examples

1. Solve $z^n = 1$. This is a simple problem because $1 = 1(\cos 0 + i\sin 0)$. Hence

   $$z_k = \cos(\tfrac{2k\pi}{n}) + i\sin(\tfrac{2k\pi}{n}) \quad k = 0, 1, 2, \ldots, n-1$$

   are the n solutions of $z^n = 1$.

2. Solve $z^2 = 1 + i\sqrt{3}$. We observe that the polar form of $1 + i\sqrt{3}$ is $2(\cos(\frac{\pi}{3}) + i\sin(\frac{\pi}{3}))$. Therefore, the roots that we are seeking stem from

   $$z_k = \sqrt{2}(\cos(\tfrac{\frac{\pi}{3}+2k\pi}{2}) + i\sin(\tfrac{\frac{\pi}{3}+2k\pi}{2})), \quad k = 0, 1.$$

   Then $z_o = \sqrt{2}(\cos\frac{\pi}{6} + i\sin\frac{\pi}{6}) = \sqrt{\frac{3}{2}} + i\frac{1}{\sqrt{2}}$, and $z_1 = \sqrt{2}(\cos(\frac{\frac{\pi}{3}+2\pi}{2}) + i\sin(\frac{\frac{\pi}{3}+2\pi}{2})) = -\sqrt{\frac{3}{2}} - i\frac{1}{\sqrt{2}}$. This same answer was achieved in the example of section 10.5.

3. Solve the equation $z^3 = -i$. We start by realizing that $-i = \cos\frac{3\pi}{2} + i\sin\frac{3\pi}{2}$, and $|-i| = 1$. Therefore, the solutions stem from

[3]The polar representation of a complex number is unique!

$$z_k = \cos(\tfrac{\frac{3\pi}{2}+2k\pi}{3}) + i\sin(\tfrac{\frac{3\pi}{2}+2k\pi}{3}), \quad k = 0, 1, 2.$$

Thus,

$$z_0 = \cos\tfrac{3\pi}{6} + i\sin\tfrac{3\pi}{6} = i,$$

$$z_1 = \cos(\tfrac{\frac{3\pi}{2}+2\pi}{3}) + i\sin(\tfrac{\frac{3\pi}{2}+2\pi}{3}) = \cos(\tfrac{7\pi}{6}) + i\sin(\tfrac{7\pi}{6}) =$$
$$\cos(\pi + \tfrac{\pi}{6}) + i\sin(\pi + \tfrac{\pi}{6}) = -\cos(\tfrac{\pi}{6}) + i\sin(\tfrac{\pi}{6}) = -\tfrac{\sqrt{3}}{2} - \tfrac{1}{2}i,$$

$$z_2 = \cos(\tfrac{\frac{3\pi}{2}+4\pi}{3}) + i\sin(\tfrac{\frac{3\pi}{2}+4\pi}{3}) = \cos(\tfrac{11\pi}{6}) + i\sin(\tfrac{11\pi}{6}) =$$
$$\cos(\tfrac{12\pi-\pi}{6}) + i\sin(\tfrac{12\pi-\pi}{6}) = \cos(2\pi - \tfrac{\pi}{6}) + i\sin(2\pi - \tfrac{\pi}{6}) = \cos\tfrac{\pi}{6} - i\sin\tfrac{\pi}{6} = \tfrac{\sqrt{3}}{2} - \tfrac{1}{2}i.$$

4. Solve $z^3 = 1 + i$. First of all, $|1 + i| = \sqrt{2}$. Moreover, since $\tan\theta = 1$, and θ lies on the first quadrant, we must have $\theta = \frac{\pi}{4}$. Therefore, the solutions stem from the formula

$$z_k = (\sqrt{2})^{1/3}(\cos(\tfrac{\frac{\pi}{4}+2k\pi}{3}) + i\sin(\tfrac{\frac{\pi}{4}+2k\pi}{3})),\ k = 0, 1, 2.$$

Thus

$$z_0 = (\sqrt{2})^{1/3}(\cos\tfrac{\pi}{12} + i\sin\tfrac{\pi}{12}),$$

$$z_1 = (\sqrt{2})^{1/3}(\cos(\tfrac{\frac{\pi}{4}+2\pi}{3}) + i\sin(\tfrac{\frac{\pi}{4}+2\pi}{3})) = (\sqrt{2})^{1/3}(\cos\tfrac{3\pi}{4} + i\sin\tfrac{3\pi}{4}),$$

$$z_2 = (\sqrt{2})^{1/3}(\cos\tfrac{\frac{\pi}{4}+4\pi}{3}) + i\sin\tfrac{\frac{\pi}{4}+4\pi}{3})) = (\sqrt{2})^{1/3}(\cos\tfrac{17\pi}{12} + i\sin(\tfrac{17\pi}{12})).$$

10.8 The Fundamental Theorem of Algebra

When studying the equation $z^n = c$, with $c = |c|(\cos\theta + i\sin\theta)$ the polar representation of the non-zero complex number c, we found that

$$z_k = |c|^{1/n}(\cos(\tfrac{\theta+2k\pi}{n}) + i\sin(\tfrac{\theta+2k\pi}{n})), \quad k = 0, 1, 2, ..., n-1$$

provides n distinct solutions. Maybe there are other solutions of $z^n = c$ that do not stem from the formula? The following result rules out this possibility.

Proposition

Every polynomial, of degree n and real coefficients, has **at most** n distinct roots.

Proof

Let us proceed by induction. A polynomial of degree 1, namely $ax+b$, has obviously just one root. Let us assume that every polynomial of degree $n-1$ has at most $n-1$ roots. Let $p(x)$ be any polynomial of degree n. If $p(x)$ has no roots we are done. Otherwise suppose that $p(x)$ has one root x_o. The division algorithm for polynomials (Hungerford, 1997) implies that there exists a polynomial of degree $n-1$, say $q(x)$, and a constant r such that $p(x) = (x-x_o)q(x)+r$. But $0 = p(x_o) = (x_o-x_o)q(x_o)+r$, so $r=0$. Consequently

$$p(x) = (x-x_o)q(x).$$

By the inductive hypothesis, $q(x)$ can have at most $n-1$ distinct roots $x_1, ..., x_{n-1}$. Right away we realize that these $n-1$ numbers are also roots of $p(x)$. Can $p(x)$ have a root s different from $x_o, x_1, ..., x_{n-1}$? Since $p(s) = 0$ and $s - x_o \neq 0$, we must have $q(s) = 0$. But this is an impossibility since the only acceptable distinct roots of $q(x)$ are $x_1, ..., x_{n-1}$. Thus $p(x)$ can have at most the n distinct roots $x_o, x_1, ..., x_{n-1}$. QED

Hence, if the polynomial $p(z) = z^n - c$ has n distinct roots, these are all the possible roots. That is, the formula described at the beginning of the section provides absolutely **all** the solutions.

The Fundamental Theorem of Algebra (FTA)[4] asserts that any polynomial with complex coefficients and positive degree has **at least** one root in C. In particular, any polynomial with real coefficients and positive degree has at least one root in C. Induction and FTA leads to the fact that given any polynomial

$$p(z) = a_n z^n + a_{n-1} z^{n-1} + ... + a_1 z + a_0 \tag{10.8}$$

where $a_i, 0 \leq i \leq n$, are complex numbers, $a_n \neq 0$, and n is a positive integer, there exist complex numbers $c_1, ..., c_n$ such that

$$p(z) = a_n(z-c_1)(z-c_2)...(z-c_n).$$

Indeed, this result is obviously true for any polynomial $a_1 z + a_o$ of degree 1 since $a_1 z + a_o = a_1(z - (-\frac{a_o}{a_1}))$. Assume, as the inductive hypothesis, that it is true for any polynomial of order $n-1$ and let $p(z)$ be an arbitrary polynomial of degree n, $p(z) = a_n z^n + a_{n-1} z^{n-1} + ... + a_1 z + a_0$. The Fundamental Theorem of Algebra asserts that $p(z)$ has a root c_1, thus the division algorithm, or more precisely a corollary of it, implies that there exists a polynomial $q(z)$ of order $n-1$, say $q(z) = b_{n-1} z^{n-1} + ... + b_o$, such that $p(z) = (z-c_1)q(z)$. By the inductive hypothesis we can assert that there exist complex numbers c_2, ..., c_n such that

[4] A proof of FTA is usually discussed in a first course on complex analysis.

$$q(z) = b_{n-1}(z - c_2)...(z - c_n).$$

Then $p(z) = b_{n-1}(z - c_1)(z - c_2)...(z - c_n)$. Since $p(z)$ has leading coefficient a_n, it follows that $b_{n-1} = a_n$ and the proof comes to a conclusion.

It should be noted that the complex numbers $c_1, ..., c_n$, which happen to be the roots of $p(z)$, do not have to be distinct. Thus, we may write

$$p(z) = a_n(z - w_1)^{b_1}...(z - w_s)^{b_s}$$

where $b_1 + ... + b_s = n$ and $w_1, ...w_s$ are the distinct numbers among $c_1, ...c_n$ (the positive integer b_i is called the 'multiplicity' corresponding to w_i).

10.9 A useful definition

For any real number θ let us define

$$e^{i\theta} = \cos\theta + i\sin\theta.$$

Without much effort we can prove that for any real numbers θ, β the following equalities hold true:

1. $e^{i\theta}e^{i\beta} = e^{i(\theta+\beta)}$
2. $\frac{e^{i\theta}}{e^{i\beta}} = e^{i(\theta-\beta)}$
3. $(e^{i\theta})^n = e^{in\theta}$, where n is any positive integer.

The third equality requires no proof because it is just another way of writing de Moivre's formula. Let us deal with a proof of the first equality:

$$e^{i\theta}e^{i\beta} = (\cos\theta + i\sin\theta)(\cos\beta + i\sin\beta) =$$
$$(\cos\theta\cos\beta - \sin\theta\sin\beta) + i(\sin\theta\cos\beta + \cos\theta\sin\beta) = \cos(\theta+\beta) + i\sin(\theta+\beta) = e^{i(\theta+\beta)}.$$

Proving the second equality is not much of a challenge either:

$$\frac{e^{i\theta}}{e^{i\beta}} = \frac{\cos\theta+i\sin\theta}{\cos\beta+i\sin\beta} = \frac{(\cos\theta+i\sin\theta)}{(\cos\beta-i\sin\beta)}\frac{(\cos\beta-i\sin\beta)}{\cos\beta+i\sin\beta)} = \frac{(\cos\theta\cos\beta+\sin\theta\sin\beta)+i(\sin\theta\cos\beta-\cos\theta\sin\beta)}{\cos^2\beta+\sin^2\beta} =$$
$$\cos(\theta-\beta) + i\sin(\theta-\beta) = e^{i(\theta-\beta)}.$$

Remark

Any complex number $c = a + bi$ can be written as $|c|e^{i\theta}$ where $0 \leq \theta < 2\pi$ is the argument and $|c|$ is the modulus. The formula for the solutions of $z^n = c$ can be written in a rather compact way:

$$z_k = |c|^{1/n} e^{i(\frac{\theta + 2k\pi}{n})}, \quad k = 0, 1, ..., n-1.$$

In particular, the roots of $z^n = 1$ stem from

$$z_k = e^{i\frac{2k\pi}{n}}, \quad k = 0, 1, ..., n-1.$$

That is, the solutions of $z^n = 1$ are 1, $e^{i\frac{2\pi}{n}}$,..., $e^{i\frac{2(n-1)\pi}{n}}$.

We should also mention that if w is a root of $z^3 = c$ then the two other roots are $we^{\frac{2\pi}{3}i}$ and $w(e^{\frac{2\pi}{3}i})^2$. Indeed

$$(we^{\frac{2\pi}{3}i})^3 = w^3(e^{\frac{2\pi}{3}i})^3 = w^3 e^{2\pi i} = w^3 \times 1 = w^3 = c,$$

and

$$(e^{\frac{4\pi}{3}i})^3 = w^3(e^{\frac{4\pi}{3}i})^3 = w^3 e^{4\pi i} = w^3 \times 1 = w^3 = c.$$

Evidently, the complex numbers w, $we^{\frac{2\pi}{3}i}$ and $we^{\frac{4\pi}{3}i}$ are distinct. Then these are all the solutions. Interestingly,

$$we^{\frac{2\pi}{3}i} = w(\cos\tfrac{2\pi}{3} + i\sin\tfrac{2\pi}{3}) = w(-\tfrac{1}{2} + i\tfrac{\sqrt{3}}{2}),$$

$$we^{\frac{4\pi}{3}i} = w(\cos\tfrac{4\pi}{3} + i\sin\tfrac{4\pi}{3}) = w(-\tfrac{1}{2} - i\tfrac{\sqrt{3}}{2}).$$

10.10 Applications

Complex numbers are widely used in many branches of mathematics and the natural sciences. We have chosen two examples from trigonometry, and an example from number theory [5]. In chapter 11 and beyond, complex numbers will be used extensively in the quest for solutions of polynomial equations of degree two, three, and four.

[5]For applications to geometry the reader can browse through a short book published by the Mathematical Association of America (Hahn, 1994).

Calculating the cosine of a particular angle

An angle of 72° corresponds to $\frac{2\pi}{5}$ radians. This angle is quite important in geometry since it is the central angle corresponding to the side of a regular pentagon inscribed in a circle. Right away we notice that the complex number

$$z = \cos\frac{2\pi}{5} + i\sin\frac{2\pi}{5}$$

has the property that $z^5 = 1$. Since $z^5 - 1 = 0$ we will have

$$(z-1)(z^4 + z^3 + z^2 + z + 1) = 0.$$

Since $z \neq 1$ it follows that $z^4 + z^3 + z^2 + z + 1 = 0$. Dividing by z^2 we get

$$(z^2 + \frac{1}{z^2}) + (z + \frac{1}{z}) + 1 = 0.$$

Then

$$(z + \frac{1}{z})^2 - 2 + (z + \frac{1}{z}) + 1 = 0.$$

That is,

$$(z + \frac{1}{z})^2 + (z + \frac{1}{z}) - 1 = 0. \tag{10.9}$$

But

$$\frac{1}{z} = \frac{1}{\cos\frac{2\pi}{5} + i\sin\frac{2\pi}{5}} = \frac{1}{e^{\frac{2\pi}{5}i}} = e^{-\frac{2\pi}{5}i} = \cos\frac{2\pi}{5} - i\sin\frac{2\pi}{5}.$$

Therefore $z + \frac{1}{z} = 2\cos\frac{2\pi}{5}$. Using (10.9) we get

$$4\cos^2\frac{2\pi}{5} + 2\cos\frac{2\pi}{5} - 1 = 0.$$

Applying the quadratic formula we arrive at $\cos\frac{2\pi}{5} = \frac{-1\pm\sqrt{5}}{4}$. Keeping in mind that the angle with measure $\frac{2\pi}{5}$ lies on the first quadrant, we have to choose a positive value for its cosine. Therefore

$$\cos\frac{2\pi}{5} = \frac{\sqrt{5}-1}{4}.$$

This is the same result obtained at the end of section 7.9 through a path that does not involve complex numbers.

Finding a closed formula

Suppose that we wish to find a closed formula for the sum $1+\cos\theta+\cos 2\theta+...+\cos(n-1)\theta$, where n is an arbitrary positive integer. With this purpose in mind, let $C=1+\cos\theta+\cos 2\theta+...+\cos(n-1)\theta$, $S=\sin\theta+\sin 2\theta+...+\sin(n-1)\theta$, and $z=\cos\theta+i\sin\theta$. Using de Moivre's formula we get

$$C+iS=1+z+z^2+...+z^{n-1}=\frac{1-z^n}{1-z}=\frac{1-(\cos n\theta+i\sin n\theta)}{1-(\cos\theta+i\sin\theta)}.$$

Multiplying the numerator and the denominator by the conjugate of the denominator we arrive at

$$C+iS=\frac{[(1-\cos n\theta)-i\sin n\theta][(1-\cos\theta)+i\sin\theta]}{[(1-\cos\theta)-i\sin\theta][(1-\cos\theta)+i\sin\theta]}=$$

$$=\frac{(1-\cos n\theta)(1-\cos\theta)+\sin n\theta\sin\theta+i[(1-\cos n\theta)\sin\theta-(1-\cos\theta)\sin n\theta]}{(1-\cos\theta)^2+\sin^2\theta}.$$

Taking the real part, it follows that

$$C=\frac{1-\cos\theta-\cos n\theta+\cos\theta\cos n\theta+\sin n\theta\sin\theta}{2-2\cos\theta}=\frac{1-\cos\theta-\cos n\theta+\cos(n-1)\theta}{2-2\cos\theta}.$$

This is the formula that we were looking for. If we were to take the imaginary part, instead of the real part, as a side benefit we would get

$$S=\frac{\sin\theta-\cos n\theta\sin\theta-\sin n\theta+\cos\theta\sin n\theta}{2-2\cos\theta}=\frac{\sin\theta-\sin n\theta+\sin(n-1)\theta}{2-2\cos\theta}.$$

A question from number theory

Given integers a, b, c, d, is it possible to find integers u, v, such that

$$(a^2+b^2)(c^2+d^2)=u^2+v^2?$$

In *Liber Quadratorum*, Leonardo de Pisa, also known as Fibonacci (circa 1175-1250), solved the problem (Burton, 2003). Using symbolic algebra, a 16th century development, we write

$$(a^2+b^2)(c^2+d^2)=a^2c^2+a^2d^2+b^2c^2+b^2d^2=a^2c^2+b^2d^2-2abcd+a^2d^2+b^2c^2+2abcd=$$

$$=(ac-bd)^2+(ad+bc)^2.$$

Thus we can choose $u=ac-bd$ and $v=ad+bc$. Interestingly, these values for u and v are not unique since

$$(a^2+b^2)(c^2+d^2)=a^2c^2+a^2d^2+b^2c^2+b^2d^2=a^2c^2+b^2d^2+2abcd+a^2d^2+b^2c^2-2abcd=$$

$$=(ac+bd)^2+(ad-bc)^2.$$

Then $u = ac + bd$ and $v = ad - bc$ also works.

Let us analyze the same problem using complex numbers. We observe that

$$(a^2+b^2)(c^2+d^2) = (a+bi)(a-bi)(c+di)(c-di) = (a+bi)(c+di)(a-bi)(c-di).$$

Define $u = Re[(a+bi)(c+di)]$, $v = Im[(a+bi)(c+di)]$. Then

$$(a+bi)(c+di) = Re[(a+bi)(c+di)] + iIm[(a+bi)(c+di)] = u+iv,$$

while

$$u - iv = \overline{u+iv} = \overline{(a+bi)(c+di)} = \overline{a+bi} \times \overline{c+di} = (a-bi)(c-di).$$

Therefore

$$(a^2+b^2)(c^2+d^2) = (u+iv)(u-iv) = u^2+v^2,$$

where $u = ac - bd$ and $v = ad + bc$. We have succeeded in giving a positive answer to the question posed at the beginning.

For instance, let us find numbers u and v such that $(3^2+5^2)(4^2+7^2) = u^2+v^2$. We know that $u = Re[(3+5i)(4+7i)] = 12-35 = -23$ and $v = Im[(3+5i)(4+7i)] = 21+20 = 41$ will do the job. Thus, we are confident that

$$(3^2+5^2)(4^2+7^2) = (-23)^2 + 41^2.$$

Remark

After analyzing the three preceding applications, the reader will surely understand the following quote from the French mathematician Jacques Hadamard (1860-1950): "The shortest path between two truths in the real domain passes through the complex domain".

Exercises

1. Find the real and imaginary parts of $(7+2i)(3-4i)$.
2. Find the real and imaginary parts of $\frac{5+i}{3-2i}$.
3. Given that $|z+9| = 3|z+1|$ find $|z|$ (Hint: Square both sides of the equality and keep in mind that $|w|^2 = w\overline{w}$ for any complex number w.)
4. Given that $|z+2i| = |z+5i|$ find the value of $z - \overline{z}$.
5. Given that $|z+ai| = |z+bi|$ (a, b real numbers, $a \neq b$) find the value of $z - \overline{z}$.

6. Find real numbers x, y such that

$$\tfrac{x}{1-i} + \tfrac{y}{1+i} = 2 + i.$$

7. Suppose that $p(z)$ is a polynomial of degree three, with real coefficients and highest coefficient 1. Furthermore, you are told that the numbers 2 and $1 + i$ are two solutions of the equation $p(z) = 0$. Find $p(z)$ explicitly, in factored and expanded form.

8. Use the polar form of $1 - i$ to calculate the solutions of $z^3 = 1 - i$.

9. Use the polar form of $1 + i$ to calculate the solutions of $z^2 = 1 + i$.

10. Find integers u, v such that $(4^2 + 6^2)(2^2 + 3^2) = u^2 + v^2$.

Chapter 11

Quadratic Equations and Vertical Parabolas

Let us start by considering the equation $x^2-5x+2=0$. Assuming that x is a solution we will have

$$x^2-5x+(\tfrac{5}{2})^2-(\tfrac{5}{2})^2+2=0.$$

Why did we add and subtract $(\frac{5}{2})^2$? Very simple, we are keeping in mind the well-known identity $(a\pm b)^2=a^2\pm 2ab+b^2$. Thus

$$(x-\tfrac{5}{2})^2-(\tfrac{5}{2})^2+2=0.$$

Consequently,

$$(x-\tfrac{5}{2})^2=\tfrac{25}{4}-\tfrac{8}{4},$$

which in turn leads to

$$(x-\tfrac{5}{2})^2=\tfrac{7}{4}=(\tfrac{\sqrt{7}}{2})^2.$$

Therefore $x-\frac{5}{2}=\pm\frac{\sqrt{7}}{2}$, that is, $x=\frac{5\pm\sqrt{7}}{2}$. The reader can check that $\frac{5+\sqrt{7}}{2}$ and $\frac{5-\sqrt{7}}{2}$ are indeed solutions of $x^2-5x+2=0$.

Another example might convince us that the method of 'completion of squares' works always when we have to solve a quadratic equation. Let us try to find the solutions of $x^2+x+1=0$. If x is a solution then

$$x^2+x+(\tfrac{1}{2})^2-(\tfrac{1}{2})^2+1=0.$$

Therefore $(x+\frac{1}{2})^2+\frac{3}{4}=0$, that is, $(x+\frac{1}{2})^2=-\frac{3}{4}$. We can see that there are no solutions in the real field. However,

$$(x+\tfrac{1}{2})^2=(\tfrac{\sqrt{3}}{2}i)^2.$$

Consequently $x+\frac{1}{2}=\pm\frac{\sqrt{3}}{2}i$, thus $x=-\frac{1}{2}\pm\frac{\sqrt{3}}{2}i$. As in the preceding example, we can check that both complex numbers are solutions. Now we are ready to consider the general case.

11.1 A formula for quadratic equations

Let us consider the equation $ax^2+bx+c=0$, $a \neq 0$, a, b, c any real numbers. Suppose that x is a solution. Then

$$x^2 + \tfrac{b}{a}x + \tfrac{c}{a} = 0.$$

Hence

$$(x + \tfrac{b}{2a})^2 - \tfrac{b^2}{4a^2} + \tfrac{c}{a} = 0.$$

That is,

$$(x + \tfrac{b}{2a})^2 = \tfrac{b^2}{4a^2} - \tfrac{c}{a}.$$

Thus

$$(x + \tfrac{b}{2a})^2 = \tfrac{b^2-4ac}{4a^2}.$$

There are two possibilities, namely $b^2 - 4ac \geq 0$ and $b^2 - 4ac < 0$.

1. Assume that $b^2 - 4ac \geq 0$. Then

   $$(x + \tfrac{b}{2a})^2 = (\tfrac{\sqrt{b^2-4ac}}{2a})^2.$$

   Therefore[1]

   $$x + \tfrac{b}{2a} = \pm\tfrac{\sqrt{b^2-4ac}}{2a},$$

   consequently

   $$x = \tfrac{-b\pm\sqrt{b^2-4ac}}{2a}.$$

   In summary, if x is a solution of the quadratic equation, and $b^2 - 4ac \geq 0$, the two possible values are the real numbers $\frac{-b+\sqrt{b^2-4ac}}{2a}$ and $\frac{-b-\sqrt{b^2-4ac}}{2a}$. Through a simple calculation we can verify that both are solutions of the given equation.

2. Assume that $b^2 - 4ac < 0$. Under these circumstances

   $$(x + \tfrac{b}{2a})^2 = (\tfrac{\sqrt{4ac-b^2}}{2a}i)^2.$$

   Hence

   $$x + \tfrac{b}{2a} = \pm\tfrac{\sqrt{4ac-b^2}}{2a}i,$$

   thus

   $$x = \tfrac{-b}{2a} \pm \tfrac{\sqrt{4ac-b^2}}{2a}i.$$

   Again, we can check that these two complex numbers are indeed solutions.

[1] In the real or complex field, $u^2 = v^2$ implies $u = \pm v$. Why? $u^2 - v^2 = 0$ is equivalent to $(u+v)(u-v) = 0$, hence $u+v=0$ or $u-v=0$.

11.2 Three problems that involve quadratics

Quadratic equations often appear in applications. We will discuss a problem from geometry and two problems that stem from elementary physics.

1. Find all rectangles of perimeter 10 and area 3.

   Denoting the length of the sides of any such rectangle by x and y we will have $x + y = 5$ and $xy = 3$. Hence $x + \frac{3}{x} = 5$, and consequently $x^2 - 5x + 3 = 0$. Thus

   $$x = \frac{5 \pm \sqrt{25-12}}{2} = \frac{5}{2} \pm \frac{1}{2}\sqrt{13}.$$

   If $x = \frac{5}{2} + \frac{1}{2}\sqrt{13}$ then $y = 5 - (\frac{5}{2} + \frac{1}{2}\sqrt{13}) = \frac{5}{2} - \frac{1}{2}\sqrt{13}$. while if $x = \frac{5}{2} - \frac{1}{2}\sqrt{13}$ then $y = \frac{5}{2} + \frac{1}{2}\sqrt{13}$. Thus, the problem has just one solution.

   If we were to analyze the system of equations $x + y = s$, $xy = b^2$, already discussed in section 7.3, from a purely algebraic point of view we would have

   $$x + \frac{b^2}{x} = s,$$

   which is equivalent to $x^2 - sx + b^2 = 0$. This quadratic equation has a real solution if and only if its discriminant $\Delta = s^2 - 4b^2$ is non-negative, that is, $s \geq 2b$ or, what is the same, $b \leq \frac{s}{2}$. No wonder that the geometrical construction did not work when $b > \frac{s}{2}$!

2. You are on the top of a 100 m. building and throw upwards a stone with an initial velocity of 2 m/s. Disregarding the resistance of air, what is its velocity when it reaches the ground?

   All the problems about kinematics of falling objects can be solved using two formulas from elementary physics, namely $v(t) = -gt + v_o$ and $s(t) = -\frac{1}{2}gt^2 + v_ot + s_o$, where g is the acceleration of gravity $(9.8m/s^2)$, v_o is the initial velocity, s_o is the initial position of the object, $v(t)$ is the velocity at any instant t, and $s(t)$ is the position of the object at any instant t (while the object is in air). It is to be noted that, in this type of problems, the positive direction is chosen to be upwards while the negative direction is downwards, and we are disregarding the resistance of air.

   In the specific problem under consideration we have $v(t) = -9.8t + 2$, and $s(t) = -4.9t^2 + 2t + 100$. The stone will hit the ground when $0 = -4.9t^2 + 2t + 100$.

Applying the quadratic formula we get[2] $t = 4.72623$ seconds. Then $v(4.72623) = -9.8 \times 4.72623 + 2 \approx$ -44.32 m/s.

3. A stone is dropped into a well and, after three seconds, one hears it hitting the bottom. Calculate d, the depth of the well, assuming that the only force acting on the stone is gravity ($g = 9.8m/s^2$) and the speed of sound is $340m/s$.

   If we do not consider the speed of sound, the solution to the problem is straightforward since $d = \frac{1}{2}9.8 \times 3^2 = 44.1m$ (we are choosing the downward direction as the positive one). But this solution is not realistic; rather, we have $3 = t_1 + t_2$, where t_1 is the time it takes the stone to reach the bottom and t_2 is the time it takes sound to travel from the bottom to the top of the well. Thus $t_2 = d/340$ and $d = \frac{1}{2}(9.8)t_1^2$. Therefore

$$3 = \sqrt{\tfrac{d}{4.9}} + \tfrac{d}{340}.$$

   Let $x = \sqrt{d}$. Then

$$0 = -3 + \tfrac{1}{\sqrt{4.9}}x + \tfrac{1}{340}x^2.$$

   Using the formula for quadratics we get $x = 6.3761$, consequently $d = x^2 = 6.3761^2 = 40.6547m$.

11.3 Some theoretical considerations

We may notice that the quadratic equation $ax^2 + bx + c = 0$ has only one root if and only if $b^2 - 4ac = 0$. The unique solution happens to be $-\frac{b}{2a}$. Customarily we will write

$$\Delta = b^2 - 4ac,$$

and call Δ the **discriminant** of the given quadratic. Thus, if $\Delta > 0$ there are two real solutions, if $\Delta = 0$ there exists a unique real solution, and if $\Delta < 0$ there are two complex solutions that happen to be conjugates of each other since the coefficients of the quadratic under scrutiny are real numbers[3].

We will prove, independently from FTA, that

$$ax^2 + bx + c = a(x - r_1)(x - r_2),$$

[2] The negative solution is not taken into account because it does not have physical meaning.
[3] Recall the property discussed at the beginning of section 10.4.

where r_1 and r_2 are the solutions of $ax^2 + bx + c = 0$. Let us assume that $\Delta \geq 0$. Then

$$r_1 = -\frac{b}{2a} + \frac{\sqrt{b^2-4ac}}{2a}, \text{ and } r_2 = -\frac{b}{2a} - \frac{\sqrt{b^2-4ac}}{2a}.$$

So $r_1 + r_2 = -\frac{b}{a}$ while

$$r_1 r_2 = (-\frac{b}{2a})^2 - \frac{b^2-4ac}{4a^2} = \frac{b^2-b^2+4ac}{4a^2} = \frac{c}{a}.$$

Therefore

$$a(x-r_1)(x-r_2) = a(x^2 - (r_1+r_2)x + r_1 r_2) = a(x^2 + \frac{b}{a}x + \frac{c}{a}) = ax^2 + bx + c.$$

The proof when $\Delta < 0$ is very similar.

It should be noted that when $\Delta = 0$ we will have

$$ax^2 + bx + c = a(x - (-\frac{b}{2a}))^2 = a(x + \frac{b}{2a})^2.$$

Remark

Suppose we have to solve the equation $x^2 + bx + c = 0$ and, by trial and error, we are able to find two numbers r, s such that $rs = c$ and $r + s = b$. Then the solutions of the equation are $-r$ and $-s$. Why is this so? Very simple:

$$x^2 + bx + c = x^2 + (r+s)x + rs = (x+r)(x+s).$$

Thus, the original equation becomes $(x+r)(x+s) = 0$. Obviously, the two solutions of this equation are $-r, -s$. This method for finding solutions is particularly useful when b and c are integers. For instance, given the equation $x^2 - 3x - 10 = 0$ we observe that $-5 \times 2 = -10$ and $-5 + 2 = -3$, thus the two roots are 5 and -2. A similar approach, although a little more laborious, can be followed when dealing with the equation $ax^2 + bx + c = 0$ (see exercise 8 at the end of the chapter).

11.4 Solving a geometrical puzzle

Suppose that a circle with radius 1 is inscribed in a right triangle with hypotenuse b. What inequality must the hypotenuse satisfy? Evidently, the hypotenuse has to be big enough in order to accommodate the inscribed circle of unit radius. Let x and y be the length of the legs (Figure 11.1). Applying proposition 7.2.2 twice we realize that $y - 1 = b - x + 1$, hence

$$y = b - x + 2.$$

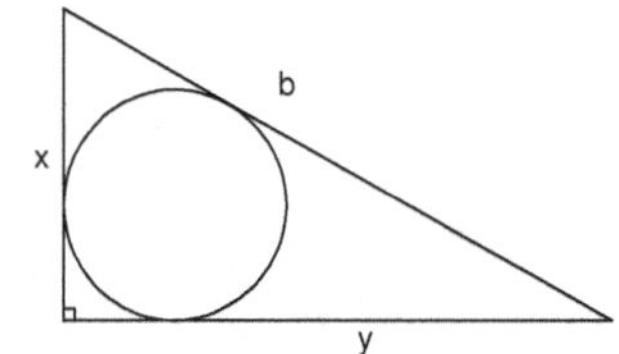

Figure 11.1: A geometrical puzzle

But in section 7.2 we found the area of a triangle in terms of the semiperimeter s and the inradius r, namely $K = s \times r$. Hence

$$\tfrac{1}{2}xy = \tfrac{1}{2}(x + y + b) \times 1.$$

That is, $xy = x + y + b$; thus, $x(b - x + 2) = x + b - x + 2 + b$. Simplifying this equation we get the quadratic equation

$$x^2 - (b + 2)x + 2b + 2 = 0.$$

The discriminant of this quadratic must be bigger than or equal to zero because otherwise the original problem has no solution. Consequently $b^2+4b+4-4(2b+2) \geq 0$, which is equivalent to $b^2 - 4b - 4 \geq 0$. Thus $(b - 2)^2 \geq 8$, hence

$$b \geq 2 + \sqrt{8}.$$

11.5 Quadratic inequalities

Suppose that we are given the inequality $x^2 - 3x + 1 < 0$ and are asked to solve it. The first idea that comes to our mind is to apply the method of completion of squares, which we used with great success when dealing with quadratic equalities. Adding and subtracting $(\frac{3}{2})^2$ we get

$$x^2 - 3x + (\tfrac{3}{2})^2 + 1 - (\tfrac{3}{2})^2 < 0.$$

Therefore

$$(x - \tfrac{3}{2})^2 < \tfrac{9}{4} - \tfrac{4}{4} = \tfrac{5}{4}.$$

Keeping in mind that $\sqrt{a^2} = |a|$ for any real number a, and extracting the square root to both sides of the inequality, we reach $|x - \frac{3}{2}| < \frac{\sqrt{5}}{2}$. Therefore

$$-\tfrac{\sqrt{5}}{2} < x - \tfrac{3}{2} < \tfrac{\sqrt{5}}{2},$$

that is, $\frac{3-\sqrt{5}}{2} < x < \frac{3+\sqrt{5}}{2}$. The solution happens to be the open interval $(\frac{3-\sqrt{5}}{2}, \frac{3+\sqrt{5}}{2})$ on the real line.

Let us try to solve the inequality $x^2 - 3x + 1 > 0$. Completing squares as in the preceding inequality, and extracting square roots, we get $|x - \frac{3}{2}| > \frac{\sqrt{5}}{2}$. Therefore

$$x - \tfrac{3}{2} > \tfrac{\sqrt{5}}{2} \text{ or } -(x - \tfrac{3}{2}) > \tfrac{\sqrt{5}}{2}.$$

Consequently

$$x > \tfrac{3+\sqrt{5}}{2} \text{ or } x < \tfrac{3-\sqrt{5}}{2}.$$

What can be said about the inequality $x^2 + x + 1 > 0$? Again, completion of squares leads to

$$x^2 + x + \tfrac{1}{4} - \tfrac{1}{4} + 1 > 0,$$

thus $(x + \frac{1}{2})^2 > -\frac{3}{4}$. This last inequality is satisfied by every real number x, hence the solution is the whole real line.

If instead of $x^2 + x + 1 > 0$ we wish to analyze $x^2 + x + 1 < 0$, right away we get $(x + \frac{1}{2})^2 < -\frac{3}{4}$. Obviously, no real number x satisfies this last inequality; thus, $x^2 + x + 1 < 0$ has no solutions in the real field.

By now the reader must have noticed that, given any quadratic inequality with real coefficients, there are only four possibilities: either the solution set is an open interval, a union of open intervals, the whole real line, or the empty set. Of course, we are looking for solutions on the real field since the complex numbers do not constitute an ordered field.

11.6 Vertical parabolas

Let us consider the point $(0, a)$, $a > 0$, and the horizontal line $y = -a$ (Figure 11.2). As the reader might remember from pre-calculus mathematics, a vertical parabola is defined as the set of all points such that its distance to $(0, a)$ is the same as the distance to the horizontal line $y = -a$. Thus, (x, y) is on the parabola if and only if

$$(x - 0)^2 + (y - a)^2 = (y + a)^2.$$

Hence

$$x^2 + y^2 - 2ay + a^2 = y^2 + 2ay + a^2,$$

so $x^2 = 4ay$. That is,

$$y = \frac{1}{4a}x^2.$$

The point $(0, a)$ is called the 'focus' and the line $y = -a$ receives the name of 'directrix'. We can see that the parabola passes through $(0, 0)$, which is called the vertex of the parabola, and opens upwards. In particular, $y = x^2$ is the equation of a verti-

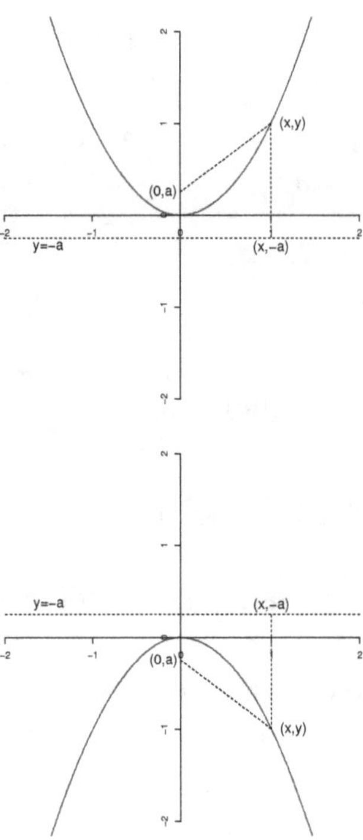

Figure 11.2: Two typical vertical parabolas with vertex at the origin

cal parabola with focus at $(0, \frac{1}{4})$. In general, $y = bx^2$ represents a vertical parabola with focus at $(0, a)$ where $\frac{1}{4a} = b$; thus, $a = \frac{1}{4b}$. The corresponding directrix is the horizontal line $y = -\frac{1}{4b}$.

In a similar fashion, if $a < 0$ then the vertical parabola with focus at $(0, a)$ and directrix $y = -a$ is given by $y = \frac{1}{4a}x^2$. This time the parabola opens downwards (Figure 11.2).

Remark

We can avoid working with the number 4 and just work with $x^2 = py$ provided that we realize that under these circumstances the focus is $(0, p/4)$ and the directrix is given by $y = -p/4$. Or, if you wish, it is possible to work with $x^2 = 2py$ provided we keep in mind that then the focus is $(0, p/2)$ and the directrix is defined by $y = -p/2$.

11.7 Translation of a parabola

Suppose that we translate, without rotation, the parabola represented by the equation $y = ax^2$ in such a way that its new vertex lies at (h, k) (Figure 11.3). The new axes x', y' are related to the original axes x, y through the expressions $x = x' + h$ and $y = y' + k$. The translated parabola will have the equation $y' = ax'^2$. Thus

$$y - k = a(x - h)^2. \tag{11.1}$$

Of course, if we were to encounter the expression (11.1), without hesitation we can assert that it represents a vertical parabola with vertex at (h, k). It opens upwards if $a > 0$, while it opens downwards if $a < 0$. Moreover, any expression of the form $y = ax^2 + bx + c$ represents a vertical parabola because we can always apply the method of completion of squares in order to transform it into $y - k = a(x - h)^2$.

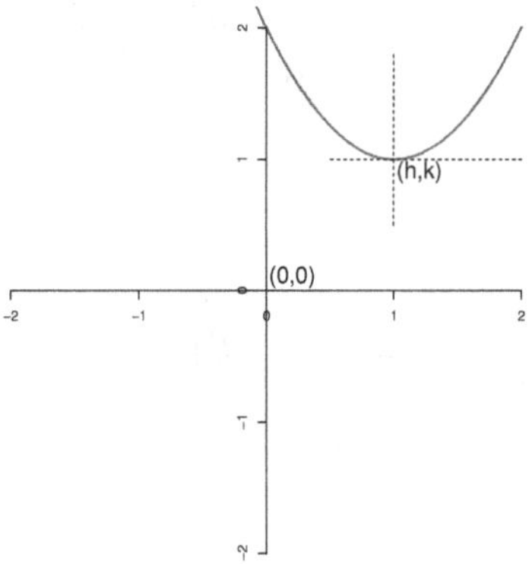

Figure 11.3: Translated vertical parabola

Examples

1. Find the vertex and focus of the parabola with equation $y = 2x^2 - 3x + 5$. Right away we use completion of squares:

$$y = 2(x^2 - \tfrac{3}{2}x + \tfrac{5}{2}) = 2((x - \tfrac{3}{4})^2 - \tfrac{9}{16} + \tfrac{40}{16}).$$

Thus

$$y = 2(x - \tfrac{3}{4})^2 + \tfrac{62}{16}.$$

Hence the vertex is at $(\frac{3}{4}, \frac{31}{8})$. Let $x' = x - \frac{3}{4}$, and $y' = y - \frac{31}{8}$. Then $y' = 2x'^2$. From $\frac{1}{4a} = 2$ we get $a = \frac{1}{8}$, consequently the focus is at $(0, \frac{1}{8})'$, which is equal to $(\frac{3}{4}, \frac{1}{8} + \frac{31}{8})$; that is, $(\frac{3}{4}, 4)$.

2. Find the vertex and focus of the parabola with equation $y = -2x^2 - 3x + 1$. First of all we note that

$$y = -2(x^2 + \tfrac{3}{2}x - \tfrac{1}{2}) = -2((x + \tfrac{3}{4})^2 - \tfrac{9}{16} - \tfrac{8}{16}) = -2(x - (-\tfrac{3}{4}))^2 + \tfrac{17}{8}.$$

Thus, the vertex is at $(-\frac{3}{4}, \frac{17}{8})$. Next, let us try to find the focus. With this purpose in mind define

$$y' = y - \tfrac{17}{8} \text{ and } x' = x + \tfrac{3}{4}.$$

Thus $y' = -2x'^2$, the equation of the parabola with respect to the new axes. Since $\frac{1}{4a} = -2$ we get $a = -\frac{1}{8}$. Therefore, the focus is $(0, -\frac{1}{8})'$. But $0 = x + \frac{3}{4}$, that is, $x = -\frac{3}{4}$. Similarly, $-\frac{1}{8} + \frac{17}{8} = y$, that is, $y = 2$. Consequently, the focus of the parabola is at $(-\frac{3}{4}, 2)$.

3. Suppose that a javelin is thrown up with an initial velocity v_o, describing an angle θ with the horizontal. According to basic laws of physics, and disregarding the resistance of air, we have

$$x = (v_o \cos\theta)t \tag{11.2}$$

$$y = (v_o \sin\theta)t - \frac{1}{2}gt^2. \tag{11.3}$$

From (11.2) it follows that $t = \frac{x}{v_o \cos\theta}$. Replacing this value in (11.3) we obtain

$$y = \frac{x(v_o \sin\theta)}{v_o \cos\theta} - \frac{1}{2}g\frac{x^2}{v_o^2 \cos^2\theta} = -\frac{g}{2v_o^2 \cos^2\theta}x^2 + x\tan\theta.$$

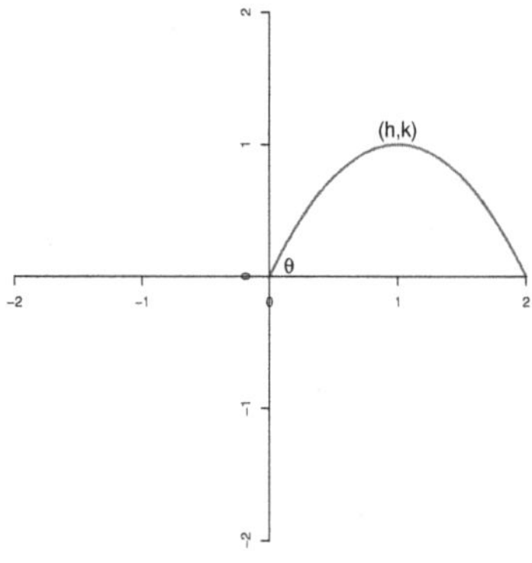

Figure 11.4: The path described by the javelin

This is a parabola that opens downwards. At what angle will the javelin reach the maximum horizontal distance from the origin? By looking at (11.3) we notice that the object will touch the ground when

$$0 = (v_o \sin\theta)t - \tfrac{1}{2}gt^2,$$

thus $t = \frac{2v_o \sin\theta}{g}$. Hence, from (11.2) it follows that

$$x = v_o \cos\theta(\tfrac{2v_o \sin\theta}{g}) = \tfrac{v_o^2}{g}\sin 2\theta.$$

The horizontal distance will be a maximum provided $\sin 2\theta = 1$. Thus $2\theta = 90$, that is, $\theta = 45$.

Using parabolas to solve inequalities

At the beginning of section 11.5 we solved the inequalities $x^2 - 3x + 1 < 0$ and $x^2 - 3x + 1 > 0$. Let us study an approach to the solutions using parabolas. Define the parabola $y = x^2 - 3x + 1$, which opens upwards since its leading coefficient is positive. The solutions of $x^2 - 3x + 1 = 0$, namely $s = \frac{3+\sqrt{5}}{2}$ and $r = \frac{3-\sqrt{5}}{2}$, are the points where the parabola crosses the horizontal axis. From Figure 11.5 we can see that the solution to the first inequality is the open interval $(\frac{3-\sqrt{5}}{2}, \frac{3+\sqrt{5}}{2})$ while the solution to the second inequality is the union of the open intervals $(-\infty, \frac{3-\sqrt{5}}{2})$ and $(\frac{3+\sqrt{5}}{2})$.

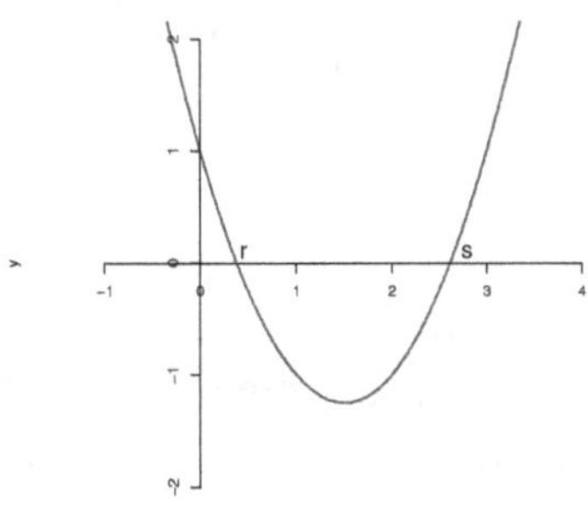

Figure 11.5: Solving a quadratic inequality using a parabola

Parabolas can help solve any quadratic inequality of the types $ax^2 + bx + c \geq 0$ or $ax^2 + bx + c \leq 0$.

11.8 Maxima and minima

If a vertical parabola opens upwards, the **minimum** is adopted at the vertex. On the other hand, if a parabola opens downwards, the **maximum** is attained at the vertex. Given the expression

$$y = ax^2 + bx + c$$

we observe that

$$y = a(x^2 + \tfrac{b}{a}x + \tfrac{c}{a}) = a[(x + \tfrac{b}{2a})^2 + \tfrac{c}{a} - \tfrac{b^2}{4a^2}].$$

Thus

$$y = a(x - (-\tfrac{b}{2a}))^2 + c - \tfrac{b^2}{4a}.$$

We can then observe that the parabola has its vertex at $(-\frac{b}{2a}, c - \frac{b^2}{4a})$. From a practical point of view, to determine the maximum or the minimum we have only to keep in mind the formula $-\frac{b}{2a}$.

Examples

1. Among all rectangles of fixed perimeter p, which one encloses the largest area? Let $s = p/2$ and denote by x and y the length of the sides of any rectangle with perimeter p. Since $x + y = s$ we can define the area function,

   $$A(x) = x(s - x) = -x^2 + sx.$$

   The minimum is adopted at $x = \frac{-s}{-2} = \frac{s}{2}$. Then the corresponding y is $y = s - \frac{s}{2} = \frac{s}{2}$; thus, the answer is to build a square with side $s/2$.

2. A farmer wants to fence a terrain in the shape of a rectangle, one of whose sides borders a river[4]. The farmer has 400 meters of wire. How should the rectangle be built in order to maximize its area? Let x denote the length of the sides perpendicular to the river and y the length of the side parallel to the river. We have $2x + y = 400$, while the area A is xy. Therefore

   $$A(x) = x(400 - 2x) = -2x^2 + 400x.$$

   We are dealing with the parabola $y = -2x^2 + 400x$, whose maximum is attained at $x = -\frac{400}{2(-2)} = 100$. Then the corresponding y happens to be $y = 400 - 2(100) = 200$. These are the 'optimal' lengths.

3. A bus is 25 meters ahead of a pedestrian, both at rest. At the same instant of time, the bus accelerates at $1m/s^2$ and the pedestrian starts running at the constant velocity of $6m/s$. Is the pedestrian able to catch the bus? If not, at what time is the distance between the bus and the pedestrian a minimum? From elementary physics we know that the distance traveled by the pedestrian is $x_p = 6t$, while the distance traveled by the bus is $x_b = 25 + \frac{1}{2}t^2$ (Figure 11.6).

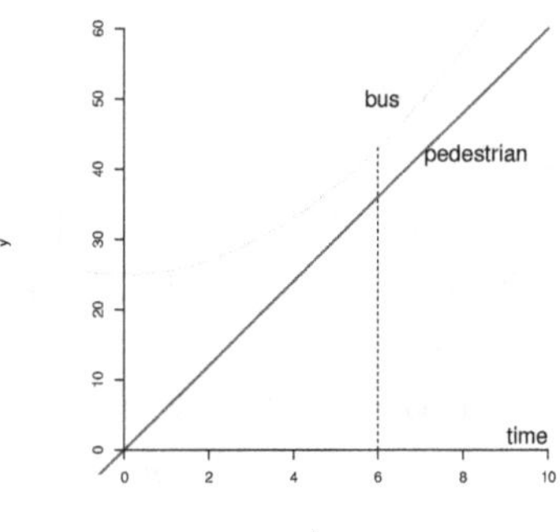

Figure 11.6: The bus and the pedestrian problem

To answer the first question we have to solve the equation $6t = 25 + \frac{1}{2}t^2$, that is,

$$t^2 - 12t + 50 = 0.$$

Since the discriminant of the equation is negative we can conclude that at no time the pedestrian will catch the bus. Let $D(t)$ denote the distance between the bus and the pedestrian, at any time t (Figure 11.7). Thus

$$D(t) = 25 + \tfrac{1}{2}t^2 - 6t = \tfrac{1}{2}(t^2 - 12t + 50) = \tfrac{1}{2}((t-6)^2 + 14) = \tfrac{1}{2}(t-6)^2 + 7.$$

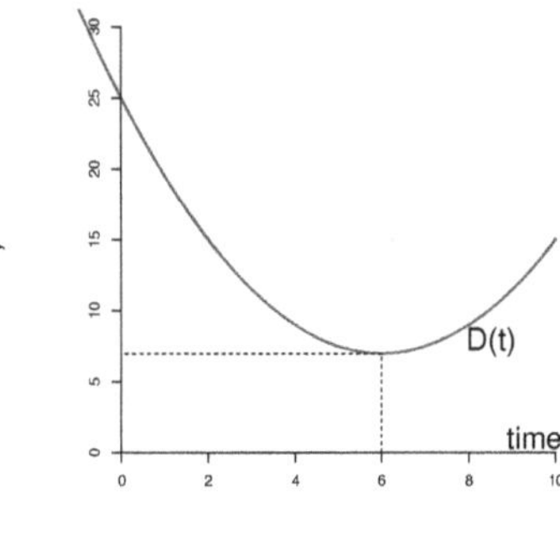

Figure 11.7: The distance between the bus and the pedestrian

Hence, this function adopts its minimum at $t = 6$. Of course, if we were to use the $\frac{-b}{2a}$ formula we would get, as expected, the same answer: $t = \frac{6}{2(1/2)} = 6$.

[4]No fence is needed on the side next to the river.

11.9 Tangent lines to parabolas

We say that a line is tangent to a parabola at a point on the curve if the line intersects, but does not cross, the parabola at the point. In other words, the tangent at a point is a line that touches the parabola only at the given point. From this definition we can obtain the equation of the tangent line T at any point (a, a^2) on the parabola (Figure 11.8) defined by the equation $y = x^2$: Let m be the slope of any line that passes through (a, a^2). The equation of this line is $y - a^2 = m(x - a)$. Since T meets the parabola at one, and only one point, it follows that the system of equations

$$y = x^2, \qquad y - a^2 = m(x - a)$$

has to have only one solution. That is to say, the quadratic equation $x^2 - a^2 = m(x - a)$ ought to have only one solution. We can rewrite this equation as

$$x^2 - mx + (ma - a^2) = 0.$$

The discriminant of this equation has to be zero, that is,

$$m^2 - 4(ma - a^2) = 0.$$

Therefore $(m - 2a)^2 = 0$, which in turn implies $m = 2a$. We have been able to show a very important fact, namely, the slope of the tangent line at any point (a, a^2) on the parabola $y = x^2$ is precisely $2a$. Hence

$$y - a^2 = 2a(x - a)$$

is the equation of the tangent line at the point (a, a^2) on the above-mentioned parabola.

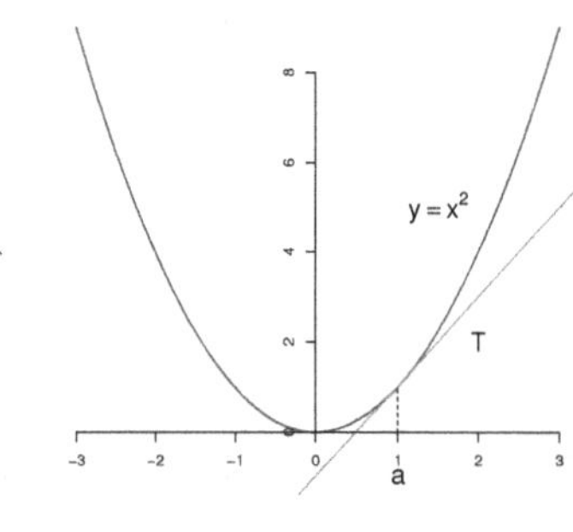

Figure 11.8: Tangent line to the parabola $y = x^2$ at $x = a$

Remark

Can we find the equation of the tangent T at any point $(a, Aa^2 + Ba + C)$ of the parabola defined by $y = Ax^2 + Bx + C$? Yes, we can! Following, step by step, the procedure presented before, it is possible to prove that $m = 2Aa + B$ is the slope of the tangent at (a, Aa^2+Ba+C). Consequently, $y-(Aa^2+Ba+C) = (2Aa+B)(x-a)$ is the equation of the straight line T. Moreover, this approach is feasible whenever we have to find the equation of the tangent to any conic (Baloglou and Helfgott, 2004).

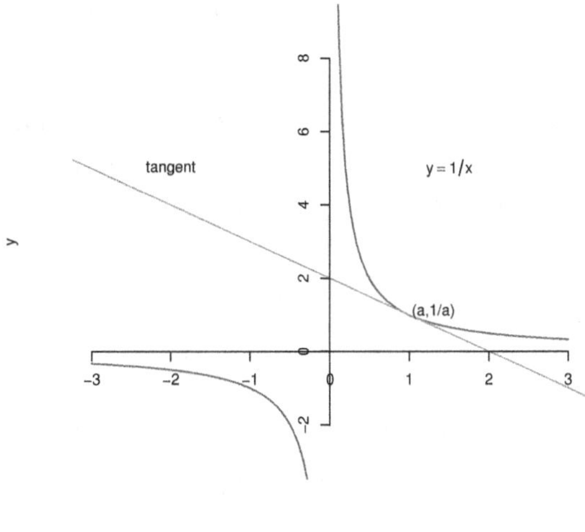

Figure 11.9: The tangent to the curve $y = \frac{1}{x}$ at a given point

For instance, let us deal with the hyperbola $y = \frac{1}{x}$ and an arbitrary point $(a, \frac{1}{a})$ on it (Figure 11.9). The equation of any line that passes through this point is $y - \frac{1}{a} = m(x-a)$. Then we have to analyze the solutions of

$$\frac{1}{x} - \frac{1}{a} = m(x-a).$$

A little bit of algebraic manipulations leads to the quadratic equation

$$mx^2 + (\tfrac{1}{a} - ma)x - 1 = 0.$$

The discriminant of it has to be zero, therefore $(\frac{1}{a} - ma)^2 + 4m = 0$. Simplifying we get to

$$(am + \tfrac{1}{a})^2 = 0,$$

thus $am + \frac{1}{a} = 0$. Then $m = -\frac{1}{a^2}$. In summary, the equation of the tangent to the hyperbola $y = \frac{1}{x}$ at $(a, \frac{1}{a})$ is

$$y - \tfrac{1}{a} = -\tfrac{1}{a^2}(x-a).$$

11.10 Area under a parabola

Let us try to find the area under the curve $y = x^2$ between 0 and $b > 0$. We start by dividing $[0, b]$ in n subintervals of equal length and build the corresponding rectangles. The area of the jth rectangle is $\frac{b}{n} \times (\frac{jb}{n})^2$ because $\frac{b}{n}$ is the width and $(\frac{jb}{n})^2$ is its height. Thereafter we add the areas of all rectangles and observe that

$$\Sigma_{j=1}^{n} \frac{b}{n}(\frac{jb}{n})^2 = \Sigma_{j=1}^{n} \frac{b^3}{n^3} j^2 = \frac{b^3}{n^3} \Sigma_{j=1}^{n} j^2 = \frac{b^3}{n^3} \Sigma_{j=1}^{n} j^2.$$

But in section 1.1 we found that $\Sigma_{j=1}^{n} j^2 = \frac{n}{6}(n+1)(2n+1)$, thus

$$\Sigma_{j=1}^{n} \frac{b}{n}(\frac{jb}{n})^2 = \frac{b^3}{6}(2 + \frac{3}{n} + \frac{1}{n^2}) \to \frac{b^3}{3},$$

where in the last step we have taken the limit when n approaches ∞. Thus, $\frac{b^3}{3}$ is the area under the curve $y = x^2$ between 0 and $b > 0$. In a similar fashion we can show that when $c < 0$ the area under the parabola $y = x^2$, between c and 0, is $-\frac{c^3}{3}$.

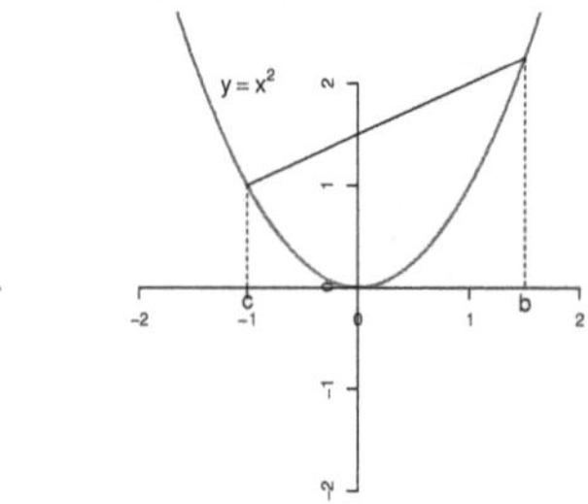

Figure 11.10: Area of a segment of a parabola

Area of a segment of a parabola

The ground is now ready to calculate the area bounded by the segment joining (c, c^2) and (b, b^2), and the curve $y = x^2$, (Figure 11.10). Such a region is often called **segment of a parabola**. We observe that

$$\text{area of segment} = \frac{(b-c)}{2}(b^2 + c^2) - (-\frac{c^3}{3} + \frac{b^3}{3}) = \frac{1}{6}(b^3 - c^3 + 3bc^2 - 3b^2c) = \frac{1}{6}(b - c)^3.$$

11.11 The reflective property of a parabola

Let us consider the curve $y = x^2$. As we already know, the tangent line to $y = x^2$ at (a, a^2) is $y - a^2 = 2a(x - a)$. Obviously, the intersection of this line with the y-axis is $(0, -a^2)$.

Suppose that a ray of light comes from a distant star and hits the mirror, whose two-dimensional projection[5] is defined by the curve $y = x^2$, at (a, a^2). We can assume that the ray is parallel to the y-axis, the axis of the parabola. The law of reflection of light asserts that the angles of incidence and reflection, measured with regard to the normal[6], are congruent (Figure 11.11). Letting α be their common measure, we can assert that $\angle LCB$ has measure 2α. Once reflected, the ray intersects the y-axis at $(0, b)$. Since the ray is parallel to the y-axis before hitting the mirror, it follows that $\angle ABC$ has measure 2α too. Let $\gamma = m\angle BAC$, $\beta = \angle BCA$. Then $\gamma + 2\alpha + \beta = 180$

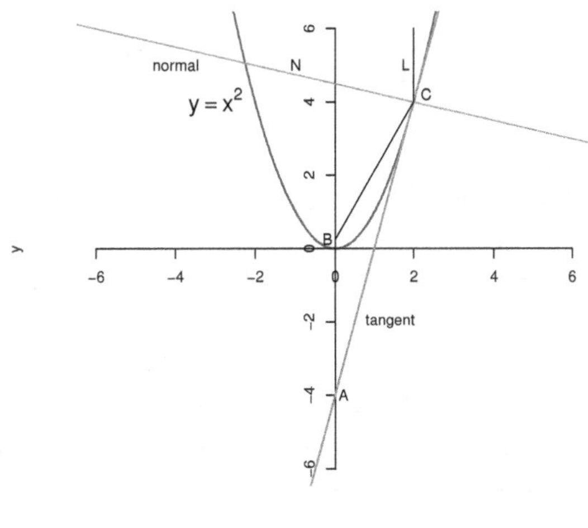

Figure 11.11: A two-dimensional simplification of a parabolic telescope

and $\alpha + \beta = 90$, thus

$$\gamma + 2\alpha + \beta = 2\alpha + 2\beta,$$

which in turn leads to $\gamma = \beta$. Hence, the triangle under consideration is isosceles. Therefore

$$\sqrt{(b + a^2)^2} = \sqrt{a^2 + (a^2 - b)^2}.$$

An elementary calculation leads to $b = 1/4$. We have then proven that any ray of light that comes from a distant star passes through the focus of the parabola.

[5]A real parabolic mirror is certainly three-dimensional. Nonetheless, we are studying a two-dimensional simplification of it.

[6]The normal being considered is the perpendicular to the tangent at (a, a^2).

Exercises

1. Find the sides of the rectangle with perimeter 12 and area 2.

2. Find the sides of a rectangle with perimeter p and area A. Under what circumstances does the problem have a solution?

3. Solve the equation $x^2 + x + 1 = 0$.

4. Use the method of completion of squares to solve the quadratic inequality $x^2 - 4x + 3 > 0$.

5. Solve the quadratic inequality $2x^2 - 6x + 1 < 0$ by any method of your choice.

6. Solve the equation $x^2 + ix - 1 = 0$.

7. A person, standing on the top of a building, drops a stone and after 4 seconds she hears it hit the ground. How tall is the building? (recall that the acceleration of gravity is 9.8 m/s^2 and the speed of sound is 340 m/s).

8. Suppose that you wish to solve the equation $ax^2 + bx + c = 0$ and, by trial and error, you are able to find numbers u, v, r, s such that $uv = a$, $rs = c$, and $vr + us = b$. Show that then $ax^2 + bx + c = (ux + r)(vx + s)$; thus, the solutions of the quadratic equation will be $-\frac{r}{u}, -\frac{s}{v}$.

9. Use the preceding exercise to solve the equation $3x^2 - 7x + 2 = 0$.

10. You wish to build a 1-inch tall rectangular box made of cardboard, without a top, such that its base is a rectangle with length 2 inches larger than its width. With this purpose in mind you cut off a 1-inch square from each corner of the rectangle and then properly fold the resulting region, which has the shape of a cross. Assuming that the intended volume of the box is 15 cubic inches, what are the dimensions of the original rectangle?

11. An individual throws a small stone upwards, from the top of a 60 m. building, with an initial velocity of 5 m/s. Disregarding the resistance of air, what is its velocity when it reaches the ground?

12. Find the vertex and focus of the parabola described by the equation $y = -x^2 + 5x - 1$.

13. At what point does the parabola $y = x^2 + 5x - 1$ adopt its minimum?

14. Find the equation of the tangent to the parabola $y = x^2 - x + 1$ at $(1, 1)$.

15. What point (or points) on the parabola $y = 2x^2 - x - 1$ has the property that the tangent to it passes through $(1, -2)$?

16. Calculate the equation of the tangent to the hyperbola $y = \frac{2}{x}$ at $(1, 2)$.

17. Solve the inequality $x^2 - 5x + 2 < 0$ by plotting the parabola $y = x^2 - 5x + 2$ and calculating the zeros[7] of it.

18. Solve the inequality $3x^2 + 2x - 1 > 0$ by plotting the parabola $y = 3x^2 + 2x - 1$ and calculating the zeros of it.

19. Consider the parabola $y = \frac{1}{2p}x^2$ and an arbitrary point $A = (a, \frac{a^2}{2p})$ on it. Let B be the point of intersection between the tangent through A and the directrix. Calculate $m\angle AFB$, where F is the focus of the parabola.

20. Let $y = \frac{1}{2p}x^2$. Through the focus of this parabola we draw a line parallel to the horizontal axis (the 'latus rectum' of the parabola). This line intersects the parabola at the points $(p, p/2)$ and $(-p, p/2)$. Then we draw the tangents, L_1 and L_2, to the parabola at both points. Show that L_1 is perpendicular to L_2 and that they intersect at a point that lies on the directrix.

21. A farmer wants to fence a terrain in the shape of a rectangle, with one of its sides next to a river (the latter side requires no fence). Assuming that the farmer has L meters of wire, how should the rectangle be built in order to maximize its area?

22. Given the parabola $y = \frac{1}{4}x^2$, show that the triangle formed by the point $(a, \frac{1}{4}a^2)$, the focus, and the intersection of the tangent line (at the above-mentioned point) with the vertical axis is an isosceles triangle.

23. Let us consider the two-dimensional analogue of the headlights of a truck, with shape $y = x^2$. Show that any ray of light that starts at the focus $(0, 1/4)$, and touches the parabolic mirror, will emerge on a path parallel to the vertical axis.

24. A bus is d meters ahead of a pedestrian, both at rest. Simultaneously, the bus accelerates at a m/s^2 and the pedestrian starts running at the constant speed of v m/s. Under what circumstances the pedestrian will not catch the bus?

25. In the preceding problem, assuming that the pedestrian does not catch the bus, at what time is the distance between the bus and the pedestrian a minimum? (Newburgh, 1997).

[7]The zeros of a parabola are the points where the curve intersects the x-axis.

26. A circle of radius r is inscribed inside a right triangle with hypotenuse of length b. What inequality does b have to satisfy?

27. Find the area under the curve $y = x^3$, $0 \leq x \leq b$, and the $x-$ axis. You are expected not to use Calculus.

Enrichment Note: The volume of a cone and of a sphere

From their earliest mathematical experience most readers might remember that the volume of a cylinder is $\pi r^2 h$, where r is the radius of the base and h is the height. Using this result, and a modicum of Cartesian geometry, we will be able to find the volume of a cone and the volume of a sphere. Interestingly, as in the calculation of the area under a parabola, again we will use the formula for $\Sigma_{j=1}^{n} j^2$ found in section 1.1.

Let us start discussing a cone with radius r and height h. If we revolve the line $y = \frac{r}{h}x$, around the x-axis, we will obtain the above-mentioned cone. Let us subdivide the interval $[0, h]$ in n subintervals of equal length. If we revolve, around the x-axis, the jth rectangle of width $\frac{h}{n}$ and corresponding height $\frac{r}{h}(\frac{jh}{n})$, that is, $\frac{rj}{n}$, we obtain a cylinder with volume $V_j = \pi(\frac{rj}{n})^2 \frac{h}{n}$. Thus

$$V_j = \frac{\pi r^2 h}{n^3} j^2.$$

Adding the volumes of the n cylinders we get

$$V_n = \frac{\pi r^2 h}{n^3} \Sigma_{j=1}^{n} j^2.$$

But $\Sigma_{j=1}^{n} j^2 = \frac{n}{6}(n+1)(2n+1)$. Hence

$$V_n = \frac{\pi r^2 h}{n^3} \frac{n}{6}(n+1)(2n+1) = \frac{\pi r^2 h}{6n^2}(2n^2 + 3n + 1) = \frac{\pi r^2 h}{6}(2 + \frac{3}{n} + \frac{1}{n^2}) \to \frac{\pi r^2 h}{3}$$

as $n \to \infty$. Therefore, the volume of the cone is simply $V = \frac{\pi r^2 h}{3}$.

Our next job is to calculate the volume of a sphere of radius r and, for this purpose, we will calculate the volume of the corresponding hemisphere. The latter is obtained by revolving the curve $y = \sqrt{r^2 - x^2}$ around the x-axis. Let us subdivide $[0, r]$ in n subintervals of equal length. The jth rectangle, with width $\frac{r}{n}$ and height $\sqrt{r^2 - (\frac{jr}{n})^2}$ generates a cylinder when it is revolved around the x-axis. Its volume is

$$V_j = \pi(\sqrt{r^2 - (\frac{jr}{n})^2})^2 \frac{r}{n} = \frac{\pi r^3}{n} - \frac{\pi r^3}{n^3} j^2.$$

Then

$V = \Sigma_{j=1}^n V_j = \Sigma_{j=1}^n \frac{\pi r^3}{n} - \frac{\pi r^3}{n^3} \Sigma_{j=1}^n j^2 = \pi r^3 - \frac{\pi r^3}{n^3} \frac{n}{6}(n+1)(2n+1) = \pi r^3 - \frac{\pi r^3}{6}(2+\frac{3}{n}+\frac{1}{n^2}).$

As $n \to \infty$, the last expression converges to $\pi r^3 - \frac{\pi r^3}{3}$, that is, $\frac{2\pi r^3}{3}$. Then, the volume of the sphere will be $\frac{4\pi r^3}{3}$.

A different approach to the volume of a sphere, using Cavalieri's Principle, can be found in the University of Chicago High School Geometry textbook (Usiskin et al., 2002). A nice heuristic derivation of the surface area of a sphere can be found there too.

Chapter 12

Biquadratic Equations and Functions

12.1 Two examples of biquadratic equations

Biquadratic equations are of the type $x^4 + ax^2 + b = 0$, where a and b are real numbers. Let us start by discussing the equation $x^4 - 1 = 0$. We define $y = x^2$, thus the equation becomes $y^2 = 1$. Hence $y^2 = 1$, which leads to $y = \pm 1$. Then $x^2 = 1$ or $x^2 = -1$. Consequently, the four solutions of the original equation are ± 1 and $\pm i$.

Next we will analyze the biquadratic $x^4 + 1 = 0$. If we use the 'natural' transformation $y = x^2$ we get $y^2 = -1$, thus $y = \pm i$. Therefore $x^2 = \pm i$. We could calculate the square root of i and $-i$, but let us choose another approach: $x^4 + 1 = 0$ can be written $(x^2+1)^2 = 2x^2$, that is, $(x^2+1)^2 = (\sqrt{2}x)^2$. Consequently

$$x^2 + 1 = \sqrt{2}x \text{ or } x^2 + 1 = -\sqrt{2}x.$$

Thus, all we have to do is solve the quadratic equations

$$x^2 - \sqrt{2}x + 1 = 0 \text{ and } x^2 + \sqrt{2}x + 1 = 0.$$

Each of them will provide us two solutions. The four of them, namely

$$\tfrac{\sqrt{2}}{2} \pm i\tfrac{\sqrt{2}}{2}, \quad -\tfrac{\sqrt{2}}{2} \pm i\tfrac{\sqrt{2}}{2},$$

are the solutions of $x^4 + 1 = 0$.

We are ready to handle the problem of solving $x^4 + b = 0$, where b is an arbitrary real number[1] . We have to consider, separately, the cases $b < 0$ and $b > 0$:

[1]We are assuming that $b \neq 0$ because the equation $x^4 = 0$ obviously has 0 as the only solution.

1. Assume that $b < 0$, that is, $-b > 0$. The equation under scrutiny is equivalent to $x^4 = -b$. Defining $y = x^2$ leads to $y^2 = -b$, hence $y = \pm\sqrt{-b}$. Therefore $x^2 = \pm\sqrt{-b}$ and consequently the four solutions of the original equation are $x = \pm\sqrt{\sqrt{-b}}$, $x = \pm i\sqrt{\sqrt{-b}}$.

2. Assume that $b > 0$, that is, $-b < 0$. The equation $x^4 + b = 0$ is equivalent to

$$((x^2 + \sqrt{b})^2 = 2\sqrt{b}x^2,$$

which in turn is equivalent to

$$(x^2 + \sqrt{b})^2 = \sqrt{2\sqrt{b}}x)^2.$$

Thus

$$x^2 + \sqrt{b} = \sqrt{2\sqrt{b}}x \text{ or } x^2 + \sqrt{b} = -\sqrt{2\sqrt{b}}x.$$

Solving these two quadratics we obtain the four solutions of $x^4 + b = 0$.

12.2 A more general setting

Let us try to solve the biquadratic equation $x^4 - 3x^2 + 2 = 0$. The transformation $y = x^2$ leads to $y^2 - 3y + 2 = 0$, that is, $(y-2)(y-1) = 0$. Thus $y = 2$ or $y = 1$. The equation $x^2 = 2$ has the two solutions $x = \pm\sqrt{2}$ while $x^2 = 1$ has the two solutions $x = \pm 1$. We have succeeded in finding the four roots of the biquadratic, namely $\pm\sqrt{2}$, ± 1.

Next let us consider the equation $x^4 + x^2 + 1 = 0$. The transformation $y = x^2$ leads to $y^2 + y + 1 = 0$, thus $y = -\frac{1}{2} \pm \frac{\sqrt{3}}{2}i$. Therefore

$$x^2 = -\tfrac{1}{2} \pm \tfrac{\sqrt{3}}{2}i.$$

To circumvent the difficulty of finding the square root of a complex number we proceed as follows: the given equation is equivalent to $x^4 + 1 = -x^2$, so

$$(x^2 + 1)^2 = 2x^2 - x^2 = x^2.$$

Thus $x^2 + 1 = x$ or $x^2 + 1 = -x$. From the quadratic equations $x^2 - x + 1 = 0$ and $x^2 + x + 1 = 0$ we get the four solutions of $x^4 + x^2 + 1 = 0$.

We are ready to discuss the general case of a biquadratic, namely $x^4 + ax^2 + b = 0$ with a and b real numbers. Two cases have to be analyzed separately, depending on whether $a^2 - 4b \geq 0$ or $a^2 - 4b < 0$:

1. Assume that $a^2 - 4b \geq 0$. Then $x^4 + ax^2 = -b$, which is equivalent to

$$(x^2 + \tfrac{a}{2})^2 = \tfrac{a^2}{4} - b.$$

That is, $(x^2 + \frac{a}{2})^2 = \frac{a^2-4b}{4}$. Therefore

$$(x^2 + \tfrac{a}{2})^2 = (\tfrac{\sqrt{a^2-4b}}{2})^2,$$

so

$$x^2 + \tfrac{a}{2} = \tfrac{\sqrt{a^2-4b}}{2} \text{ or } x^2 + \tfrac{a}{2} = -\tfrac{\sqrt{a^2-4b}}{2}.$$

From these two quadratics we get the four solutions of $x^4 + ax^2 + b = 0$, which can be real or complex. Of course, we could have defined $y = x^2$ and proceeded to the solutions in a rather smooth way.

2. Assume that $a^2 - 4b < 0$. The biquadratic can be written $x^4 + b = -ax^2$, which is equivalent to

$$(x^2 + \sqrt{b})^2 = 2\sqrt{b}x^2 - ax^2. \quad (12.1)$$

How do we know that $b \geq 0$? It follows from the inequality $a^2 < 4b$. On the other hand, $a^2 < 4b$ implies $a \leq |a| < 2\sqrt{b}$, thus $2\sqrt{b} - a > 0$. Then from (12.1) we get

$$(x^2 + \sqrt{b})^2 = (x\sqrt{2\sqrt{b} - a})^2,$$

so

$$x^2 + \sqrt{b} = (\sqrt{2\sqrt{b} - a})x \text{ or } x^2 + \sqrt{b} = -(\sqrt{2\sqrt{b} - a})x.$$

Solving these two quadratics we get the four roots, real or complex, of $x^4 + ax^2 + b = 0$.

12.3 Biquadratic functions

As we know , any function of the type $f(x) = ax^2 + bx + c$, $a \neq 0$, attains its minimum at $x = -\frac{b}{2a}$ when $a > 0$, and attains its maximum at $x = -\frac{b}{2a}$ when $a < 0$. Such a result is obtained by observing that

$$f(x) = a(x^2 + \tfrac{b}{a}x + \tfrac{c}{a}) = a((x + \tfrac{b}{2a})^2 + \tfrac{c}{a} - \tfrac{b^2}{4a^2}) = a(x + \tfrac{b}{2a})^2 + (c - \tfrac{b^2}{4a}).$$

The point $(-\frac{b}{2a}, c - \frac{b^2}{4a})$ happens to be the vertex of the parabola given by the expression $y = ax^2+bx+c$. The '$-\frac{b}{2a}$ formula' helps to solve several interesting optimization problems without the recourse of calculus, a topic that was discussed in the previous chapter.

Biquadratic functions are of the form $f(x) = ax^4+bx^2+c$, $a \neq 0$. We will analyze them closely, especially with regard to the problem of finding where they adopt either a maximum or a minimum. As we will see later in the chapter, several optimization questions in geometry lead to biquadratic functions.

First of all, we observe that biquadratic functions are symmetric with respect to the y-axis since $f(-x) = f(x)$. Let us analyze the function $f(x) = x^4 + x^2 + 1$. Obviously, the function adopts its minimum at $x = 0$, namely 1, and thereafter it increases for $x \geq 0$; the above-mentioned symmetry implies that the function is decreasing for $x \leq 0$ (Figure 12.1). A similar observation leads to the conclusion that the function $g(x) = -x^4 - x^2 + 1$ attains its maximum when $x = 0$ (Figure 12.1).

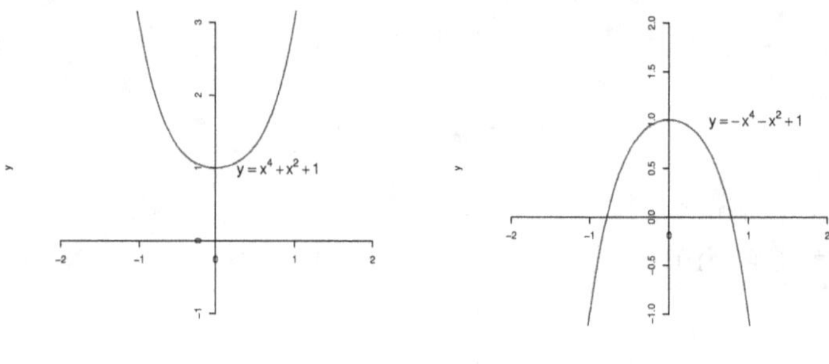

Figure 12.1: Two typical biquadratic functions

Next let us consider $h(x) = x^4 - x^2 + 1$. Using the method of completion of squares we get

$$h(x) = (x^2 - \tfrac{1}{2})^2 + \tfrac{3}{4}.$$

Thus $h(x)$ attains its minimum, namely $3/4$, whenever $x^2 = 1/2$. Then the minimum is adopted at $x = \pm 1/\sqrt{2}$ (Figure 12.2). In a similar fashion, given the function $k(x) = -x^4 + x^2 + 1$ we realize that

$$k(x) = -(x^2 - \tfrac{1}{2})^2 + \tfrac{5}{4}.$$

Thus, the function attains its maximum, namely $5/4$, at $x = \pm 1/\sqrt{2}$ (Figure 12.2).

Interestingly, the graph of any biquadratic function will be of one of the four types shown in the four figures of this section. Of course, whether they intersect the x-axis, or how many times they do so, will depend on the nature of the solutions of the

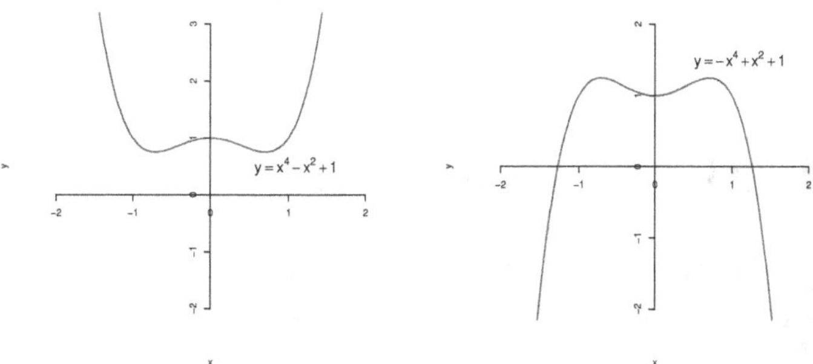

Figure 12.2: Two more typical biquadratic functions

corresponding biquadratic equation. Indeed, let $f(x) = ax^4 + bx^2 + c$. Two cases need to be analyzed separately, depending on whether $a > 0$ or $a < 0$:

1. Assume that $a > 0$. If $b \geq 0$, then it is obvious that f will attain its minimum at $x = 0$. So there is no loss of generality in assuming that $b < 0$. By completion of squares we get

   $$f(x) = a(x^4 + \tfrac{b}{a}x^2 + \tfrac{c}{a}) = a(x^4 + \tfrac{b}{a}x^2 + \tfrac{b^2}{4a^2} + \tfrac{c}{a} - \tfrac{b^2}{4a^2}) = a((x^2 + \tfrac{b}{2a})^2 + \tfrac{4ac-b^2}{4a^2}).$$

   Thus
   $$f(x) = a(x^2 + \frac{b}{2a})^2 + \frac{4ac - b^2}{4a}. \tag{12.2}$$

   for all x. We note that the function will attain its minimum $\frac{4ac-b^2}{4a}$ when $x^2 = -\frac{b}{2a}$, so the minimum will be adopted at the two real numbers

   $$\boxed{\pm\sqrt{-\frac{b}{2a}}}.$$

2. Assume that $a < 0$. If $b \leq 0$ we can observe, just by looking at $f(x) = ax^4 + bx^2 + c$, that the maximum c is attained at $x = 0$. Otherwise, if $b > 0$, the expression (6.2) allows us to conclude that the maximum $\frac{4ac-b^2}{4a}$ is attained when $x^2 + \frac{b}{2a} = 0$; that is to say, at the real numbers $\pm\sqrt{-\frac{b}{2a}}$. This is called the formula for extrema of biquadratic functions.

We can summarize the behavior of biquadratic functions with regard to extrema:

Suppose that $a > 0$. If $b \geq 0$ then the minimum is adopted at $x = 0$, while if $b < 0$ then the minimum is attained at $\pm\sqrt{\frac{-b}{2a}}$. On the other hand, assume that $a < 0$. If

$b > 0$ then the maximum is adopted at $\pm\sqrt{\frac{-b}{2a}}$, while if $b \leq 0$ then the maximum is attained at $x = 0$.

12.4 Applications to geometry

1. Among all right triangles of fixed hypotenuse a, which one encloses the greatest area? Let x and y denote the lengths of the legs. Then $x^2 + y^2 = a^2$ while the area is $A = xy/2$, so

$$A(x) = \tfrac{1}{2}x\sqrt{a^2 - x^2}.$$

We square the area function and obtain

$$f(x) = A^2(x) = -\tfrac{1}{4}x^4 + \tfrac{1}{4}a^2x^2,$$

which is a biquadratic function. Evidently, $f(x)$ will attain its maximum at the same point where $A(x)$ attains its maximum. According to the formula for extrema of biquadratics, the maximum will be adopted at

$$x = \sqrt{\tfrac{-a^2/4}{2(-1/4)}} = \tfrac{a}{\sqrt{2}}.$$

The corresponding length of the other leg will be

$$y = \sqrt{a^2 - (\tfrac{a}{\sqrt{2}})^2} = \tfrac{a}{\sqrt{2}}.$$

Thus, the solution is to build an isosceles right triangle. There is an intriguing relationship between optimization and symmetry!

2. Suppose that we want to build a 10 ft. long tent, with lateral walls of length 3 ft. (Figure 12.3). What should be the measure of the angle at the top of the isosceles triangle if we wish to maximize the volume of the tent? Since the length is a fixed number, namely 10, we need to maximize the area $A(x)$ of the triangular wall, where x is the length of the base.

We have

$$A(x) = \tfrac{x}{2}\sqrt{9 - \tfrac{x^2}{4}},$$

thus

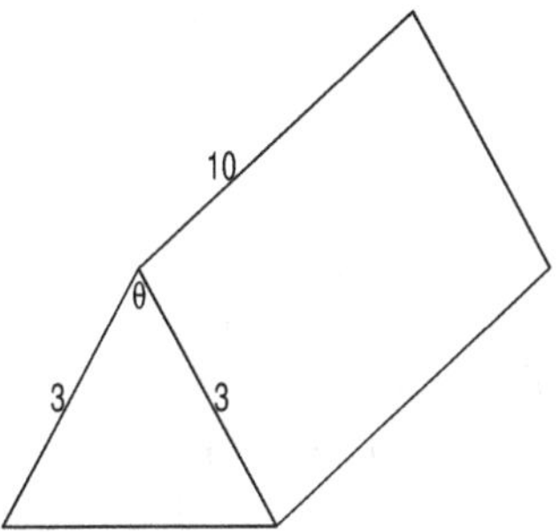

Figure 12.3: The optimal tent problem

$$f(x) = A^2(x) = -\tfrac{1}{16}x^4 + \tfrac{9}{4}x^2.$$

The function $f(x)$ attains its maximum at

$$x = \sqrt{\tfrac{-9/4}{2(-1/16)}} = \sqrt{18} = 3\sqrt{2}.$$

Since $A(x)$ and $f(x)$ adopt their maximum at the same point, the best we can do is to choose $x = 3\sqrt{2}$. We observe that $3^2 + 3^2 = (3\sqrt{2})^2$, hence the converse of the Pythagorean proposition implies that the angle at the top has to be a right angle. A different approach to the solution of the tent problem, using trigonometry, was used in section 6.4.

3. Of all cones with the same lateral surface s, which one has the greatest volume? Let x denote the radius of the base and y the length of the slant side of any cone with the given lateral surface. We have $s = \pi x^2 + \pi xy$, thus $y = \frac{s}{\pi x} - x$. The height of the cone is found by applying the Pythagorean proposition, thus

$$h = \sqrt{(\tfrac{s}{\pi x} - x)^2 - x^2} = \sqrt{\tfrac{s^2}{\pi^2 x^2} - \tfrac{2s}{\pi}}.$$

Therefore,

$$V(x) = \tfrac{\pi x^2}{3}\sqrt{\tfrac{s^2}{\pi^2 x^2} - \tfrac{2s}{\pi}}.$$

The square of the volume will be a biquadratic function, namely

$$V^2(x) = -\tfrac{2\pi s}{9}x^4 + \tfrac{s^2}{9}x^2.$$

Its maximum is attained at

$$x = \sqrt{\frac{-s^2/9}{2(-2\pi s/9)}} = \sqrt{\tfrac{s}{4\pi}},$$

exactly the same value of x at which the maximum of $V(x)$ is attained. The corresponding optimal value of y will be

$$\frac{s - \frac{\pi s}{4\pi}}{\pi\sqrt{\frac{s}{4\pi}}},$$

that is, $y = 3\sqrt{\frac{s}{4\pi}}$; three times the optimal value of x.

A different set of optimization problems, related to cylinders inscribed in spheres or cones, can be found in (Natanson, 1963).

Exercises

1. Use formula (10.7), which involves the polar form of a complex number, to find the four solutions of $x^4 = 1$.

2. Use formula (10.7) to find the four solutions of $x^4 = -1$. Compare this approach with the one used at the beginning of this chapter.

3. Calculate the four solutions of $x^4 + 2x^2 + 6 = 0$.

4. Calculate the four solutions of $x^4 - x^2 - 1 = 0$.

5. Calculate the four solutions of $2x^4 + 4x^2 + 8 = 0$.

6. Where does the function $f(x) = -3x^4 + x^2 + 1$ adopt its maximum?

7. Where does the function $f(x) = 2x^4 - x^2 + 1$ attain its minimum?

8. Where does the function $f(x) = -x^4 + 3x^2 - 1$ adopt its maximum?

9. Among all rectangles inscribed in a circle of radius r, which one encloses the greatest area? Find the dimensions of this optimal rectangle.

10. Among all cylinders inscribed in a sphere of radius r, which one has the greatest lateral surface area? What are the dimensions of the optimal cylinder?

Chapter 13

Third Degree Polynomial Equations

13.1 Two problems from geometry

Third order polynomial equations do appear when solving geometrical problems. Let us discuss in detail two examples of this nature.

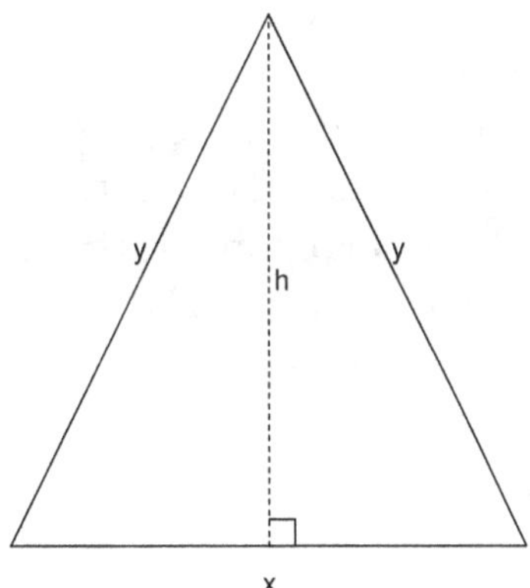

Figure 13.1: The problem of an isosceles triangle of given area and perimeter

1. Suppose we wish to determine all isosceles triangles of area 3 and perimeter 10. Let x be the length of the base and y the length of the two congruent sides

(Figure 13.1). Then the height of the isosceles triangle that we have in mind is equal to $\sqrt{y^2 - x^2/4}$. Therefore,

$$x + 2y = 10 \text{ and } \tfrac{x}{2}\sqrt{y^2 - \tfrac{x^2}{4}} = 3.$$

Hence $\frac{x}{2}\sqrt{(5-\frac{x}{2})^2 - \frac{x^2}{4}} = 3$, which in turn leads to

$$5x^3 - 25x^2 + 36 = 0.$$

2. Assume that, given a right triangle, its area is 2 and the hypotenuse is 2 units bigger than one of the legs. What are the lengths of the three sides? Let x and y denote the lengths of the legs. Then the length of the hypotenuse will be $\sqrt{x^2 + y^2}$ and we will have $xy = 4$, $\sqrt{x^2 + y^2} = x + 2$. Squaring the last expression we get $x^2 + y^2 = x^2 + 4x + 4$, therefore

   $$x^2 + \tfrac{16}{x^2} = x^2 + 4x + 4.$$

   Simplifying this equation and multiplying by x^2 leads to

   $$x^3 + x^2 - 4 = 0.$$

Surprisingly, compared to quadratic or biquadratic equations, it is considerably more difficult to solve cubic equations. So much so that at the end of the 15th century mathematicians thought that it might be impossible to find an algebraic procedure to solve all third degree polynomial equations. But, in the early 16th century Scipione del Ferro (1465-1526) and in 1535 Niccolo Fontana-Tartaglia (1499-1557), independently of each other, found a way to solve cubics of the type $x^3 + px = q$, where p and q are positive numbers. The procedure appeared in print, for the first time (1545), in *Ars Magna* by Girolamo Cardano (1501-1576)[1]. Let us illustrate the method, from a modern perspective, through an example.

13.2 A particular equation

Suppose we wish to find a solution of $x^3 + x = 1$. Taking into consideration that

$$(u + v)^3 = u^3 + 3u^2v + 3uv^2 + v^3 \tag{13.1}$$

we may write

[1] *Ars Magna* is also known as *The Great Art*. The latter is the English translation of the title.

$$((u+v)^3 - 3uv(u+v) = u^3 + v^3.$$

Thus, if we could solve the system of equations

$$-3uv = 1, u^3 + v^3 = 1 \tag{13.2}$$

then $u+v$ would be a solution of $x^3 + x = 1$. Hence, let us concentrate our attention on the above-mentioned system. Right away we note that $v = -\frac{1}{3u}$, so

$$u^3 - \frac{1}{27u^3} = 1,$$

and multiplying by u^3 we get

$$u^6 - u^3 - \frac{1}{27} = 0.$$

Let $z = u^3$. Then

$$z^2 - z - \frac{1}{27} = 0.$$

This is a quadratic equation, whose solutions are given by $z = \frac{1}{2} \pm \frac{1}{2}\sqrt{1 + \frac{4}{27}}$. That is

$$u^3 = \frac{1}{2} \pm \frac{1}{2}\sqrt{1 + \frac{4}{27}}.$$

Let us choose one of these two options, say $u^3 = \frac{1}{2} + \frac{1}{2}\sqrt{1 + \frac{4}{27}}$. Then

$$u = \sqrt[3]{\frac{1}{2} + \frac{1}{2}\sqrt{1 + \frac{4}{27}}},$$

which in turn leads to

$$v^3 = 1 - (\frac{1}{2} + \sqrt{\frac{1}{4} + \frac{1}{27}}) = \frac{1}{2} - \sqrt{\frac{1}{4} + \frac{1}{27}}.$$

Hence

$$v = \sqrt[3]{\frac{1}{2} - \sqrt{\frac{1}{4} + \frac{1}{27}}}.$$

Therefore, it is to be expected that

$$x = u + v = \sqrt[3]{\frac{1}{2} + \sqrt{\frac{1}{4} + \frac{1}{27}}} + \sqrt[3]{\frac{1}{2} - \sqrt{\frac{1}{4} + \frac{1}{27}}}$$

is a solution of the original cubic equation. With a little bit of patience the reader can verify that this real number is indeed a solution. By the way, if we had chosen to work with

$$u^3 = \tfrac{1}{2} - \tfrac{1}{2}\sqrt{1 + \tfrac{4}{27}}$$

then $v^3 = \frac{1}{2} + \frac{1}{2}\sqrt{1 + \frac{4}{27}}$ and we get the same solution

$$r_1 = \sqrt[3]{\tfrac{1}{2} + \sqrt{\tfrac{1}{4} + \tfrac{1}{27}}} + \sqrt[3]{\tfrac{1}{2} - \sqrt{\tfrac{1}{4} + \tfrac{1}{27}}}.$$

Using the division algorithm for polynomials, we can find real numbers a and b such that

$$x^3 + x - 1 = (x - r_1)(x^2 + ax + b).$$

Since quadratic equations can be solved easily, the important task is to find one real root of the given cubic. It should be noted that if instead of starting with (13.1) we would have preferred to work with the identity

$$(u - v)^3 = u^3 - 3u^2v + 3uv^2 - v^3,$$

then the same root r_1 is found.

13.3 A general method

Does the method used to find a solution of $x^3+x = 1$ work for any equation $x^3+px = q$ where $p, q > 0$? Indeed it does, as we will see next. Keeping in mind that

$$x^3 = -px + q \tag{13.3}$$

and

$$(u + v)^3 = 3uv(u + v) + u^3 + v^3,$$

if we were to be able to solve the system of equations

$$u^3 + v^3 = q$$

$$3uv = -p$$

then $x = u + v$ would be a solution of (13.3). Since $v = -\frac{p}{3u}$ we will have

$$u^3 - \tfrac{p^3}{27u^3} = q,$$

so

$$u^6 - qu^3 - \tfrac{p^3}{27} = 0.$$

Defining $z = u^3$ we get the quadratic equation $z^2 - qz - \frac{p^3}{27} = 0$, whose solutions are

$$z = \frac{q}{2} \pm \sqrt{\frac{q^2}{4} + \frac{p^3}{27}}.$$

If we choose to work with $z = \frac{q}{2} + \sqrt{\frac{q^2}{4} + \frac{p^3}{27}}$, then

$$u = \sqrt[3]{\frac{q}{2} + \sqrt{\frac{q^2}{4} + \frac{p^3}{27}}}.$$

By the same token $v^3 = q - (\frac{q}{2} + \sqrt{\frac{q^2}{4} + \frac{p^3}{27}})$, thus

$$v = \sqrt[3]{\frac{q}{2} - \sqrt{\frac{q^2}{4} + \frac{p^3}{27}}}.$$

We can then conclude that

$$x = \sqrt[3]{\frac{q}{2} + \sqrt{\frac{q^2}{4} + \frac{p^3}{27}}} + \sqrt[3]{\frac{q}{2} - \sqrt{\frac{q^2}{4} + \frac{p^3}{27}}}$$

is a solution of the original problem. Of course, the reader can verify that this number, which happens to be a real number, is indeed a solution of $x^3 + px = q$. The process that led to the solution is called the Cardano-Tartaglia approach. Maybe it should be called the del Ferro-Tartaglia-Cardano approach.

13.4 An example with historical significance

Emboldened with their success when dealing with the equation $x^3 + px = q$, $(p, q > 0)$, 16th century mathematicians tried to apply Cardano-Tartaglia's method to the equation $x^3 = 15x + 4$ and found an apparent paradox. Let us discuss what happens when searching for a real solution under the above-mentioned methodology. Since $(u+v)^3 = 3uv(u+v) + u^3 + v^3$, our job is to solve the system

$$u^3 + v^3 = 4,$$

$$3uv = 15$$

in order to claim that $x = u + v$ is a solution of $x^3 = 15x + 4$. Right away we have $v = \frac{5}{u}$, consequently

$$u^3 + \frac{125}{u^3} = 4.$$

Thus $u^6 - 4u^3 + 125 = 0$. Letting $z = u^3$ we get $z^2 - 4z + 125 = 0$, therefore

$$z = \frac{4 \pm \sqrt{16-500}}{2} = 2 \pm \sqrt{-121} = 2 \pm 11i.$$

The big surprise is that we have to deal with the square root of a negative number. That is to say, complex numbers appear even though by simple inspection we can conclude that the real number $x = 4$ is a solution of $x^3 = 15x + 4$. Neither Cardano nor Tartaglia could accept that it made any sense to take the square root of a negative number. One generation after them, Rafael Bombelli (1526-1572) started to work, hesitantly, with the symbol $\sqrt{-1}$. With the benefit of hindsight, and knowledgeable about complex numbers, we can proceed to choose the solution $z = 2 + 11i$ without any qualms. Then $u^3 = 2 + 11i$, so $u = \sqrt[3]{2 + 11i}$. The challenge upon us is to find a number $a + bi$ such that $(a + bi)^3 = 2 + 11i$. That is to say, we need to calculate a cubic root of $2 + 11i$. Since

$$a^3 + 3a^2bi + 3a(-b^2) - b^3i = 2 + 11i,$$

we will have $a(a^2 - 3b^2) = 2$ and $b(3a^2 - b^2) = 11$. By trial an error we can find a simple solution of this system, namely $a = 2$, $b = 1$. Thus $u = 2 + i$, which in turn leads to

$$v = \frac{5}{2+i} = \frac{5(2-i)}{(2+i)(2-i)} = \frac{10-5i}{4+1} = 2 - i.$$

Hence $u + v = 2 + i + 2 - i = 4$, precisely the same solution of $x^3 = 15x + 4$ that we found before by simple inspection.

13.5 A wider perspective

In the previous section we found that the Cardano-Tartaglia approach also leads to a real solution when dealing with a particular example of an equation of the type $x^3 = px + q$, $(p, q > 0)$. Is this true in general? That is, do we always get a real solution when dealing with

$$x^3 + px + q = 0 \tag{13.4}$$

where p and q are *any* real numbers? The Cardano-Tartaglia method leads to the solution

$$x = \sqrt[3]{-\frac{q}{2} + \sqrt{\Delta}} + \sqrt[3]{-\frac{q}{2} - \sqrt{\Delta}}, \tag{13.5}$$

where $\Delta = \frac{q^2}{4} + \frac{p^3}{27}$.

If $\Delta > 0$, then we get a real solution and we do not have to worry anymore. What happens if $\Delta = 0$? Under these circumstances, obviously $x = 2\sqrt[3]{-\frac{q}{2}}$ is a real solution. The most interesting case, from a mathematical point of view, takes place when $\Delta < 0$. Could we also obtain a real solution? We can write the solution (13.5) as

$$x = \sqrt[3]{a+bi} + \sqrt[3]{a-bi},$$

where $a = -\frac{q}{2}$ and $b = \sqrt{-\frac{q^2}{4} - \frac{p^3}{27}}$. The complex number $c = a + bi$, expressed in polar form, is $|c|e^{i\theta}$. Then $|c|^{1/3}e^{i\theta/3}$ is one cubic root of $a + bi$. Similarly, $|c|^{1/3}e^{-i\theta/3}$ is another cubic root of $a - bi$. Therefore,

$$x = |c|^{1/3}e^{i\theta/3} + |c|^{1/3}e^{-i\theta/3} = 2|c|^{1/3}\cos\tfrac{\theta}{3}$$

is a real solution of $x^3 + px + q = 0$ whenever $\Delta < 0$. In an enrichment note, at the end of the chapter, we will discuss the nature of *all* the solutions of $x^3 + px + q = 0$, where p, q are any real numbers. We will see that when $\Delta > 0$ the equation under consideration has only one real root (the other two are complex numbers), precisely the number defined in (13.5). If $\Delta < 0$, unexpectedly there will be three real solutions (one of them is $2|c|^{1/3}\cos\frac{\theta}{3}$). Finally, if $\Delta = 0$ and $q = 0$ evidently the only solution is zero (with multiplicity three) while if $\Delta = 0$ and $q \neq 0$, then the three roots are real (one of them is $2\sqrt[3]{-q/2}$).

13.6 Cardano's transformation

So far we have only analyzed cubic equations with no quadratic term. Can we solve an equation like

$$x^3 + px^2 + qx + r = 0\ ?$$

Let $y = x + \frac{p}{3}$. Then

$$(y - \tfrac{p}{3})^3 + p(y - \tfrac{p}{3})^2 + q(y - \tfrac{p}{3}) + r = 0,$$

that is,

$$y^3 - y^2p + y\tfrac{p^2}{3} - \tfrac{p^3}{27} + py^2 - 2\tfrac{p^2}{3}y + \tfrac{p^3}{9} + qy - \tfrac{qp}{3} + r = 0.$$

Thus, thanks to the transformation $y = x + \frac{p}{3}$, called Cardanos's transformation, we have to deal with the equation

$$y^3 + (\tfrac{p^2}{3} - 2\tfrac{p^2}{3} + q)y + (r - \tfrac{p^3}{27} - \tfrac{pq}{3}) = 0,$$

which has no quadratic term, so is amenable to the Cardano-Tartaglia approach.

Examples

1. In section 13.1, when analyzing a geometrical problem linked to right triangles, we found the equation $x^3 + x^2 - 4 = 0$. Firstly, we define Cardano's transformation, namely $y = x + \frac{1}{3}$. Then

   $$(y - \tfrac{1}{3})^3 + (y - \tfrac{1}{3})^2 - 4 = 0.$$

   Thus,

   $$y^3 - 3y^2\tfrac{1}{3} + 3y\tfrac{1}{9} - \tfrac{1}{27} + y^2 - \tfrac{2}{3}y + \tfrac{1}{9} - 4 = 0.$$

   That is,

   $$y^3 - \frac{1}{3}y - \frac{106}{27} = 0. \tag{13.6}$$

   Since

   $$\Delta = \tfrac{(106/27)^2}{4} - \tfrac{\frac{1}{27}}{27} = 3.85185$$

   we can conclude that one solution of (13.6) is

   $$y = \sqrt[3]{\tfrac{106}{54} + \sqrt{3.85185}} + \sqrt[3]{\tfrac{106}{54} - \sqrt{3.85185}} = 1.64796.$$

   Hence $x = 1.64796 - \frac{1}{3} = 1.31463$ is a solution. Since $\Delta > 0$ we cannot expect any other real roots; thus, the geometrical problem under consideration has only one solution.

2. The problem of finding all isosceles triangles of area 3 and perimeter 10, developed in section 13.1, led to the cubic equation

   $$x^3 - 5x^2 + \tfrac{36}{5} = 0.$$

   Let $y = x - \frac{5}{3}$, the corresponding Cardano transformation. Then

   $$(y + \tfrac{5}{3})^3 - 5(y + \tfrac{5}{3})^2 + \tfrac{36}{5} = 0,$$

   so

   $$y^3 + 3y^2\tfrac{5}{3} + 3y\tfrac{25}{9} + \tfrac{125}{27} - 5(y^2 + \tfrac{10}{3}y + \tfrac{25}{9}) + \tfrac{36}{5} = 0.$$

   That is,

$$y^3 - \frac{25}{3}y - \frac{278}{135} = 0.$$

Since

$$\Delta = \frac{(\frac{278}{135})^2}{4} - \frac{(\frac{25}{3})^3}{27} = -20.3733 < 0$$

we should expect three real solutions. One of them will be $2|c|^{1/3}\cos\frac{\theta}{3}$, where $c = a + bi$, $a = \frac{278}{270} = 1.02963$, $b = \sqrt{\frac{-(278/135)^2}{4} + \frac{(25/3)^3}{27}} = 4.51368$. Moreover,

$$\theta = \arctan(\tfrac{b}{a}) = \arctan\frac{4.51368}{1.02963} = 1.34652 \text{ radians.}$$

Hence

$$y = 2|c|^{1/3}\cos\tfrac{\theta}{3} = 2(1.02963^2 + 4.51368^2)^{1/6} \times \cos(\tfrac{1.34652}{3}) = 3.00317.$$

Therefore $x = 3.00317 + \frac{5}{3} = 4.66984$. What are the two other real solutions?

Applying the division algorithm for polynomials we get

$$x^3 - 5x^2 + \tfrac{36}{5} = (x - 4.66984)(x^2 - 0.33016x - 1.54179).$$

The quadratic equation $x^2 - 0.33016x - 1.54179 = 0$ has two solutions, namely $x = 1.41769$ and $x = -1.08753$. Thus, the geometrical problem under consideration has two solutions, namely $x = 4.66984$ and $x = 1.41769$. Their corresponding y-values are $y = 5 - \frac{4.66984}{2} = 2.66508$ and $y = 5 - \frac{1.41769}{2} = 4.29116$. That is to say, there are exactly two isosceles triangles with area 3 and perimeter 10, namely one with base 4.66984 and side of common measure 2.66508, and the other with base 1.41769 and side of common measure 4.29116. The geometrical problem under consideration is the starting point of a wider search of relationships between angles, area, and perimeter of a triangle (Baloglou and Helfgott, 2008).

Using Newton's method (Helfgott, 2012), a well-known tool from Calculus, we could easily find the solutions of $x^3 - 5x^2 + \frac{36}{5} = 0$. This is precisely what a graphing calculator does when we use the *solve* command and write Solve$(x^3 - 5x^2 + 36/5 = 0, x)$, making sure that we are on the approximate mode. The rationale behind the algebraic approach to cubics, vis-a-vis the Calculus approach, has to do with historical and methodological factors. For close to one hundred and fifty years, roughly between 1545 and 1690, Cardano-Tartaglia's method was the only available tool. And the algebraic approach allows us to understand under what circumstances the solutions will be real or complex.

13.7 Using geometry to solve a cubic

Islamic mathematicians from the 11th century used geometry to solve the cubic equation $x^3 + Ax = B$, where A and B are positive numbers. Let us take advantage of a Cartesian approach to simplify the calculations[2]. Define $a = \sqrt{A}$ and $b = B/A$. Then the equation under study is

$$x^3 + a^2x = ba^2.$$

Let us pay attention to the parabola $y = \frac{1}{a}x^2$. Then we draw the semi-circle with diameter of length b, whose endpoints are $(0,0)$ and $(R,0)$ (Figure 13.2). Let P be the point of intersection between the parabola and the semi-circle. From P we drop the perpendicular to the horizontal axis and label Q their point of intersection. We claim that OQ, which we will denote with the letter c, is a solution of the given cubic.

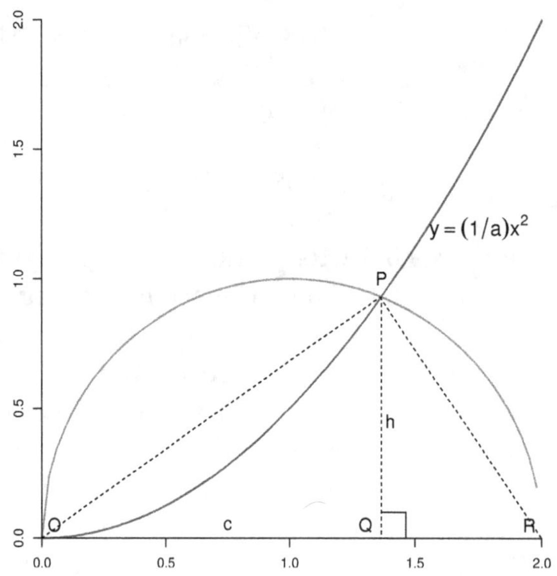

Figure 13.2: Solving a cubic through geometry

Let us provide a justification. We start by building triangle OPR. Since $\angle OPR$ is an inscribed angle that subtends a diameter, we can conclude that $m\angle OPR = 90$. Then

$$h^2 = c(b - c),$$

where $h = PQ$ (exercise 6, chapter 4). On the other hand, $h = \frac{1}{a}c^2$ since the pair (c, h) belongs to the parabola. Thus, $h^2 = \frac{c^4}{a^2}$, and consequently

[2]Let us recall that Cartesian geometry is a 17th century development.

$$\frac{c^4}{a^2} = cb - c^2.$$

So $c^3 + a^2c = ba^2$, that is, $c^3 + Ac = B$.

Exercises

1. Find the dimensions of all isosceles triangles with area $\sqrt{3}$ and perimeter 6.

2. Find the dimensions of all isosceles triangles of area 4 and perimeter 12.

3. Solve the equation $x^3 + x + 1 = 0$.

4. Use the Cardano transformation to eliminate the quadratic term of the equation $x^3 + x^2 + x + 1 = 0$. Then solve the resulting equation.

5. We have a cube of side x and a rectangular prism of base area 4 and height x. Calculate x assuming that the volume of the cube is equal to the volume of the prism plus 2 units (Hint: Keep in mind that the volume of the prism under consideration is $4x$.)

6. Solve the equation $x^3 + 3x = 2$.

7. You have a hollow sphere made of metal that has a uniform thickness of 1 cm. Given that the volume of the shell is equal to the volume of the hollow sphere, calculate the radius of the latter.

8. Assume that, given a right triangle, the product of its legs is 6 and the hypotenuse is 1 unit bigger than one of the legs. Calculate the length of the hypotenuse.

9. Find two solutions, real or complex, of the equation $z^6 - z^2 + 1 = 0$.

10. Can the equation $z^6 + z^2 + 1 = 0$ have a real solution?

Enrichment Note: The nature of the solutions of a cubic equation

Let us recall that given a cubic equation $x^3 + px + q = 0$, with p, q real numbers, the Cardano-Tartaglia method leads to the solution

$$x = \sqrt[3]{-\tfrac{q}{2} + \sqrt{\Delta}} + \sqrt[3]{-\tfrac{q}{2} - \sqrt{\Delta}}$$

where $\Delta = \frac{q^2}{4} + \frac{p^3}{27}$. Let us consider three cases separately, depending on the sign of Δ.

1. Let us assume that $\Delta > 0$. Then the above-mentioned solution is a real number. How about the other two solutions? Let

$$u_1 = \sqrt[3]{-\tfrac{q}{2} + \sqrt{\Delta}}, \quad v_1 = \sqrt[3]{-\tfrac{q}{2} - \sqrt{\Delta}}.$$

Then[3] u_1, $u_2 = u_1 e^{2\pi i/3}$, and $u_3 = u_1(e^{2\pi i/3})^3$ are the solutions of $u^3 = -\frac{q}{2}+\sqrt{\Delta}$. Similarly, v_1, $v_2 = v_1 e^{2\pi i/3}$, and $v_3 = v_1(e^{2\pi i/3})^2$ are the solutions of $v^3 = -\frac{q}{2} + \sqrt{\Delta}$. We claim that $u_2 + v_3$ is a solution of $x^3 + px + q = 0$.

Indeed,

$$u_2^3 + v_3^3 = u_1^3 + v_1^3 = -q$$

and

$$u_2 v_3 = u_1 e^{2\pi i/3} v_1 e^{4\pi i/3} = u_1 v_1 = -\tfrac{p}{3}.$$

In a similar fashion we have that $u_3 + v_2$ is a solution of $x^3 + px + q = 0$. Thus,

$$u_1 e^{2\pi i/3} + v_1 e^{4\pi i/3}$$

and

$$u_1 e^{4\pi i/3} + v_1 e^{2\pi i/3}$$

are solutions of $x^3 + px + q = 0$. That is, a complex solution happens to be

$$u_1(\cos\tfrac{2\pi}{3} + i\sin\tfrac{2\pi}{3}) + v_1(\cos\tfrac{4\pi}{3} + i\sin\tfrac{4\pi}{3}) = u_1(-\tfrac{1}{2} + \tfrac{\sqrt{3}}{2}i) + v_1(-\tfrac{1}{2} - \tfrac{\sqrt{3}}{2}i) = (-\tfrac{1}{2}u_1 - \tfrac{1}{2}v_1) + i(\tfrac{\sqrt{3}}{2}u_1 - \tfrac{\sqrt{3}}{2}v_1).$$

On the other hand,

$$u_1(\cos\tfrac{4\pi}{3} + i\sin\tfrac{4\pi}{3}) + v_1(\cos\tfrac{2\pi}{3} + i\sin\tfrac{2\pi}{3}) = u_1(-\tfrac{1}{2} - \tfrac{\sqrt{3}}{2}i) + v_1(-\tfrac{1}{2} + \tfrac{\sqrt{3}}{2}i) = (-\tfrac{1}{2}u_1 - \tfrac{1}{2}v_1) + i(-\tfrac{\sqrt{3}}{2}u_1 + \tfrac{\sqrt{3}}{2}v_1)$$

[3] We are using a simple fact, proven at the end of section 10.9: If w is a solution of $z^3 = d$ then $we^{2\pi i/3}$ and $we^{2\pi i/3}$ are the two other solutions.

is also a complex solution, distinct from the preceding complex solution. We have found three distinct solutions of $x^3 + px + q = 0$ (recall that $u_1 + v_1$ is a real solution), and there are no other solutions.

2. Let us assume that $\Delta < 0$. Under these circumstances we know that

$$x = \sqrt[3]{a + bi} + \sqrt[3]{a - bi},$$

where $a = -\frac{q}{2}$ and $b = \sqrt{-\frac{q^2}{4} - \frac{p^3}{27}}$, is a solution of $x^3 + px + q = 0$. We have to keep in mind that $\sqrt[3]{a + bi}$ has three different values and that $\sqrt[3]{a - bi}$ has three different values too. Let $c = a + bi$.

The three possible values of $\sqrt[3]{a + bi}$ are $|c|^{1/3}e^{i\theta/3}$, $|c|^{1/3}e^{i\theta/3}e^{i2\pi/3}$, and $|c|^{1/3}e^{i\theta/3}e^{i4\pi/3}$, where θ is the argument of $c = a - bi$. Similarly, the three possible values of $\sqrt[3]{a - bi}$ are

$$|c|^{1/3}e^{-i\theta/3},\ |c|^{1/3}e^{-i\theta/3}e^{i2\pi/3},\ \text{and } |c|^{1/3}e^{-i\theta/3}e^{i4\pi/3}.$$

We notice that only

$$|c|^{1/3}e^{i\theta/3} + |c|^{1/3}e^{-i\theta/3},\ |c|^{1/3}e^{i\theta/3}e^{i2\pi/3} + |c|^{1/3}e^{-i\theta/3}e^{i4\pi/3},\ \text{and}$$
$$|c|^{1/3}e^{i\theta/3}e^{i4\pi/3} + |c|^{1/3}e^{-i\theta/3}e^{i2\pi/3}$$

have the property that $u^3 + v^3 = -q$ and $uv = -p/3$. For instance,

$$(|c|^{1/3}e^{i\theta/3})^3 + (|c|^{1/3}e^{-i\theta/3})^3 = |c|e^{i\theta} + |c|e^{-i\theta} = 2|c|\cos\theta = 2a = -q$$

and

$$|c|^{1/3}e^{i\theta/3} \times |c|^{1/3}e^{-i\theta/3} = |c|^{2/3} = (a^2 + b^2)^{1/3} = (\tfrac{q^2}{4} - \tfrac{q^2}{4} - \tfrac{p^3}{27})^{1/3} = -\tfrac{p}{3}.$$

The six other possibilities do not have this property (their product happens to be a complex number instead of the real number $-p/3$). Thus, the three solutions are

$$2|c|^{1/3}\cos\tfrac{\theta}{3},\ |c|^{1/3}(-\cos\tfrac{\theta}{3} - \sqrt{3}\sin\tfrac{\theta}{3}),\ \text{and } |c|^{1/3}(-\cos\tfrac{\theta}{3} + \sqrt{3}\sin\tfrac{\theta}{3}).$$

Hence, we have three real roots!

Table 13.1: The solutions of $x^3 + px + q = 0$, with $\Delta = \frac{q^2}{4} + \frac{p^3}{27}$

Δ	solutions
$\Delta > 0$	One real solution and two complex solutions
$\Delta < 0$	Three real solutions
$\Delta = 0$ and $q \neq 0$	Two real solutions (one of them with multiplicity 2)
$\Delta = 0$ and $q = 0$	Only the solution 0 (with multiplicity 3)

3. Suppose that $\Delta = 0$. If $q = 0$ then $0 = \Delta = p^3/27$, so $p = 0$. The equation becomes $x^3 = 0$, whose only solution is 0 (with multiplicity 3). We may well assume that $q \neq 0$. Then

$$x = \sqrt[3]{-\tfrac{q}{2}} + \sqrt[3]{-\tfrac{q}{2}} = 2\sqrt[3]{-\tfrac{q}{2}}$$

is a real solution. Using the division algorithm for polynomials we get

$$x^3 + px + q = (x - 2\sqrt[3]{-\tfrac{q}{2}})(x^2 + 2\sqrt[3]{-\tfrac{q}{2}}x + (\tfrac{q}{2})^{2/3}).$$

The equation $x^2 + 2\sqrt[3]{-\frac{q}{2}}x + (\frac{q}{2})^{2/3} = 0$ has only one solution, namely $\sqrt[3]{\frac{q}{2}}$. Therefore, the two solutions (when $\Delta = 0$ and $q \neq 0$) are $2\sqrt[3]{-\frac{q}{2}}$ and $\sqrt[3]{\frac{q}{2}}$ (with multiplicity 2). We can summarize everything in a table (Table 13.1).

Chapter 14

Fourth Degree Polynomial Equations

14.1 The Chinese Castle Problem

We have been able to study a particular family of fourth degree polynomials, namely biquadratic equations. This time we wish to discuss equations of the type $x^4 + ax^3 + bx^2 + cx + d = 0$, where a, b, c, d are real numbers. Let us consider the following problem, which was stated and solved by 13th century Chinese mathematicians (Swetz, 2004): *A circular castle of unknown radius has doors on the north side and south side, and a large tree lies 3 miles above the north door (Figure 14.1). A man starts from the south door and walks toward the east, and sees the tree for the first time at point B after 9 miles, that is,* $BC = 9$. *What is the radius of the castle?*

From the nature of the problem we can assert that $\overline{BD}$ is tangent to the circle (at a certain point A) and that $\overline{OC}$ is perpendicular to $\overline{BC}$. Since the tangents to a circle from an exterior point are congruent, it follows that $AB = 9$. Then

$$AD = \sqrt{(3+r)^2 - r^2} = \sqrt{9+6r}.$$

But $BD^2 = (3+2r)^2 + 81$. Hence $(9 + \sqrt{9+6r})^2 = (3+2r)^2 + 81$. Thus $81 + 9 + 6r + 18\sqrt{9+6r} = 9 + 4r^2 + 12r + 81$, that is, $4r^2 + 6r = 18\sqrt{9+6r}$. Therefore

$$16r^4 + 36r^2 + 48r^3 = 324(9+6r).$$

Simplifying we get $16r^4 + 48r^3 + 36r^2 = 2916 + 1944r$, that is,

$$16r^4 + 48r^3 + 36r^2 - 1944r - 2916 = 0. \tag{14.1}$$

We have reached a quartic polynomial. In the last chapter, when discussing the phenomenon of refraction we will find a quartic too:

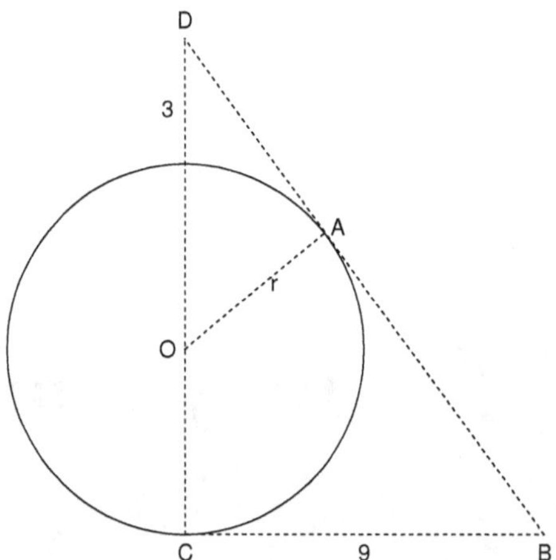

Figure 14.1: The castle problem

$$9(R-4)^2((7-R)^2+4)-16(7-R^2)((R-4)^2+9)=0$$

Thus, quartics sometimes appear when discussing problems that stem from geometry or physics.

14.2 Ferrari's technique

Let us recall how we solved the biquadratic equation $x^4+x^2+1=0$. We started by rewriting the equation as $x^4+1=-x^2$, hence $(x^2+1)^2=2x^2-x^2$, that is $(x^2+1)^2=x^2$. Then $x^2+1=x$ or $x^2+1=-x$. We have reached two quadratic equations, which can be easily solved. The crux of the matter was to reach the equation $(x^2+1)^2=x^2$, wherein two perfect squares appear at each side of the equality sign.

Lodovico Ferrari (1522-1565), a disciple of Cardano, invented a systematic procedure to solve all quartics without a cubic term, that is, equations of the type

$$x^4+ax^2+bx+c=0$$

and he is duly recognized in *Ars Magna* for this important achievement. Let us illustrate Ferrari's method through two particular examples from *Ars Magna*, which we will analyze from a modern perspective.

1. Problem XII, chapter 39 from *Ars Magna*, deals with the equation $x^4+3=12x$, which is equivalent to $(x^2)^2 = 12x - 3$. For any z we have

$$(x^2+z)^2 = 2zx^2 + 12x + (z^2-3). \tag{14.2}$$

We wish to find a value of z that will make a perfect square from the expression to the right of the equality sign. This can be accomplished by having the discriminant equal to zero, that is, $0 = 144 - 4(2z)(z^2-3)$. Thus, we reach the cubic

$$z^3 - 3z - 18 = 0,$$

which is called the 'resolvent cubic'. The rational root test (Hungerford, 1997) implies that any rational solution of the preceding cubic has to be an integer that is a divisor of 18. The only divisors of 18 are $\pm1, \pm2, \pm3, \pm6, \pm9, \pm18$. By simple inspection we find that $z = 3$ is a solution of the resolvent cubic. Replacing this value in (14.2) we get

$$(x^2+3)^2 = 6x^2 + 12x + (3^2-3),$$

or, what is the same,

$$(x^2+3)^2 = 6(x^2+2x+1).$$

The last equation is equivalent to

$$(x^2+3)^2 = (\sqrt{6}(x+1))^2.$$

Therefore $x^2+3 = \sqrt{6}(x+1)$ or $x^2+3 = -\sqrt{6}(x+1)$. These are quadratic equations with solutions

$$\sqrt{\tfrac{3}{2}} \pm \sqrt{-\tfrac{3}{2}+\sqrt{6}}, \quad -\tfrac{\sqrt{6}}{2} \pm \tfrac{i}{2}\sqrt{6-4\sqrt{6}},$$

respectively. These are precisely the four solutions of the original quartic. We should mention that Cardano only presents the two real solutions since complex numbers were unknown when *Ars Magna* was published.

2. Problem IX, chapter 39 from *Ars Magna*, deals with the quartic

$$x^4 + 4x + 8 = 10x^2,$$

which is equivalent to $x^4 = 10x^2 - 4x - 8$. Then for any z we have

$$(x^2 + z)^2 = (10 + 2z)x^2 - 4x + (z^2 - 8). \tag{14.3}$$

We wish to find a real value of z that will make the expression to the right of the equal sign of (14.3) a perfect square. This can be accomplished by having the discriminant equal to zero. That is,

$$0 = 16 - 4(2z + 10)(z^2 - 8).$$

This equation is equivalent to

$$z^3 + 5z^2 - 8z - 42 = 0,$$

which is the resolvent cubic of the given quartic. With the help of the rational root test we can reach a solution of it, namely $z = -3$. Replacing this value in (14.3) we obtain

$$(x^2 - 3)^2 = 4x^2 - 4x + 1,$$

which is equivalent to $(x^2 - 3)^2 = (2x - 1)^2$. Hence $x^2 - 3 = 2x - 1$ or $x^2 - 3 = -(2x - 1)$. The solutions of these quadratics are $1 \pm \sqrt{3}$ and $-1 \pm \sqrt{5}$ respectively. These are, then, the four solutions of the given quartic.

The two quartics that we have just solved have something in common: it was relatively easy to find a real root of the corresponding cubic, a fact that allowed us to concentrate on showing how Ferrari's technique works. Usually we are not so lucky. Let us consider the equation

$$y^4 + y^2 - y - 2 = 0,$$

which is equivalent to $y^4 = -y^2 + y + 2$. For any z we have

$$(y^2 + z)^2 = 2y^2 z + z^2 - y^2 + y + 2.$$

That is,

$$(y^2 + z)^2 = (2z - 1)y^2 + y + z^2 + 2. \tag{14.4}$$

The left side is already a perfect square, so our task is to find z such that the expression to the right of the equal sign is a perfect square too. This can be achieved by making its discriminant equal to zero. Thus

$$0 = 1 - 4(2z - 1)(z^2 + 2).$$

Multiplying and grouping coefficients of equal powers we get the cubic equation

$$z^3 - \tfrac{1}{2}z^2 + 2z - \tfrac{9}{8} = 0.$$

Cardano's transformation $u = z - \frac{1}{6}$ leads to the reduced cubic $u^3 + \frac{23}{12}u - \frac{173}{216} = 0$. Using Cardano-Tartaglia's method we get a solution $u = 0.387513$; thus, $z = 0.387513 + \frac{1}{6} = 0.55418$. Then we replace this value in (14.4) and obtain

$$(y^2 + 0.55418)^2 = 0.10836y^2 + y + 2.30712,$$

which is equivalent[1] to

$$(y^2 + 0.55418)^2 = 0.10836(y + \tfrac{1}{2\times 0.10836})^2.$$

Therefore

$$(y^2 + 0.55418)^2 = (\sqrt{0.10836}(y + \tfrac{1}{0.21672}))^2.$$

Thus

$$y^2 + 0.55418 = \sqrt{0.10836}(y + \tfrac{1}{0.21672}) \text{ or } y^2 + 0.55418 = -\sqrt{0.10836}(y + \tfrac{1}{0.21672}).$$

The two solutions of the first quadratic equation are 1.1605 and -0.831317. The solutions of the second quadratic are two complex conjugate roots.

We had to spend a lot of work to reach the two real roots. Once we realize that Ferrari's method works for all quartics (see enrichment note at the end of the chapter), Newton's method, a calculus technique, is frequently the tool of choice. For instance, it would be an onerous task to find the solutions of (14.1) employing Ferrari's technique. Using a graphing calculator we get $Solve(16r^4 + 48r^3 + 36r^2 - 1944r - 2916 = 0, r) = 4.5$ (the negative value $-3/2$ has to be discarded). What happens is that all graphing calculators have incorporated Newton's method, or a variation of it, as part of their software.

Remark

Let $p(x) = x^4 + ax^3 + bx^2 + cx + d$, where a, b, c, d are real numbers. A well-known consequence of FTA (section 10.8) implies that there exist complex numbers $r_1, ..., r_4$ such that $p(x) = (x - r_1)(x - r_2)(x - r_3)(x - r_4)$. The r_i, $1 \leq i \leq n$, are precisely the solutions of $x^4 + ax^3 + bx^2 + cx + d = 0$. Then

$$p(x) = x^4 - (r_1 + r_2 + r_3 + r_4)x^3 + ... + r_1r_2r_3r_4.$$

In particular $r_1 + r_2 + r_3 + r_4 = -a$. This simple fact will be used in the next section.

[1] Recall the formula from section 11.3, written just before a remark.

14.3 Descartes' technique

Some ninety years after the publication of *Ars Magna*, René Descartes (1600-1653) presented a new method to solve quartics. Let us illustrate his method analyzing the equation $x^4 + 4x + 8 = 10x^2$, which is none other but Problem IX (chapter 39) of Cardano's masterpiece.

The main idea is to express $x^4 - 10x^2 + 4x + 8$ as a product of two quadratic factors for certain undetermined coefficients a, b, c, d, say

$$x^4 - 10x^2 + 4x + 8 = (x^2 + ax + b)(x^2 + dx + c).$$

Let r_1, r_2 be the solutions of $x^2 + ax + b = 0$ and let r_3, r_4 be the solutions of $x^2 + dx + c = 0$. Then $r_1 + r_2 = -a$ and $r_3 + r_4 = -d$. But r_1, r_2, r_3, r_4 are solutions of a quartic without cubic term, hence $r_1 + r_2 + r_3 + r_4 = 0$. Therefore $-a - d = 0$, that is, $d = -a$. The original problem is thus reduced to finding a, b, c such that

$$x^4 - 10x^2 + 4x + 8 = (x^2 + ax + b)(x^2 - ax + c).$$

Hence

$$x^4 - 10x^2 + 4x + 8 = x^4 - ax^3 + cx^2 + ax^3 - a^2x^2 + acx + bx^2 - abx + bc =$$
$$x^4 + (c - a^2 + b)x^2 + (ac - ab)x + bc.$$

So

$$b + c - a^2 = -10,$$

$$ac - ab = 4,$$

$$bc = 8.$$

We have a system of three equations with three unknowns, which we can rewrite as follows:

$$b + c = a^2 - 10,$$

$$c - b = \tfrac{4}{a},$$

$$bc = 8.$$

Adding and subtracting the first two equalities we obtain $c = \frac{1}{2}(a^2 - 10 + \frac{4}{a})$, $b = \frac{1}{2}(a^2 - 10 - \frac{4}{a})$. Then

$$8 = bc = \tfrac{1}{4}[(a^2 - 10)^2 - \tfrac{16}{a^2}],$$

that is, $32 = a^4 - 20a^2 + 100 - \frac{16}{a^2}$. Therefore

$$a^6 - 20a^4 + 68a^2 - 16 = 0.$$

Letting $u = a^2$ we transform it into

$$u^3 - 20u^2 + 68u - 16 = 0.$$

By simple inspection we note that $u = 4$ is a solution of it. Hence $a = 2$, $b = -4$, and $c = -2$. Consequently

$$x^4 - 10x^2 + 4x + 8 = (x^2 + 2x - 4)(x^2 - 2x - 2).$$

The solutions of the quadratic equation $x^2 + 2x - 4 = 0$ are $1 \pm \sqrt{3}$ while the solutions of the quadratic equation $x^2 - 2x - 2 = 0$ are $-1 \pm \sqrt{5}$. Thus, these four numbers are the solutions of the original quartic.

Remark

Using advanced mathematics it can be proven that any algebraic method to solve general quartics will involve cubics. In other words, no matter how much we struggle in quest of a method to solve general quartics, we cannot avoid the need to solve a cubic. That is why cubics appear when we use either Ferrari's method or Descartes' method.

14.4 Quasisymmetric equations

Let us consider the equation

$$x^4 + x^3 + x^2 + 6x + 36 = 0.$$

Since $x = 0$ is not a solution of it we may well divide the equation by x^2, following the path we took in section 10.10 (first example) when solving the equation $x^4 + x^3 + x^2 + x + 1 = 0$. We get the equivalent equation

$$x^2 + x + 1 + \tfrac{6}{x} + \tfrac{36}{x^2} = 0,$$

that is, $x^2 + \frac{36}{x^2} + x + \frac{6}{x} + 1 = 0$. In turn, this equation is equivalent to

$$(x + \tfrac{6}{x})^2 - 12 + (x + \tfrac{6}{x}) + 1 = 0.$$

Defining

$$z = x + \frac{6}{x} \tag{14.5}$$

we get $z^2 + z - 11 = 0$. This is a quadratic with solutions $z_1 = \frac{-1+3\sqrt{5}}{2}$, $z_2 - \frac{1+3\sqrt{5}}{2}$. From (14.5) it follows that

$$x^2 - zx + 6 = 0.$$

Thus, the four solutions of the original equation stem from the quadratics $x^2 - z_1x + 6 = 0$ and $x^2 - z_2x + 6 = 0$.

The example that we have just discussed is a particular case of the family of equations

$$a_0x^4 + a_1x^3 + a_2x^2 + a_1mx + a_0m^2 = 0,$$

called quasisymmetric equations. They are equivalent to

$$a_0x^2 + a_1x + a_2 + \frac{a_1m}{x} + \frac{a_0m^2}{x^2} = 0,$$

that is,

$$a_0(x + \frac{m}{x})^2 - 2ma_0 + a_1(x + \frac{m}{x}) + a_2 = 0.$$

Defining $z = x + \frac{m}{x}$ we get the quadratic

$$a_0z^2 + a_1z + a_2 - 2ma_0,$$

which has two solutions, z_1 and z_2. Then the four solutions of the original equation stem from $x^2 - z_1x + m = 0$ and $x^2 - z_2x + m = 0$.

14.5 Quintics and beyond

Having solved the most general cubic and quartic, mathematicians thought that they might be able to solve the quintic and higher order polynomial equations in the sense that a formula, involving only the four arithmetic operations (addition, subtraction, multiplication, and division) and radicals, could be found. We can certainly find the solutions of some quintics, for instance, $x^5 - 13x^4 + 49x^3 - 43x^2 - 50x + 56 = 0$ has the numbers $1, -1, 2, 4, 7$ as the solutions. But the question posed is to solve *all* quintics.

At the end of the 18th century, people engaged in mathematical research started to think that it might be impossible to find a formula to solve all quintics. Their suspicions were confirmed in 1824 when Niels Abel (1802-1829) proved that, indeed, it is not possible to solve all polynomial equations of degree five or higher through the four arithmetic operations and radicals. Shortly thereafter, Evariste Galois (1811-1832) found under what conditions a quintic, or higher order polynomial equation, is solvable. Galois' work appeared in print in 1846, twelve years after the death of the author, and revolutionized algebra.

Exercises

1. Use Descartes' method to find the four solutions of $x^4 + 3 = 12x$.

2. Show that Cardano's transformation $y = x + \frac{d_1}{4}$, applied to the general quartic

$$x^4 + d_1x^3 + d_2x^2 + d_3x + d_4 = 0,$$

   leads to an equation $y^4 + ay^2 + by + c = 0$ with no cubic term. While solving the equation $x^4 + 4x + 8 = 10x^2$ (problem IX, chapter 39, of *Ars Magna*) we found its resolvent cubic $z^3 + 5z^2 - 8z - 42 = 0$ and noticed, by simple inspection, that $z = -3$ is one solution. The use of the division algorithm for polynomials leads to

$$(z + 3)(z^2 + 2z - 4) = 0.$$

   Thus, $\sqrt{5} - 1$ is another solution of the resolvent cubic. Use this solution to find the four solutions of $x^4 + 4x + 8 = 10x^2$.

3. Find the resolvent cubic of $x^4 + 6x^2 + 36 = 60x$ and calculate a four-decimal approximation of the above-mentioned cubic.

4. Use the decimal approximation, found in the previous exercise, to find four-decimal approximations to the solutions of $x^4 + 6x^2 + 36 = 60x$.

5. Find the four solutions of $x^4 + 6x^3 = 64$ (problem VII, chapter 39, of *Ars Magna*).

6. Calculate the four solutions of $x^4 + x = 0$ (Hint: A simple factorization will lead to the answer; neither Ferrari's method nor Descartes' method are needed.)

7. Suppose that we need to find the four solutions of $x^4 - x^2 + x = 1$. A simple inspection will make us realize that $x = 1$ is a solution of it. Then the division algorithm for polynomials leads to $(x - 1)(x^3 + x^2 + 1) = 0$. Thus, our task now is to solve $x^3 + x^2 + 1 = 0$.

8. Apply Ferrari's method at the very beginning, assuming that you had not noticed that $x = 1$ is a solution of the quartic from exercise 7. Is the resolvent cubic any simpler than $x^3 + x^2 + 1 = 0$? Use both paths in the quest for the solutions of $x^4 - x^2 + x = 1$.

9. Solve the quartic $x^4 - 6x^2 + 12 = 0$ using Ferrari's method. How does this method compare with the standard way of solving biquadratics? (Hint: The number $2\sqrt{3}$ happens to be a root of the corresponding resolvent cubic.)

10. Given the quartic $x^4 + 6x^3 - 64 = 0$, use the transformation $y = \frac{4}{x}$ to obtain an equation without a cubic term. Then solve this new equation using Ferrari's method. Further question: Would it have been more difficult to use Cardano's transformation at the very beginning?

Enrichment Note: Ferrari's Method in a General Setting

As we know, the most general quartic that we are studying is $x^4 + d_1x^3 + d_2x^2 + d_3x + d_4 = 0$ where the coefficients are real numbers. Using Cardano's transformation $y = x + \frac{d_1}{4}$ we can convert it into an equation of the type

$$y^4 + ay^2 + by + c = 0.$$

Next we will apply Ferrari's technique to this equation. Indeed, $y^4 = -ay^2 - by - c$. Thus

$$(y^2 + z)^2 = 2y^2z + z^2 - ay^2 - by - c$$

for any z. Rearranging terms we get

$$(y^2 + z)^2 = (2z - a)y^2 - by + (z^2 - c).$$

Our immediate goal is to find a real number z such that the right hand side of the last equality becomes a perfect square. This will happen provided that its discriminant is zero; that is,

$$b^2 - 4(2z - a)(z^2 - c) = 0. \tag{14.6}$$

We are now dealing with a cubic equation, customarily called the 'resolvent cubic.' Either by simple inspection or using Cardano-Tartaglia's method we can find a real solution z_1 of it. Then

$$(y^2 + z_1)^2 = (2z_1 - a)y^2 - by + z^2 - c = (2z_1 - a)(y - \tfrac{b}{2(2z_1-a)})^2$$

provided that $2z_1 - a \neq 0$.

1. Assume that $2z_1 - a > 0$. Then

$$(y^2 + z_1)^2 = (\sqrt{2z_1 - a}(y - \tfrac{b}{2(2z_1-a)}))^2.$$

Consequently

$$y^2 + z_1 = \sqrt{2z_1 - a}(y - \tfrac{b}{2(2z_1-a)}) \text{ or } y^2 + z_1 = -\sqrt{2z_1 - a}(y - \tfrac{b}{2(2z_1-a)}).$$

Solving these quadratics we get the four roots of $y^4+ay^2+by+c=0$, which in turn lead to the four roots of the original quartic through the use of Cardano's transformation.

2. Assume that $2z_1 - a < 0$. Under these circumstances the expression

$$(y^2+z_1)^2 = (2z_1-a)(y - \tfrac{b}{2(2z_1-a)})^2$$

becomes

$$(y^2+z_1)^2 = (i\sqrt{a-2z_1})^2(y - \tfrac{b}{2(2z_1-a)})^2.$$

Then

$$y^2+z_1 = i\sqrt{a-2z_1}(y - \tfrac{b}{2(2z_1-a)}) \text{ or } y^2+z_1 = -i\sqrt{a-2z_1}(y - \tfrac{b}{2(2z_1-a)}).$$

These are two quadratic equations with complex coefficients. Their four roots are the solutions of $y^4+ay^2+by+c=0$. The solutions of the original quartic are obtained using Cardano's transformation.

We still have to address the possibility that $2z_1 - a = 0$. From (14.6) we get $b^2 = 0$, hence $b = 0$. Therefore

$$(y^2+z_1)^2 = z_1^2 - c.$$

1. Assume that $z_1^2 - c > 0$. Then $(y^2+z_1)^2 = (\sqrt{z_1^2-c})^2$. So $y^2+z_1 = \sqrt{z_1^2-c}$ or $y^2+z_1 = -\sqrt{z_1^2-c}$. From these two equations we obtain the four solutions of $y^4+ay^2+by+c=0$.

2. Assume that $z_1^2 - c < 0$. We will have $(y^2+z_1)^2 = z_1^2 - c$, therefore $(y^2+z_1)^2 = (i\sqrt{c-z_1^2})^2$. Thus

$$y^2+z_1 = i\sqrt{c-z_1^2} \text{ or } y^2+z_1 = -i\sqrt{c-z_1^2}.$$

From these two equations we get the four needed solutions.

3. We have to consider the possibility that $z_1^2 - c = 0$. Under these circumstances $y^2+z_1 = 0$, hence $y^2 = -z_1$. Then $y = \pm\sqrt{-z_1}$ if $-z_1 \geq 0$, $y = \pm i\sqrt{z_1}$ if $-z_1 < 0$. There will be two solutions, each of them with multiplicity two. Of course, if $z_1 = 0$ then $y^4+ay^2+by+c=0$ has only 0 as a solution (with multiplicity 4).

Chapter 15

Some Uses of Technology

15.1 Approximations of π

In chapter 7, on the geometry of the circle, we found the recursive system

$$s_{n+1} = \sqrt{2 - \sqrt{4 - s_n^2}}, \quad n = 0, 1, 2, \dots .$$

$$s_0 = \sqrt{3}.$$

Let us recall that s_n is the length of the regular polygon of 3×2^n sides inscribed in a unit circle. Then

$$p_n = 3 \times 2^n \times s_n$$

is the perimeter of the above-mentioned polygon. Since $p_n \to p = 2\pi$ we can assert that $\pi \approx \frac{p_n}{2}$ when n is large. Let us build a program using a TI-83 (or any similar graphing calculator):

Program: Archimedes 1
Input "n", n
$\sqrt{3} \to s$
For(I,1,n)
$\sqrt{2 - \sqrt{4 - s^2}} \to s$
End
$3 * 2^n * s \to p$
Disp $p/2$

Comment

The *For* command replaces $\sqrt{3}$ in the formula $\sqrt{2-\sqrt{4-s^2}}$ in order to obtain a new value for s and keeps repeating this process n times. Exactly what we are supposed to do with a recursive system.

Finding the first six decimals of π

By choosing $n = 5$ (a regular polygon with $3 * 2^5$, or 96 sides) we see on the screen the value 3.141031951. Using $n = 6$ (a regular polygon with 192 sides) the value on the screen happens to be 3.141452472, while for $n = 10$ (a regular polygon with 3072 sides) the output is 3.141592106. Taking into consideration that a ten-decimal approximation of π is 3.1415926535 (Beckmann, 1977), we can conclude that for $n = 10$ we have found the first six 'true' decimals of π.

The phenomenon of cancellation

Could we find more 'true' decimals by taking a larger value of n, say $n = 15$? Surprisingly, for $n = 15$ we get 3.141616513. We have lost four 'true' decimals of π. What has happened? When n is large, s becomes very small; thus, in the expression $\sqrt{2-\sqrt{4-s^2}}$ we are subtracting two numbers that are very close to each other, namely 2 and $\sqrt{4-s^2}$. This fact leads to a computational problem called 'cancellation' (Lotspeich, 1988), which determines a loss of significant digits.

To avoid 'cancellation' we start by doing a little bit of algebra:

$$\sqrt{2-\sqrt{4-s^2}} = \frac{\sqrt{2-\sqrt{4-s^2}}\sqrt{2+\sqrt{4-s^2}}}{\sqrt{2+\sqrt{4-s^2}}} = \frac{\sqrt{4-(4-s^2)}}{\sqrt{2+\sqrt{4-s^2}}} = \frac{s}{\sqrt{2+\sqrt{4-s^2}}}.$$

The following modified program works well for any n:

Program: Archimedes 2

```
Input "n?", n
√3 → s
For(I, 1, n)
s/√(2+√(4−s²)) → s
End
3 * 2^n * s → p
Disp p/2
```

For instance, $n = 20$ leads to 3.141592654. Now we have eight 'true' decimals of π (our calculator is set to provide nine decimals on the screen, but the last is usually not reliable because it is an approximation). Higher values of n lead to the same number on the screen. We cannot do any better insofar as the calculator settings are not changed.

There are many other ways of calculating π, especially using series (Halpin and Helfgott, 2003).

15.2 The birthday problem

Suppose there are n students in a classroom and we wish to find out what is the probability that at least two were born on the same day. The sample space S will be the set of all n-tuples $(a_1, ..., a_n)$ where a_i can be any positive integer between 1 and 365 (we do not consider leap years). Thus, $\#(S)=365^n$. That is to say, the number of elements in S is precisely 365^n.

Let E be the event that at least two among n people were born on the same day. We wish to determine $\#(E)$, the number of elements in E. It will be easier to calculate $\#(E^c)$ (the number of elements in the complement of E) and then use the fact that $\#(E) + \#(E^c) = \#(S) = 365^n$.

How do we calculate $\#(E^c)$? For the first person we have 365 possibilities, while for the second person the number of possibilities has been reduced to 364. Continuing in this fashion we can assert that for the nth person we have $365 - n + 1$ different possibilities. By the multiplication principle it follows that

$$\#(E^c) = 365 \times 364 \times ... \times (365 - n + 1).$$

Thus

$$\#(E) = 365^n - 365 \times 364 \times ... \times (365 - n + 1).$$

Each of the 365^n simple events has the same probability, namely $\frac{1}{365^n}$. Then the probability of E, namely $p(E)$, is

$$p(E) = \frac{\#(E)}{\#(S)} = \frac{\#(E)}{365^n} = 1 - \frac{365 \times 364 \times ... \times (365 - n + 1)}{365^n}$$

The next challenge is to build a program to calculate $p(E)$ for any n.

Program: Birthday 1, TI83

```
Input "n?", n
365 → a
For(I,2,n)
a * (365 − I + 1) → a
End
Disp 1 − a/365^n
```

Table 15.1: Birthday problem probabilities

n	p
20	0.41
30	0.71
40	0.89
50	0.97
60	0.99

The program Birthday 1 works well up to $n = 39$. Thereafter, for $n \geq 40$, it displays the words ERROR(OVERFLOW). To avoid this difficulty let us make a small modification to the program.

Program: Birthday 2, TI83

Input "n?", n
$1 \to a$
For (I,1,n)
$\frac{a*(365-I+1)}{365} \to a$
End
Disp $1 - a$

The reader can check that the program Birthday 2 works for any n, without restrictions. It should be noted that a slight modification leads to a program with a TI-89 or similar calculators:

Program: Birthday 2, TI89

birthday( )
Prgm
Input "n", n
$1 \to a$
For i, 1, n
$\frac{a*(365-i+1)}{365} \to a$
EndFor
Disp $1 - a$
EndPrgm

15.3 The phenomenon of refraction

Fermat's Principle of Least Time asserts that when a ray of light goes from one point to another, it does so in the least amount of time. Accepting this principle, a cornerstone of optics, we will provide evidence in the sense that Snell's law of refraction is valid, that is,

$$\frac{\sin\alpha}{\sin\beta} = \frac{v_1}{v_2}. \tag{15.1}$$

It is to be noted that α is the angle of incidence and β is the angle of refraction, while v_1 is the speed of light in the upper medium and v_2 is the speed of light in the lower medium. Willebrord Snell (1580-1626) was a Dutch scientist.

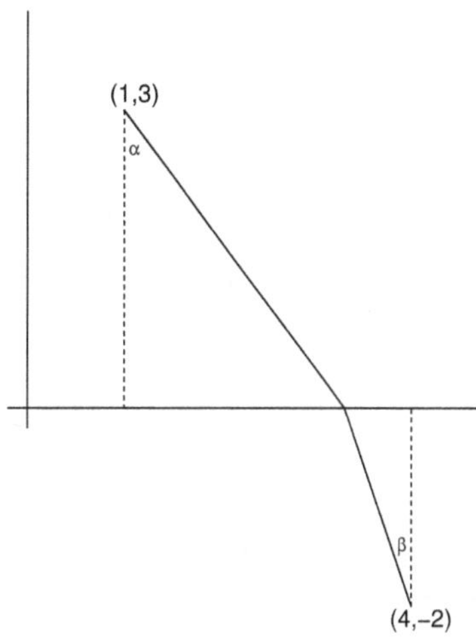

Figure 15.1: A particular example of the phenomenon of refraction

Suppose that a ray of light goes from $(1, 3)$, in air, down to $(4, -2)$ in a block of glass. The x-axis is set at the border of both media (Figure 15.1). The speed of light in air is $v_1 = 3$ cm/ns while we assume that the speed of light inside the block of glass is $v_2 = 2$ cm/ns, where 1 ns=10^{-10} seconds. Since time = distance/speed, the time function will be

$$f(x) = \tfrac{1}{3}\sqrt{(x-1)^2 + 3^2} + \tfrac{1}{2}\sqrt{(4-x)^2 + 2^2}, \quad 1 \le x \le 4.$$

Graphing this function and using the *min* option from a TI-83 (found under *2nd Calc*) we get $x = 3.1554852$. Then, using the Pythagorean proposition and the definition of sine, we get $\sin\alpha = 0.5834994463$ and $\sin\beta = 0.3889997186$. Hence

$$\tfrac{\sin\alpha}{\sin\beta} = 1.499999662 \approx 1.5 = \tfrac{v_1}{v_2}.$$

To avoid repeated calculations we may as well write a program keeping in mind Figure 15.2 (Helfgott, 1998).

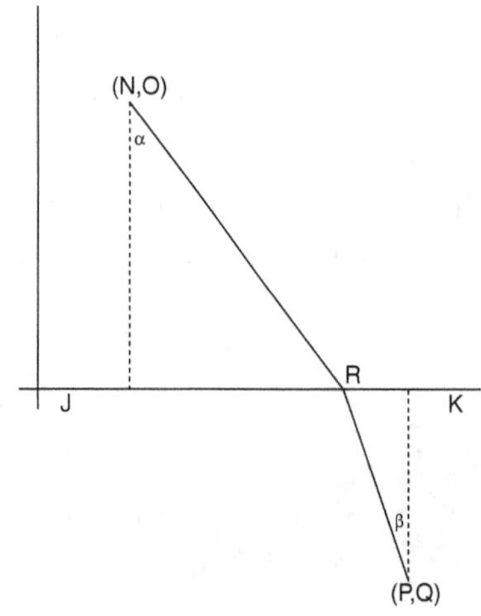

Figure 15.2: The phenomenon of refraction

Program: Refrac

Input "left bound?", J

Input "right bound?", K

Input "$x1$?", N

Input "$y1$?", O

Input "$x2$?", P

Input "$y2$?", Q

fMin$((1/3)\sqrt{((x-N)^2+O^2)}+(1/2)\sqrt{((P-x)^2+Q^2)}, x, J, K) \to R$

$R - N \to S$

$P - R \to U$

$\sqrt{S^2+O^2} \to M$

$\sqrt{U^2+Q^2} \to W$

$S/M \to A$

$U/W \to B$

Disp A/B

Remark

The left bound (J) and the right bound (K) need to be chosen in order that the command *fMin* works. The numbers N, O, P, Q are subjected to the following inequalities: $J \le N \le K$, $J \le P \le K$, $O > 0$, $Q < 0$.

Running the program

Once the program is understood, students can choose several points in air and in glass and verify that Snell's Law holds if we accept Fermat's Principle. For instance, choosing $J = -1$ and $K = 10$, for the points $(2, 7)$ and $(8, -1)$ the program provides the number 1.499994133, while for $(3, 21)$, $(6, -10)$ the program provides the number 1.499991053. In all cases we get a number very close to 1.5, which is the quotient v_1/v_2. This way we garner evidence in the sense that Fermat Principle of Least Time implies Snell's law of refraction. Is it possible to provide a mathematical proof? It can be done without using Calculus (Helfgott and Helfgott, 2002), but it is quite a laborious task. If we were to use Calculus, the proof is rather straightforward. It should be noted that Gottfried Leibniz (1646-1716) dealt with this problem in the first ever published paper on Calculus (1684). Does Snell's law imply Fermat's Principle of Least Time? The answer is yes (Golomb, 1964).

A problem from optics

Having settled the question about the equivalence between Fermat's Principle and Snell's law, it is time to solve a specific problem from the realm of optics. Suppose we have air on the first medium and water on the second medium (the speed of light in water is, approximately, 2.25 cm/ns). Assume that a ray of light starts at $(4, 3)$ and we wish to find the measure of angle α, in degrees, so that the ray hits the point $(7, -2)$ (Figure 15.3). We define the pertinent time function, namely

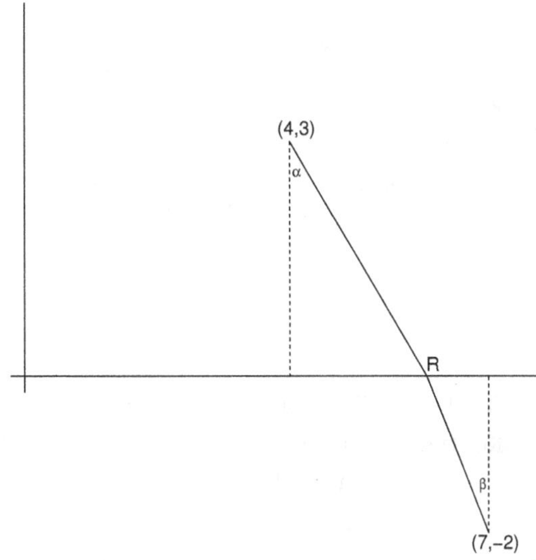

Figure 15.3: A specific problem of refraction

$$f(x) = \frac{1}{3}\sqrt{(x-4)^2+3^2} + \frac{1}{2.25}\sqrt{(7-x)^2+2^2}.$$

Using a graphing calculator we can find where it attains its minimum: 6.061593727. Then

$$\tan\alpha = \frac{6.061593727-4}{3} = \frac{2.061593727}{3} = 0.687197909.$$

Thus

$$\alpha = \tan^{-1}(0.687197909) = 34.49676805^o.$$

An alternative approach uses Snell's law: If α is the angle of incidence and β is the angle of refraction, then

$$\frac{\sin\alpha}{\sin\beta} = \frac{v_1}{v_2} = \frac{3}{2.25} = \frac{4}{3}.$$

Let $(R, 0)$ be the point where the ray of light touches the surface of separation between air (on the top) and water (on the bottom). Since

$$\sin\alpha = \frac{R-4}{\sqrt{(R-4)^2+3^2}}$$

and

$$\sin\beta = \frac{7-R}{\sqrt{(7-R)^2+2^2}}$$

we have that

$$\frac{(R-4)\sqrt{(7-R)^2+4}}{(7-R)\sqrt{(R-4)^2+9}} = \frac{4}{3}.$$

Squaring the preceding equality, we obtain

$$\frac{(R-4)^2[(7-R)^2+4]}{(7-R)^2[(R-4)^2+9]} = \frac{16}{9}.$$

That is,

$$9(R-4)^2((7-R)^2+4) - 16(7-R)^2((R-4)^2+9) = 0.$$

This is a quartic equation. Using a graphing calculator we get two real solutions: 6.06159 and 8.61793. We keep only the first one since the other has no physical sense (the ray crosses the interface at a point R, $4 < R < 7$). As expected, $R = 6.06159$ is exactly the same value we found when Fermat's Principle was employed. Then, to finish the problem we follow the same steps used before:

$$\tan\alpha = \frac{6.0159-4}{3} = 0.6871966667.$$

Hence $\alpha = 34.4967197^o$.

15.4 The ship and ambulance problem

A ship is 30 miles from a point that is 65 miles down-shore from a hospital (Figure 15.4). You are the captain and need to order an ambulance, which starts from the hospital, to meet the ship at a point on the shoreline that will minimize the time needed to bring an ill passenger to the hospital. Assuming that the ship travels at the constant speed of 20 miles/hour and the ambulance travels at the constant speed of 50 miles/hour, where should the ambulance meet the ship?

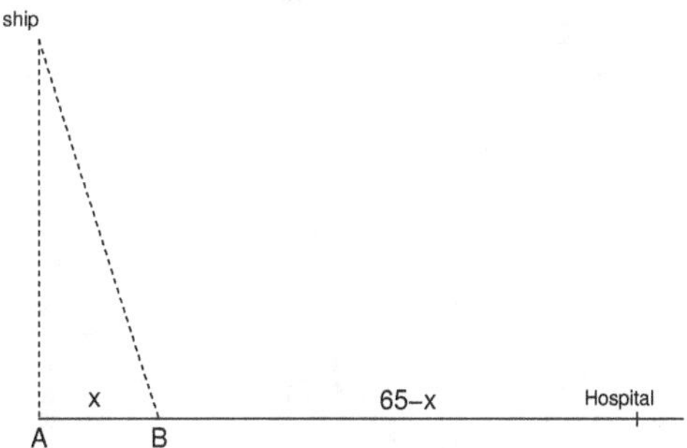

Figure 15.4: Taking a sick person to the hospital in the shortest amount of time

We have to deal with the function

$$f(x) = \frac{\sqrt{30^2+x^2}}{20} + \frac{65-x}{50}, \quad 0 \leq x \leq 65.$$

This function provides the time it takes the boat to reach a point B on the shore, which is at a distance x from A, and from there to the hospital. First of all let us check whether the ambulance will arrive at any point, between A and the hospital, before the ship. Indeed,

$$\frac{65-x}{50} \leq \frac{65}{50} = \frac{13}{10} < \frac{3}{2} = \frac{30}{20} = \frac{\sqrt{30^2+0^2}}{20} \leq \frac{\sqrt{30^2+x^2}}{20}.$$

Thus,

$$\frac{65-x}{50} < \frac{\sqrt{30^2+x^2}}{20}.$$

Next we graph the function $f(x)$ and choose the *minimum* option to find where the minimum is adopted. On the screen we get the number 13.093082 (the corresponding y-coordinate is 2.6747727, which is the minimum time). In summary, the best that the captain can do is to land 13.09 miles from A. The total time needed to bring the ill passenger to the hospital is 2.67 hours.

There is an alternative approach to the problem that uses the fact that Snell's law of refraction is equivalent to Fermat's Principle of Least Time, in the sense that one implies the other. When β, the angle of refraction, gets closer and closer to 90° (Figure 15.5) the segment $\overline{OQ}$ is rotated toward the surface of separation. In the limit, the law of refraction asserts that

$$\frac{\sin\alpha}{\sin 90} = \frac{20}{50},$$

hence $\sin\alpha = 0.4$. But $\sin\alpha = \frac{x}{\sqrt{x^2+30^2}}$, therefore $\frac{x^2}{x^2+900} = 0.4^2$.

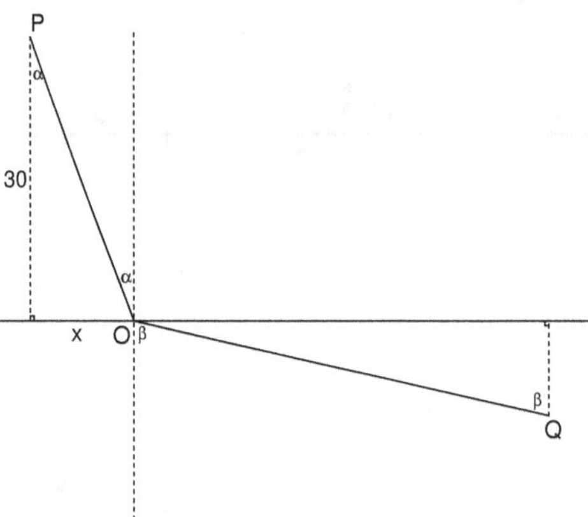

Figure 15.5: An alternative solution to the ship and ambulance problem

Solving this quadratic equation we get $x = 13.09$, exactly the same answer found before. The reader may have noticed that the optimal value, 13.09 miles, is independent from the distance d between A and the hospital, provided that we make sure that the ambulance arrives first. This will happen if $d < 75$ miles.

Extensions to the ship and ambulance problem can be found in the literature (Helfgott and Lutz, 2002).

15.5 Annuities

Across the book we have often discussed applications, either to basic physics (mainly kinematics and optics) or to problems with underlying geometrical interest. They help to provide a clear motivation for several topics that otherwise might seem far removed from our experience.The world of finance is a rich source of applications too, especially the topic of annuities. The study of the latter requires very little background and its usefulness cannot be contested.

The formula for annuities

Let us recall the basic formula for compound interest, namely $A = P(1+i)^n$ where P is the 'principal,' that is, the amount deposited, i is the monthly interest rate and n is the number of months (it works as well if i stands for the yearly interest rate and n denotes the number of years). How could we justify the above-mentioned formula? We deposit P dollars and at the end of the first month we will have $P + iP$ dollars, that is, $P(1+i)$. At the end of the second month we will have $P(1+i) + iP(1+i)$ dollars. This last expression is none other than $P(1+i)(1+i)$, that is, $P(1+i)^2$. It is then evident that at the end of n months our deposit amounts to $P(1+i)^n$.

Let us now look at a different problem: If i is the monthly interest rate, what capital, P_1, should be deposited in a bank in order to draw R dollars at the end of one month? Applying the basic formula for compound interest we will have $R = P_1(1+i)$; thus, $P_1 = R(1+i)^{-1}$. Similarly, if we wish to draw R dollars at the end of two months we should have deposited P_2 dollars at the beginning, so that $R = P_2(1+i)^2$. Hence $P_2 = R(1+i)^2$. In a similar fashion, if we wish to draw R dollars from our account at the end of the nth month, we should have deposited P_n dollars at the beginning, where $P_n = R(1+i)^{-n}$. Therefore, if we wish to draw R dollars from the account after the first, second,...,n^{th} month, we should have deposited in the bank $P = P_1 + ... + P_n$ dollars. That is,

$$P = R(1+i)^{-1} + ... + R(1+i)^{-n}.$$

The reader can check that, for any $a \neq 1$,

$$1 + a + ... + a^n = \tfrac{1-a^{n+1}}{1-a}.$$

Hence

$$a + a^2 + ... + a^n = \tfrac{a-a^{n+1}}{1-a} = \tfrac{1-a^n}{\frac{1}{a}-1}.$$

Letting $a = \frac{1}{1+i}$ we get

$$P = R\tfrac{1-(1+i)^{-n}}{1+i-1}.$$

Therefore

$$P = \frac{R}{i}[1 - (1+i)^{-n}] \tag{15.2}$$

This is the *fundamental formula for annuities.* Observe that

$$\lim_{n\to\infty} \frac{R}{i}[1 - (1+i)^{-n}] = \frac{R}{i}. \tag{15.3}$$

That is to say, a person has to deposit $\frac{R}{i}$ dollars in a bank (that pays a monthly interest i) if he/she wishes to draw R dollars each month every single month, that is, indefinitely.

Examples

We will discuss four examples related to annuities. The first three require a pretty standard use of graphing calculators, but the fourth requires a careful choice of windows.

1. John wishes to buy a car from an automobile enterprise. He is given a loan of $10,000 to be repaid in 60 equal monthly installments. If the bank charges 1% monthly interest, how much is each installment? We just need to apply the fundamental formula:

   $$10,000 = \tfrac{R}{0.01}[1 - 1.01^{-60}].$$

   Therefore $R = \frac{100}{1-1.01^{-60}} = 222.44$ dollars.

2. Mary deposits $15,000 in a bank and wishes to draw $400 at the end of each month. Taking into consideration the fact that the bank pays 1% of interest per month, for how many months can Mary draw 400 dollars? From the fundamental formula for annuities we have

   $$15,000 = \tfrac{400}{0.01}[1 - 1.01^{-n}].$$

   Then $\frac{150}{400} = 1 - 1.01^{-n}$, thus $1.01^{-n} = 1 - \frac{3}{8} = \frac{5}{8}$. Consequently $1.01^{n} = \frac{8}{5}$. Taking logarithms we get $n \log 1.01 = \log \frac{8}{5}$, hence $n = 47.2349$.

   Suppose that Mary makes 48 payments. Evidently, her last payment should not be $400 but a lesser amount.Indeed, if she were to make 48 payments of $400 each, the original deposit would be

   $$P = \tfrac{400}{0.01}[1 - 1.01^{-48}] = 15,189.5837974.$$

This should have been the loan at the very beginning. Since the loan amounted to only \$15,000, the difference of \$189.5837974 had to be in a separate account earning interest for 48 months (there should be an agreement between the bank and Mary). Applying the formula for compound interest we get $(189.5837974)1.01^{48} = 305.65$. Thus, the last payment should be $400 - 305.65 = 94.35$.

3. Josh deposits \$100,000 in a bank that pays 10% on a yearly basis. He wants his heirs to draw \$5,000 at the end of each year, toward tuition. For how many years can his heirs draw the above-mentioned amount? Applying the fundamental formula for annuities we get

$$100,000 = \tfrac{5,000}{0.1}[1 - 1.1^{-n}].$$

Therefore $2 = \frac{10}{5} = 1 - 1.1^{-n}$, which in turn leads to $-1 = 1.1^n$. We have reached an apparent paradox because 1.1^n can never be negative. What has happened?

We have to remember that a deposit of R/i dollars provides yearly drawings of R dollars indefinitely. Thus, Josh should have deposited, at the beginning, only $5,000/0.1$ dollars. That is, 50,000 dollars instead of 100,000 dollars. The other 50,000 dollars should be deposited in another account in order to gain interest.

4. Carmen is given a loan of \$20,000, to be repaid in 60 months, each installment valued at \$4000. What monthly interest is the bank charging? From the fundamental formula for annuities we get

$$20,000 = \tfrac{400}{i}[1 - (1+i)^{-60}].$$

Therefore $50i = 1 - (1+i)^{-60}$. Simplifying we reach the equation $50i(1+i)^{60} - (1+i)^{60} + 1 = 0$. We might be tempted to use the powerful *Solve* command, but a standard calculator is not able to provide an answer to

$$\mathit{Solve}(50x \times (1+x)^{60} - (1+x)^{60} + 1 = 0, x).$$

Rather, let us use a graphing calculator (with windows $x_{min} = 0$, $x_{max} = 0.015$, $y_{min} = -0.1$, $y_{max} = 0.1$) to graph the function

$$f(x) = 50x(1+x)^{60} - (1+x)^{60} + 1.$$

The *zero* command leads to the number 0.006183. Thus, the answer is $i = 0.6183\%$, the monthly interest rate. The challenge in this problem is to find an appropriate window!

15.6 The optimal box

Suppose that we have a rectangular closed box whose base is a square. The surface area of the box is a given number L. What are the dimensions of the box of maximum volume? Let x be the side of the base and y the height of the box. Then $4xy+2x^2 = L$, so $y = \frac{L-2x^2}{4x}$. Thus

$$V(x) = \tfrac{x}{4}(L - 2x^2).$$

Next we choose a value for L, say $L = 5$, and use our graphing calculator to graph $V(x)$. The *calc* command will provide the approximate point where this function attains its maximum, namely 0.91287202. Then we calculate the corresponding height and find out that the side of the base and the height are practically the same. In other words, the cube with side of length 0.9128 has the maximum volume among all closed boxes with $L = 5$. Evidently, if the answer is a cube then the side of this cube has to be $\sqrt{5/6}$ since $4x^2 + 2x^2 = 5$. A further step leads us to the construction of a simple four-step program to test our conjecture.

Program: MaxVol
Input "surface?", L
fmax$(0.25x * (L - 2x^2), x, 0, \sqrt{L/2}) \to R$
$(L - 2R^2)/(4R) \to S$
Disp R, S

Remark

We chose $\sqrt{L/2}$ as the right bound inside the *fmax* command since $L - 2x^2 > 0$ implies $x < \sqrt{L/2}$; any number bigger than $\sqrt{L/2}$ will work as well.

Running the program

For instance, if $L = 10$ we will get on the screen the two numbers 1.290993139 and 1.290997068 once we run the program. These numbers are pretty close to each other, as predicted by our conjecture. Experimenting with many different values for L we may conclude that it is highly likely that the height of the box of maximum volume should be equal to the length of the side of its base. That is to say, a cube of side $\sqrt{L/6}$ seems to be the answer. The examples on refraction of light and a closed box show how a systematic quasi-empirical approach may be used as a first step in the analysis of a problem.

Searching for a mathematical proof

Is it possible to find an acceptable mathematical proof that does not use calculus? The answer is yes, provided that we apply the inequality between the arithmetic and geometric means for any three positive numbers, a topic that was discussed in the first chapter. Let us recall that, given $a, b, c > 0$, $\sqrt[3]{abc} \le \frac{1}{3}(a+b+c)$, where equality holds if and only if $a = b = c$. Applying the AGM inequality to the problem of the closed box we get

$$V^2 = \tfrac{1}{16}x^2(L-2x^2)(L-2x^2) = x^2(\tfrac{L}{4}-\tfrac{x^2}{2})(\tfrac{L}{4}-\tfrac{x^2}{2}) \le [\tfrac{1}{3}(x^2+\tfrac{L}{4}-\tfrac{x^2}{2}+\tfrac{L}{4}-\tfrac{x^2}{2})]^3 = (\tfrac{1}{3}\tfrac{L}{2})^3.$$

The function $V^2(x)$ will attain its maximum when $x^2 = \frac{L}{4} - \frac{x^2}{2}$, that is, when $x = \sqrt{L/6}$. This is the same point where $V(x)$ will attain its maximum. Corresponding to $x = \sqrt{L/6}$ we will have

$$y = \tfrac{L-2(L/6)}{4\sqrt{L/6}} = \sqrt{\tfrac{L}{6}}.$$

Thus, the answer is to build a cube. Undoubtedly, the use of Calculus would lead to the solution of the problem right away because $V(x)$ happens to be a simple polynomial of degree 3. However, students will probably appreciate better the power of Calculus if they are first exposed to a purely algebraic approach to the problem.

Exercises

1. Suppose we have air on the top and water at the bottom. A ray of light starts at (3,2). What should be the angle α so that the ray hits the point (5,-3)? The velocity of light in air is 300,000 Km/s while the velocity of light in water is 225,000 Km/s. Solve this problem using Fermat's Principle of Least Time.

2. Solve the preceding problem using Snell's law.

3. Build a program with your graphing calculator in order to explore whether there are triangles whose sides are three consecutive numbers and their area is a natural number too. Find two such triangles.

4. Given a right cylinder with fixed volume 100, what should be the height and the radius of the base in order that the surface S be minimal? Use a graphing calculator to solve this problem, once you realize that $S(x) = 2\pi x^2 + \frac{200}{x}$, where x is the radius. It seems that the height has to be twice as big as the radius. Test this conjecture building an appropriate program.

5. Use your graphing calculator to solve the ship and ambulance problem if the distance downshore from the hospital is 70 miles. All other parameters remain the same (ship velocity: 20 mph; ambulance velocity: 50 mph; distance from the ship to the shore: 30 miles.)

6. Solve the preceding problem using the equivalence between Fermat's Principle of Least Time and the law of refraction of light.

7. Keep all parameters as in problem 5, but assume that the hospital is not located on the shore but 2 miles inland and the ambulance can drive in a straight line to the pickup point. Where should the ambulance wait for the ship?

8. You deposit \$30,000 in a bank that pays 5% on a yearly basis and wish to draw \$3,000 at the end of each year. For how many years can you retrieve money from the bank?

9. John is given a loan of \$10,000 to be repaid in 48 months, each installment amounting to \$300. What monthly interest is the bank charging the client?

10. Mary is given a loan of \$5,000 to be repaid in 60 equal monthly installments. If the bank charges 1.2% interest per month, how much will be each payment?

Bibliography

[1] Baloglou, G. and Helfgott, M. (2004), Finding Equations of Tangents to Conics, *The AMATYC Review*, Vol. 25, No. 2, pp. 35-45.

[2] Baloglou, G. and Helfgott, M. (2008), Angles, Area, and Perimeter Caught in a Cubic, *Forum Geometricorum*, Volume 8, pp. 13-25.

[3] Beckmann, P. (1977), *A History of* π, The Golem Press, 4th edition.

[4] Berele, A. and Goldman, J. (2001), *Geometry: Theorems and Constructions*, Prentice Hall.

[5] Birkhoff, G.D. (1932), A Set of Postulates for Plane Geometry, Based on Scale and Protractor, *Annals of Mathematics*, Vol. 33, No. 2, pp. 329-345.

[6] Burton, D.M. (2003), *The History of Mathematics*, 5th edition, McGraw Hill.

[7] Fetisov, A.I. (1963), *Proof in Geometry*, D.C. Heath and Company.

[8] Gilbert, G. and MacDonnell, D. (1963), The Steiner-Lehmus Theorem, *American Mathematical Monthly*, Vol. 70, pp. 79-80.

[9] Grunbaum, B. (2012), Is Napoleon's Theorem *Really* Napoleon's Theorem?, *American Mathematical Monthly*, pp. 495-501.

[10] Golomb, S.(1964), Elementary Proofs for the Equivalence of Fermat's Principle and Snell's Law, *American Mathematical Monthly*, Vol. 71, pp. 541-543.

[11] Halpin, P. and Helfgott, M (2004). Programming Recursive Algorithms and Approximations of π, *New York State Mathematics Teacher's Journal*, Vol. 54, No. 1, pp. 28-32.

[12] Hahn, L-S. (1996), *Complex Numbers in Geometry*, Mathematical Association of America.

[13] Helfgott, M. (2012), *Calculus of One Variable: An Eclectic Approach*, ISBN 978-1477633878.

[14] Helfgott, M. (1998), Computer Technologies and the Phenomenon of Refraction, *The Physics Teacher*, pp. 236-238.

[15] Helfgott, H. and Helfgott, M. (2009), A Modern Vision of the Work of Cardano and Ferrari on Quartics, *Convergence*, Mathematical Association of America.

[16] Helfgott, H. and Helfgott, M. (2002), A noncalculus proof that Fermat's principle of least time implies the law of refraction, *American Journal of Physics*, Vol. 70 (12), pp. 1224-1225.

[17] Helfgott, M. and Lutz, M. (2002), The Boat and Ambulance Problem Revisited, *Mathematics Teacher*, Vol. 95, No. 6, pp. 270-274.

[18] Hungerford, T. (1997), *Abstract Algebra*, 2nd. edition, Saunders College Publishing.

[19] Isaacs, I.M. (2001), *Geometry for College Students*, pp. 70-71, Brooks/Cole.

[20] Lotspeich, R. (1988), Archimedes' Pi - an Introduction to Iteration, *Mathematics Teacher*, March 1988, 208-210.

[21] Moise, E. (1963), *Elementary Geometry from an Advanced Point of View*, Addison Wesley.

[22] McShane, E.J. (1941), The Addition Formulas for the Sine and Cosine, *American Mathematical Monthly*, Vol. 48, pp. 688-689.

[23] Natanson, I.P. (1960), *Simplest Maxima and Minima Problems*, D.C. Heath and Company.

[24] Newburgh, R. (1996), Real, Imaginary, and Complex Numbers: Where does the Physics Hide?, *The Physics Teacher*, Vol. 34, pp. 23-25.

[25] Oliver, B.M. (1993), Heron's Remarkable Triangle Area Formula, *The Mathematics Teacher*, Vol. 86, No. 2, pp. 161-163.

[26] Schaumberger, N. (1962), A Simple Proof of the Formula for $\sin(A+B)$, *Mathematics Magazine*, Vol. 35, p. 229.

[27] Swetz, F. (2004), Using Problems from the History of Mathematics, *Convergence*, Mathematical Association of America.

[28] Taback, S. (1990), Coordinate Geometry: A Powerful Tool for Solving Problems, *Mathematics Teacher*, pp. 264-268.

[29] Usiskin, Z., Hirschhorn, D., and Coxford, A. (2002), *Geometry*, University of Chicago Mathematics Project.

[30] Xenos, T. (2006), *Selected Topics in Mathematics*, pp. 248-249, Ziti Publishers, Thessaloniki, Greece (in Greek).

Appendix A

A.1 List of enrichment notes

1. Saccheri-Legendre Theorem (Chapter 2).
2. Legendre's Theorem (Chapter 3).
3. Euclid's proof of Pythagoras' Theorem (Chapter 4)
4. Euler Line (Chapter 5).
5. Napoleon's Theorem (Chapter 6).
6. Proving a basic equality (Chapter 7).
7. Yet another proof of the addition formula (Chapter 8)
8. Volume of a cone and a sphere (Chapter 11).
9. The nature of the solutions of a cubic equation (Chapter 13).
10. Ferrari's method in a general setting (Chapter 14).

A.2 A set of axioms for neutral geometry

As undefined terms we choose the notions of point, line, and plane. All other terms are defined in the context of set theory[1]. The first three axioms are as follows:

1. Given two points A, B there is one and only one line L that passes through both points (we write $L = \overleftrightarrow{AB} = \overleftrightarrow{BA}$.)

2. For any different points P and Q there is a real positive number called *distance*, denoted PQ or QP.

3. Every line can be put in a one-to-one correspondence with the real numbers (this correspondence is called a 'coordinate system' for the line; thus, we can measure the distance between any two points using a ruler[2]).

Before stating the next axiom we need some definitions. Given three points A, B, and C, lying on a line, B is between A and C if $AB + BC = AC$ (commonly written $A - B - C$). The *segment* $\overline{AB}$ denotes the set that contains A, B and every point in between. Besides, a set Ω is 'convex' if, for any P, Q in Ω, $\overline{PQ} \subseteq \Omega$. Having defined the notions of *segment* and *convex set* we can state the fourth axiom, namely:

4. Given an arbitrary line L, the points on the plane that do not belong to L are divided into two disjoint convex sets (called half-planes, H_1 and H_2). Moreover, if P belongs to H_1 and Q belongs to H_2 then $\overline{PQ}$ intersects the line (Figure A1). This axiom has a name: **Axiom of Separation of the Plane**.

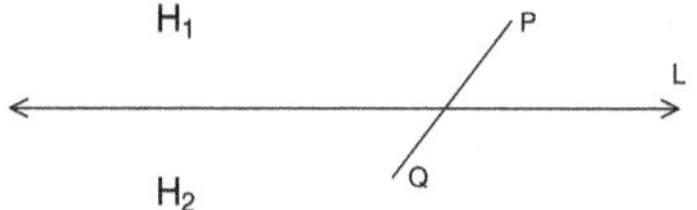

Figure A.1: The Axiom of Separation of the Plane

The second group of axioms, which describe what a protractor does, requires the notion of *ray* and *angle*. Given any two points A, B, the ray $\overrightarrow{AB}$ is the set of all points C in $\overleftrightarrow{AB}$ such that A is not between C and B. That is, $\overrightarrow{AB} = \overline{AB} \cup \{C : A - B - C\}$. Next let us consider the rays $\overrightarrow{OB}$ and $\overrightarrow{OC}$, which do not lie on the same line. We define $\angle BOC = \overrightarrow{OB} \cup \overrightarrow{OC}$. The rays $\overrightarrow{OB}$ and $\overrightarrow{OC}$ are called the 'sides' of $\angle BOC$ (also denoted $\angle COB$) while O is called the 'vertex'. Is it possible to provide a

[1] Moise (1963).

[2] A professional mathematician would say that given any line there exists a bijection $f : line \to \Re$ such that $PQ = |x_1 - x_2|$, where $x_1 = f(P)$, $x_2 = f(Q)$.

precise definition of the intuitively clear idea of interior of an angle? Indeed we can: The interior of $\angle BOC$ is the intersection of the half-plane determined by $\overleftrightarrow{OB}$ and containing C, with the half-plane determined by $\overleftrightarrow{OC}$ and containing B. The terrain is now ready to state four more axioms:

5. Given any angle, $\angle AOB$, there is a real number r $(0 < r < 180)$, called the measure of $\angle AOB$ and denoted by $m\angle AOB$.

6. Given any real number α, $0 < \alpha < 180$, and a ray $\overrightarrow{OA}$ let H_1 be one of the two half-planes determined by $\overleftrightarrow{OA}$. Then there exists a unique ray $\overrightarrow{OB_1}$ such that B_1 lies on H_1 and $m\angle AOB_1 = \alpha$ (Figure A2). Similarly, if H_2 is the other half-plane determined by $\overleftrightarrow{OA}$, there exists a unique ray $\overrightarrow{OB_2}$ such that B_2 lies on H_2 and $m\angle AOB_2 = \alpha$.

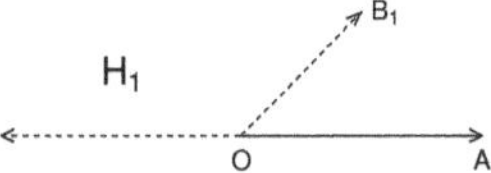

Figure A.2: The Axiom of Construction of Angles

7. Given $\angle AOB$ and a point C in its interior, we have $m\angle AOB = m\angle AOC + m\angle COB$.

8. Let A, O, and C lie on a line, with O in between A and C, and let B be a fourth point not on the above-mentioned line. Then we say that $\angle AOB$ and $\angle COB$ form a linear pair (Figure A3). The eighth axiom asserts that $m\angle AOB + m\angle BOC = 180$.

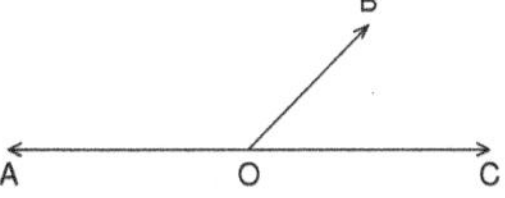

Figure A.3: The Linear Pair Axiom

The last axiom, needed to develop neutral geometry, has to do with triangles. So, we might as well define the well-known notion of a 'triangle'. Given any three points A, B, C, which do not lie on the same line, ΔABC (the triangle with vertices A, B, C and sides $\overline{AB}$, $\overline{BC}$, and $\overline{AC}$) is simply the union of the three segments $\overline{AB}$, $\overline{BC}$, and $\overline{AC}$. Furthermore, we will say that ΔABC and ΔDEF are under a SAS correspondence if $AB = DE, BC = EF$ and $m\angle ABC = m\angle DEF$ (also written $\angle ABC \cong \angle DEF$). Moreover, congruence between triangles is defined in the usual way: $\Delta ABC \cong \Delta DEF$ if and only if $AB = DE, AC = DF, BC = EF$, $\angle ABC \cong \angle DEF$, $\angle BCA \cong \angle EFD$, and $\angle CAB \cong \angle FDE$.

9. If ΔABC and ΔDEF are under a SAS correspondence then $\Delta ABC \cong \Delta DEF$ (in other words, the three corresponding sides have equal length and the three corresponding angles have equal measure).

Appendix B

B.1 A proof of the AGM inequality for $n = 3$

We need to show that, given any three positive numbers a, b, c,

1. $\sqrt[3]{abc} \leq \frac{a+b+c}{3}$.
2. $\sqrt[3]{abc} = \frac{a+b+c}{3}$ if and only if $a = b = c$.

Although it might seem odd, we will first prove AGM for $n = 4$ and then we will use this result to prove AGM for $n = 3$. Thus, we have to prove, at the beginning, that for any four positive numbers a, b, c, d,

1. $\frac{a+b+c+d}{4} \geq \sqrt[4]{abcd}$.
2. $\frac{a+b+c+d}{4} = \sqrt[4]{abcd}$ if and only if $a = b = c = d$.

Let us try to prove the inequality. Using AGM ($n = 2$) we get $a + b \geq 2\sqrt{ab}$ and $c + d \geq 2\sqrt{cd}$. Then

$$\frac{a+b+c+d}{2} \geq \sqrt{ab} + \sqrt{cd}.$$

Once more we apply AGM ($n = 2$), this time to $\sqrt{ab}$ and $\sqrt{cd}$:

$$\sqrt{ab} + \sqrt{cd} \geq 2\sqrt{\sqrt{ab}\sqrt{cd}} = 2\sqrt[4]{abcd},$$

then $\frac{a+b+c+d}{2} \geq 2\sqrt[4]{abcd}$; that is, $\frac{a+b+c+d}{4} \geq \sqrt[4]{abcd}$.The inequality has been proven.

Next suppose that $a = b = c = d$. Then $\sqrt[4]{abcd} = \sqrt[4]{a^4} = \frac{a+b+c+d}{4}$. More of a challenge will be to prove that if $\frac{a+b+c+d}{4} = \sqrt[4]{abcd}$ then $a = b = c = d$. We will provide a proof of the logical contrapositive of this implication, that is,

$$\text{if it is false that } a = b = c = d \text{ then } \frac{a+b+c+d}{4} \neq \sqrt[4]{abcd}.$$

Let us assume the antecedent, that is, suppose that it is false that $a = b = c = d$. Without loss of generality we may assume that $a \neq b$. Then, thanks to AGM ($n = 2$) we get $a + b > 2\sqrt{ab}$ and $c + d \geq 2\sqrt{cd}$. Therefore

$$\tfrac{a+b+c+d}{2} > \sqrt{ab} + \sqrt{cd}.$$

But we proved before that $\sqrt{ab} + \sqrt{cd} \geq 2\sqrt[4]{abcd}$. Hence

$$\tfrac{a+b+c+d}{2} > 2\sqrt[4]{abcd},$$

that is, $\frac{a+b+c+d}{4} > \sqrt[4]{abcd}$. Consequently $\frac{a+b+c+d}{4} \neq \sqrt[4]{abcd}$. The proof of AGM ($n = 4$) has come to an end.

Our task now is to provide a proof of AGM($n = 3$). Let us first prove that $\frac{a+b+c}{3} \geq \sqrt[3]{abc}$. With this purpose in mind we apply AGM ($n = 4$) and conclude that

$$\tfrac{a+b+c+(abc)^{1/3}}{4} \geq \left(abc(abc)^{1/3}\right)^{1/4} = \left((abc)^{4/3}\right)^{1/4} = (abc)^{1/3}.$$

Then $a + b + c + (abc)^{1/3} \geq 4(abc)^{1/3}$, hence $a + b + c \geq 3(abc)^{1/3}$. Therefore $\frac{a+b+c}{3} \geq (abc)^{1/3}$.

It only remains to prove that $\frac{a+b+c}{3} = (abc)^{1/3}$ if and only if $a = b = c$. Well, if $a = b = c$ then

$$\tfrac{a+b+c}{3} = \tfrac{3a}{3} = a = (a^3)^{1/3} = (aaa)^{1/3} = (abc)^{1/3}.$$

How about proving that if $\frac{a+b+c}{3} = (abc)^{1/3}$ then $a = b = c$? As we did when working with the case $n = 4$, we will prove the logical contrapositive of this implication, namely, if it is false that $a = b = c$ then $\frac{a+b+c}{3} \neq (abc)^{1/3}$. So, let us assume that it is false that $a = b = c$. Without loss of generality we can suppose that $a \neq b$. Using AGM ($n = 4$) it follows that

$$\tfrac{a+b+c+(abc)^{1/3}}{4} > \left(abc(abc)^{1/3}\right)^{1/4} = (abc)^{1/3}.$$

Therefore $a + b + c + (abc)^{1/3} > 4(abc)^{1/3}$; hence $a + b + c > 3(abc)^{1/3}$, which in turn implies $\frac{a+b+c}{3} > (abc)^{1/3}$. Consequently

$$\tfrac{a+b+c}{3} \neq (abc)^{1/3}.$$

QED

Index

www.ingramcontent.com/pod-product-compliance
Lightning Source LLC
Chambersburg PA
CBHW081432170526
45166CB00008B/2177

* 9 7 8 1 4 9 2 7 9 8 8 9 7 *